DEC 1 9 2016

NAPA COUNTY LIBRARY
580 COOMBS STREET
NAPA, CA 94559

PROTECTING THE PLANET

PROTECTING THE PLANET

Environmental Champions from **Conservation** to **Climate Change**

BUDD TITLOW & MARIAH TINGER

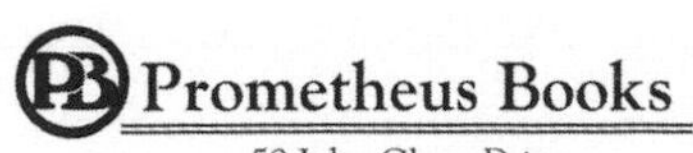

59 John Glenn Drive
Amherst, New York 14228

Published 2016 by Prometheus Books

Protecting the Planet: Environmental Champions from Conservation to Climate Change. Copyright © 2016 by Budd Titlow and Mariah Tinger. All rights reserved. No part of this publication may be reproduced, stored in a retrieval system, or transmitted in any form or by any means, digital, electronic, mechanical, photocopying, recording, or otherwise, or conveyed via the Internet or a website without prior written permission of the publisher, except in the case of brief quotations embodied in critical articles and reviews.

Trademarked names appear throughout this book. Prometheus Books recognizes all registered trademarks, trademarks, and service marks mentioned in the text.

Every attempt has been made to trace accurate ownership of copyrighted material in this book. Errors and omissions will be corrected in subsequent editions, provided that notification is sent to the publisher.

The Internet addresses listed in the text were accurate at the time of publication. The inclusion of a website does not indicate an endorsement by the author(s) or by Prometheus Books, and Prometheus Books does not guarantee the accuracy of the information presented at these sites.

Cover image of globe © NPeter/Shutterstock
Cover design by Jeff Schaller
Cover design © Prometheus Books

Inquiries should be addressed to
Prometheus Books
59 John Glenn Drive
Amherst, New York 14228
VOICE: 716–691–0133 • FAX: 716–691–0137
WWW.PROMETHEUSBOOKS.COM

20 19 18 17 16 5 4 3 2 1

Library of Congress Cataloging-in-Publication Data

Names: Titlow, Budd, author. | Tinger, Mariah, author.
Title: Protecting the planet : environmental champions from conservation to climate change / by Budd Titlow & Mariah Tinger.
Description: Amherst, New York : Prometheus Books, [2016] | Includes bibliographical references and index.
Identifiers: LCCN 2016017840 (print) | LCCN 2016031100 (ebook) | ISBN 9781633882256 (hardcover) | ISBN 9781633882263 (ebook)
Subjects: LCSH: Environmental protection—United States—History. | Environmentalists—United States—Biography. | Environmental policy—United States—History. | Climatic changes. | Global environmental change.
Classification: LCC GE180 .T57 2016 (print) | LCC GE180 (ebook) | DDC 363.70092/273—dc23
LC record available at https://lccn.loc.gov/2016017840

Printed in the United States of America

To a sustainable and livable planet for all future generations, most especially my own children—Mariah (my remarkable coauthor), Merisa, and David—and grandchildren—Harrison, Sierra, Max, and Wyeth. Also to my wife, Debby, and my aunt Betty, without whose love and support this book would have not been possible.

—Budd Titlow

I dedicate this book to all of the heroes who share our love of nature and have sacrificed greatly in its honor, but most importantly to my father and coauthor, Budd, who has been my environmental hero since I was old enough to say "out(side)."

—Mariah Tinger

CONTENTS

PART THREE: THE US ENVIRONMENTAL MOVEMENT: A STORY FOR THE AGES: 1969 THROUGH 2016

INTRODUCTION

FROM MARIAH'S PERSPECTIVE

While writing this book, I escaped to a monastery in Petersham, Massachusetts, for peace, inspiration, and focus. The volleyed chanting of the monks and sisters during the services eased my mind to a gentle place of introspection. Under the angelic voice of Sister Mary Frances, I recalled the moments where I have felt closest to God. Without exception, these moments transpire in a beautiful spot in nature. One happened on a run in a cemetery near Hartford, Connecticut, where I attended Trinity College. The sun filtered through the trees—leaves gold due to autumn's chill. The light infused the cemetery with an ethereality that pulled my heart toward the surface of my skin. The glow I felt in the cemetery that day echoed the golden light of the trees and sun. My feet floated down the path.

Another distinct moment came when I was a naturalist in Yosemite National Park. John Muir had described luminescent rainbows that appear on Yosemite Falls under full moonlight. Striking off solo, with only celestial light to illuminate the path, I endeavored to find these mythical moonbows. The air on the trail was ponderous and silent. The silver-blue light of the moon glancing off the snow softened my movements and even my breath on the hike. When I finally saw Yosemite Falls, even it seemed tamed by the quiet force of the full moon. The ice spray on the sides of the waterfall often cracks and crashes with a noise that fills the valley. That night, however, the cold air preserved the spray softly on the granite wall—lacy and ethereal. I did not find any moon rainbows shimmering over the cliffs that night, but I did find an incredible oneness with this pulchritudinous world of which we are blessed to be a part.

With nature as a divine force and light for goodness in my life, it strikes a special chord in my heart when I see it being destroyed ruthlessly, recklessly, and without consciousness for the long-term ramifications. In my graduate studies at Harvard, the more I learned about the dangers of climate change, the more puzzled I became about why we were not taking deep and drastic actions to stop releasing greenhouse gas pollution. In lecture after lecture, I learned the ways that climate change would affect every aspect of our existence. For example, the distinguished public-health expert Paul Epstein, who left us too early, warned of increasing disease vectors. I think of him now when I see how warmer winters have dramatically increased the deer tick population—and therefore the risk of Lyme disease—in my backyard. Paul Kirshen, civil and environmental engineer, told us how ill-prepared Boston's infrastructure is to absorb the sea-level rise coming our way. Bill Moomaw, IPCC author and Nobel Prize winner, explained that climate change–induced drought in Syria was a primary factor in the disintegration of that region and the rise of ISIS. Tim Weiskel, social anthropologist and historian, spoke movingly about the legions of climate refugees from island nations that were slowly being swallowed by the sea. Mark Leighton, rainforest ecologist, told us of the massive loss of species due to climate change—including many of my favorite butterfly species. The list goes on and on—climate change could potentially touch sundry areas of our lives in frightening ways. We cover the aforementioned topics in detail in this book, but, rest assured, it is an optimistic tome.

I embarked on this project as a remedy to sleepless nights induced by this overwhelming climate-change information. I wanted to find a positive way to participate in problem-solving. Not just for my kids (Budd's grandchildren), but for every citizen of the global community. This is not tomorrow's problem—in fact, today's elderly are most at risk from climate change–induced temperature increases in the United States. We are already seeing summers with highs so extreme that folks without a robust constitution cannot survive. No generation is safe from its ravages—it is here and now.

Fortunately, the research and conversations that we have had with our heroes have provided an about-face for us. The heroic actions performed by these leaders have led to incredible solutions for a transition away from fossil fuels. The future no longer looks hopeless. In fact, it is entirely sanguine, in large part due to their efforts.

—Mariah Tinger

FROM BUDD'S PERSPECTIVE

As a baby boomer, I came to understand climate change in a completely different manner than Mariah. In fact, I believe this intergenerational contrast represents—in many ways—the essence of the climate-change issue. While I had been a staunch environmentalist for my entire forty-year professional career, I tended to focus more on immediate impacts—water pollution, wetland filling, loss of wildlife habitat, and protection of endangered species. I was, of course, aware of the climate-change controversy, but—like many of my generation—I believed that it involved something that might happen long after my time on Earth. It was not until a detailed discussion I had with Mariah after she started her graduate coursework at Harvard in 2009 that my mind was really opened up to both the immediacy and severity of climate change.

But fifteen years before my personal epiphany on the subject, my life started being dramatically altered by climate change—I just didn't realize it at the time.

With my family, I moved to Southborough, Massachusetts—twenty-five miles west of Boston—in 1994. Southborough lies on the eastern edge of what is often locally called the Worcester Snow Belt (WSB). According to many people who have lived in east/central Massachusetts for a long time, the WSB is an oblong-shaped, northwest-southeast trending polygon that roughly centers on the seven hills that the city of Worcester—like Rome—is built upon.

Due to a combination of factors—including locally high topographic elevations and locations on the leading northwest edge of nor'easter storm rotations—where the heaviest precipitation tends to be produced—the WSB receives significantly more snow than the city of Boston. This includes the landmark 1997 April Fools' Day Blizzard—which was an ill-appreciated joke to those of us living there—when the area received an all-time record of thirty-eight inches of snow in twenty-four hours. Depending on which records you review, during the ten years (1994 to 2003) that we lived in Southborough, the WSB received five of the heaviest annual snowfall totals recorded in the more than one hundred years of Massachusetts weather history. Of course, at the time, I didn't connect these extreme winter weather events with climate change.

What I did know for sure, however, was that I was sick of continuously blowing snow off the driveway, knocking snow and icicles off the eaves of our 1840s farmhouse, and positioning gallon buckets inside the house to collect the water from ceiling leaks for six months—November through April—every year. Plus, since I needed two full hip replacements, I couldn't keep up with the physical demands of six hours of snow and ice removal drudgery each time we had a major snowfall. Moreover, I could no longer participate in cross-country skiing or any of the other endeavors that make living in "snow country" enjoyable. While we all really loved Boston—certainly one of the greatest cities in the world (Symphony Hall, Fenway Park, the Boston Marathon, the Common and Duck Pond)—it was time for me to think about moving south ... into some location that actually experiences the season of spring. So, early in 2004, I resigned my job as a professional wildlife biologist and private environmental consultant in the city of Worcester, sold our Southborough home, and moved, with my wife, Debby, to the warmer climes of North Carolina.

To paraphrase the late, great radio newsman Paul Harvey, "and now for the rest of the story": I'm currently living even further south, in the city of Tallahassee, Florida, where I'm pursuing a completely different lifestyle—coordinating agricultural food safety audits, teaching courses at Florida State University (FSU), writing newspaper columns for the *Tallahassee Democrat*, and authoring books.

It has been more than ten years since I fled the vagaries of winter in the WSB and—thanks to Mariah—I now fully understand that climate change is to blame for my leaving. Warmer than normal weather actually favors snowstorms.[1] Very simply put (we'll get into more detail on this later in the book), as the weather warms throughout the world, more moisture—evaporating from both the land and the oceans—is trapped in the atmosphere. The more moisture that is trapped in the atmosphere, the greater the occurrence of extreme weather events throughout the world.

Since storms primarily happen when warm, moisture-laden air masses meet cold, dry air masses, this explains how climate change produces more numerous—and often with heavier snow accumulations—snowstorms and blizzards. So my experience with living through five of east/central Massachusetts's snowiest winters was *not* just some anomaly, fluke, or normal cyclical occurrence. No, if I had understood the science of climate change at the time, I would have fully anticipated—and actually expected—that this would occur.

I next had the authenticity of climate change driven home clearly to me during a 2014 fall ecology course I was teaching on the subject at FSU. Professor William Landing—a guest speaker from FSU's Department of Oceanography—presented some results from his summer of work aboard an Antarctic research vessel. He showed us that ice cores drilled from deep within the Antarctic ice sheet contained bubbles of carbon dioxide (CO_2) gas that were at least 800,000 years old.[2] The Scripps Institution of Oceanography collects current CO_2 readings on the big island of Hawaii, at the Mauna Loa Observatory. Out of all of the CO_2 readings taken from these ancient glacial ice cores, none were higher than the CO_2 readings collected in 2014.[3]

So not only were the levels of CO_2 in our atmosphere the highest recorded in more than 800,000 years, they were also in excess of the 350 parts per million (ppm) "cautionary threshold" put forth by Dr. James Hansen and other climate scientists in 2008.[4] In fact, CO_2 measurements on Mauna Loa in the spring of 2014 exceeded the "extreme danger" climate-change threshold of 400 ppm from which some scientists have said that *recovery is not possible.*[5] Then, in the fall of 2015, CO_2 measurements taken at Mauna Loa stayed above the 400 ppm level for an entire month, leading some climate scientists—like Ralph Keeling, Peter Gleick, and Michael Mann—to speculate that we may not see readings below 400 ppm again for quite some time—if ever.[6]

This information made me realize that if we don't take decisive action soon we are facing the reality of fomenting the dire consequences associated with a 2.0 C (3.6 F) global increase in air temperature. Such things as (1) the flooding loss of entire Pacific island nations and heavily populated coastal cities, (2) the relocation of millions, if not billions, of people, (3) the acidification and loss of most of our coral reefs, (4) the inundation of more than half of the world's coastal wetland systems—including the bulk of our tidal salt marsh oceanic nursery habitats, (5) the salt water contamination of hundreds of municipal groundwater supplies, (6) mega-droughts that last for more than thirty years, and (7) political revolutions and the formation of new terrorist groups—are all now imminent possibilities. As bestselling author and climate-change activist, Naomi Klein aptly titled her most recent book, *This Changes Everything*—the suite of 2014 and 2015 Mauna Loa CO_2 measurements exceeding 400 ppm should definitely put us all in a this-changes-everything mode of thinking![7] In fact, the overall debate has now shifted from "Is climate change really occurring and how fast is it happening" to "Okay, climate change is definitely happening right now so what can we do about it?" That's where this book comes in.

In part 1, we begin by reviewing the history of the climate-change crisis. We tell you how scientists first identified the phenomenon in the 1800s. Then we move through the proof that it is actually happening and conclude with when and where ongoing impacts are currently occurring.

In part 2 and part 3, we take a look at the history of the US environmental movement. But we don't just talk about the environmental elements as stand-alone events. Instead, we weave a story that covers everything from the first settlement by Native American tribes and colonization by "dominion over nature" European settlers through Manifest Destiny and westward expansion, the Industrial Age, the rebellious sixties, and the heyday of regulatory policy.

As you'll read, the evolution of the United States from a pioneering, scrapping, win-at-all-costs collection of colonies into one of the world's most powerful nations and leaders in natural resource protection is a phenomenon unto itself. How did we make this remarkable transition? Who were the people most responsible for keeping us heading in the right direction each step of the way? How did they achieve their goals? In answering these questions, we also provide insights into the strategies that worked best for our past environmental heroes and discuss how their methods can be applied to dealing with our current climate-change crisis.

Then, in part 4, we describe our current climate-change heroes, telling you exactly who they are, what they have done, and how and why they think the ongoing climate-change crisis can best be stopped in its tracks. In this section, we include a broad range of opinions from an array of distinguished scientists, authors, activists, grassroots organizers, politicians, businesspeople, and celebrities.

Finally, in part 5, Mariah and I tell you how and why we think a gigantic push by humanity will result in worldwide solutions for all future generations to mitigate climate change and keep world temperatures as we know them today, if not lower. We're convinced that by working together for the same goal we can get this done. After all, this is the United States. We are the people who corrected the abominations of slavery, gave all people equal rights under our Constitution, endured the Great Depression and the Dust Bowl, defeated the scourge of the Nazis, put men on the moon, and severely degraded Osama Bin Laden's Al Qaeda management hierarchy.

In summary, we believe the time for action is now. We must heed the clarion call of our new climate-change heroes who—by working in tribute to the accomplishments of our past environmental heroes—can come forward and save us all from the pending ravages of climate change.

One Earth—there is no Planet B!

—Budd Titlow

Part One

CLIMATE CHANGE

Now Humanity's Greatest Environmental Challenge

Chapter 1

WHAT EXACTLY IS CLIMATE CHANGE?

Climate change is just what the name implies—the climate on Earth is changing based on the fact that the average global temperature is steadily increasing. While there are some natural phenomena involved in this process, the primary driver of the worldwide temperature increase is the emission of Greenhouse Gases (GHG)—primarily carbon dioxide (CO_2)—that is associated with human activities, especially the burning of fossil fuels. The problem is that our climate is changing at an accelerated rate—as currently verified by almost one hundred percent of the world's climate scientists—something that has never before happened on Earth. How do we know this?

Let's begin by clearing up two common misconceptions. First, it's certainly true that Earth's climate has been changing over hundreds of millennia. In fact, climate scientists generally recognize at least five distinct "ice ages" in Earth's history. And, in between these ice ages, gradual—very long term—and natural—not human influenced—warming and cooling trends have occurred. But these gradual and natural changes are *not* what we are talking about here.

The climate change we are currently experiencing on Earth is *not* gradual—at least not in geologic terms. In fact, when compared to the epochs between the ice ages, our present climate-change crisis is just a fleeting moment in time—the veritable beat of a gnat's eyelash. Plus, our climate-change crisis is *very much human influenced*, being caused by the billions of metric tons of GHG—most notably CO_2 emissions—that we are spewing forth into our atmosphere every year.

Here's how it works: The sunlight that enters our Earth's atmosphere is largely visible light, meaning that we can see it. We all know the buoyant feeling we get from being outside and seeing brilliant warm rays of sunlight streaming down on Earth's surface. The energy in visible light passes down through our atmosphere without being absorbed by atmospheric gases.

But the light that is reflected back up into our atmosphere from Earth's surface is changed from visible light energy into infrared light energy. And a certain percentage of this reflected infrared light energy is—always has been—absorbed by gas molecules in our atmosphere. In fact, this atmospheric heat absorption is what makes it possible for us to live on Earth. If our atmosphere did not absorb any heat energy from infrared light, Earth would be a huge snowball, and life would not be present or even possible.[1]

Now, here's the rub. Up until the start of the Industrial Revolution—which we'll discuss in detail in part 2—Earth's temperature had been rising very gradually, fitting the pattern of cyclical warming trends that had occurred every few epochs—between ice ages, if you will. But as soon as we started burning fossil fuels—wood, coal, and oil—to run our newfangled steam engines and other machines, we started artificially increasing the amount of CO_2 and other GHG in our atmosphere.

Throughout the 1800s, the burgeoning use of fossil fuels to power our ever-improving and expanding mechanical technology had a pronounced dual effect. The spread of machines was rapidly transforming our society from largely rural agrarian to urban industrial. Factory work provided many Americans with dependable, fixed incomes—something they had never had before—and more comfortable lives, while also creating our first wave of incredibly wealthy corporate entrepreneurs. But with this dramatic upturn in our still-young nation's economic production and individual riches came the dark side of the environmental equation. At the start of the Industrial Revolution, atmospheric CO_2 was around 280 parts per million (ppm).[2] With every passing year, the clouds of pollution and atmospheric gases unleashed by our flourishing technology caused the regression line of global air temperature rise to become steeper and steeper.

Now, today, we have arrived at the point where the CO_2 contamination of our atmosphere has exceeded the alarming level of more than 400 ppm. So, in order to continue life as we know it, we must figure out how to stop this now-dramatic increase in global temperatures and keep them from rising an additional 1.5 C (2.7 F) (from 2016 levels).

The other commonly held—but erroneous—belief is that climate change (global warming) can't really be occurring because we are still having cold weather—like descending "polar vortexes" and record-setting blizzards. Anyone remember in February 2015 when Senator Jim Inhofe threw a snowball at the sitting Senate president and claimed it was evidence of a global-warming hoax? Here's the simple way to understand that cold weather and blizzards do not prove that the globe is not warming: First, it's a scientific fact that warm air holds more water vapor than cold air. Next, the warming global air temperatures are causing the world's glaciers and ice sheets to melt at unprecedented rates—also a scientifically documented fact. Finally, this melting ice over land causes ocean and sea levels throughout the world to warm and rise.

So—based on the standard water cycle that we all studied in school—it makes sense that greater amounts of water vapor evaporate into our atmosphere each year from the surfaces of our warming oceans and seas. In other words, the amount of water condensed into clouds in our skies is constantly increasing as our Earth warms.

Now think about the conditions that must be present to create a storm of any type. Simply put, stormy weather typically occurs when a front of cold, dry air collides with a front of warm, wet air. When this happens, the clouds are like a sponge being squeezed—the strength of the resulting precipitation is dependent on the total amount of moisture trapped in the air. If this happens in the spring, summer, or fall, the colliding fronts could provide a "climate-change assist" to extreme weather events like tornados and cyclones.[3] (Also, higher sea levels will exacerbate damage caused by storm surges associated with Atlantic hurricanes, Pacific typhoons, and other oceanic storms.) But when moisture-laden fronts come together in the winter, record-setting snowfalls and even blizzards can and do occur. It makes practical sense: the greater the amount of moisture trapped in our atmosphere, the greater likelihood of an extreme weather event—no matter what time of year it is.

The key point to emphasize here is that the severity of *all* weather events—including snowstorms—is being augmented by climate change. So the often-heard denial argument that Earth can't really be warming because we still have cold, snowy winters, really doesn't—excuse the pun—hold any water. In fact, snowier winters should actually be expected to occur due to climate change.

Across the globe, the warming atmosphere is also exacerbating existing drought conditions and creating severe droughts where none have occurred in the past. If the air temperature is increasing on a daily, monthly, and annual basis—as is actually happening for all three time intervals—then the surface of Earth has to be drying out more and more with each passing day.

When the soil becomes too dry to support crop growth—and irrigation water is no longer available—the crops fail and the ground no longer has any deep-rooted plant cover to protect the topsoil. The topsoil then dries up and either washes away during intense rainstorms or blows away during windstorms. Afterward, the ground surface is left barren and is no longer capable of supporting crops of any type.

On the flip side of the coin—since sea level is rising—flooding is becoming more prevalent and severe in low-lying areas of the world. Land at the southern tip of Florida, Louisiana's Mississippi River Delta, and Vietnam's Mekong Delta is being severely eroded by encroaching surf. Plus the southern lowlands of Bangladesh—which are the outlets for three of the world's major river systems—and many Pacific archipelagos—like the Marshall Islands and the Maldives—are all fast disappearing beneath the rising tides. We'll describe the specific flooding effects of sea-level rise plus many other severe environmental effects of climate change in greater detail later in this section.

Chapter 2

WHAT IS THE HISTORY OF CLIMATE CHANGE?

The early history (1820 to 1970) of climate change is reasonably well known. We can begin with a fellow named Jean Baptiste Joseph Fourier, a French physicist who, in the 1820s, began to wonder how a planet of Earth's size could maintain its warmth at the vast distance of 94 million miles away from the sun's energy source.

The more he thought about it, the more Fourier realized that some other process beyond just straight solar radiation had to be keeping Earth's temperature so steady. He then postulated that gases in Earth's atmosphere were trapping and holding the sun's radiation after it reflected off Earth's surface. While he had no mechanism to test his hypothesis, Fourier envisioned this as a similar process to heat being trapped in a glass-covered box. In so doing, he was the first scientist to describe greenhouse gases (GHG).[1]

Next, in the 1860s, Irish physicist and Alpine climbing pioneer John Tyndall noted an interesting fact when reviewing the geologic records of his day. Much of northern Europe was—at one time—covered by ice sheets. Wondering how Earth's climate could change so dramatically that the ice sheets melted, Tyndall devised a series of experiments showing that water vapor and carbon dioxide (CO_2) had excellent heat-trapping capabilities—despite the fact that they were both just trace gases.

As a result of this work, Tyndall supported Fourier's metaphoric image of atmospheric gases forming an "insulating blanket of warmth around the Earth."[2] So, while Fourier and Tyndall both provided an inside understanding of what was going on to keep Earth's atmosphere warm, neither of them mentioned any thoughts about human activities having a part in this global process.

Fast forward to the year 1896, when Swedish physicist/chemist and Nobel Prize winner Svante Arrhenius used estimates of coal burning to determine that human-driven activities involving fossil fuels were affecting Earth's atmosphere by making it warmer. His research showed that Earth's average temperature was about 15 C (59 F), primarily because of the capacities of water vapor and CO_2 to absorb infrared radiation from the sun. Arrhenius also used his calculations to surmise that doubling the CO_2 level in Earth's atmosphere would raise the planet's temperature by 5.0 C to 6.0 C (9.0 F to 10.8 F).[3] Aha—so now we had it . . . the cat was finally out of the bag and the civilized world was about to know that human lifestyles were creating a potential future disaster for our planet.

But not so fast there, friends. Many scientists believed that Arrhenius's calculations were too simple and that he had failed to include many factors—such as the potential ability of Earth's vast bodies of saltwater to soak up huge volumes (as much as five-sixths) of the CO_2 that was being generated. This led to the assumption that absorption by the oceans would significantly reduce the GHG concentrations in the atmosphere.[4] In turn, this would then prevent any measurable increases in the planet's annual average temperatures—at least for the next few thousand years or so!

After cold water was thrown on Arrhenius's findings, the historic trail of the greenhouse effect hypothesis went silent for a few decades, until 1931, when American physicist Edward O. Hulburt took an interest. Hulburt revisited Arrhenius's work, again testing what doubling the CO_2 concentration would do to the temperature of Earth's atmosphere. His calculations came up with a predicted increase of around 4 C (7.2 F) of global warming.[5]

In the process, Hulburt discovered that the interception of reflected energy in infrared radiation was of key importance.[6] Hulburt's results received minimal attention from other scientists who were studying the climate. The prevailing thought remained: Earth's climate system used some sort of natural balancing process to maintain itself.

The year 1938 brought the first measurements indicating that global warming was actually occurring. Using data showing that CO_2 concentrations had risen over the previous century, Guy Callendar, a British engineer, illustrated that global air temperatures had also risen over the same period. His correlation of increasing CO_2 levels with rising atmospheric temperatures was labeled the Callendar Effect and was—for the most part—summarily dismissed by meteorologists and the general public.[7]

Several events significant to the history of climate change occurred in the 1950s. First, in 1955, climate researcher Gilbert Plass performed detailed modeling analyses—using a newfangled device called "the computer"—to show that doubling CO_2 concentrations in the atmosphere would indeed increase global temperatures by 3.0 C to 4.0 C (5.4 F to 7.2 F).[8] In a break with scientific thought at the time, Plass argued that water vapor absorption would not significantly alter this predicted CO_2 effect.[9]

In fact, Plass also believed that the world's oceans would only be able to absorb a minimal amount of the carbon that was being generated by human activities. He calculated that consumption of all of Earth's fossil-fuel resources over the next millennium could result in a surface temperature increase of 7.0 C (12.6 F).[10]

Plass's work was pivotal in the establishment of the central role of CO_2 in climate change, and in the danger that anthropogenic (human-induced) carbon emissions posed to Earth's climate system. Because of these findings, Plass is often referred to as the "Father of Modern Greenhouse-Gas Theory."[11]

Next, in 1957, a pair of scientists—chemist Hans Suess and oceanographer Roger Revelle—supported Plass's work by proving that seawater would not, as previously believed, absorb all CO_2 that enters the atmosphere. In fact, portending the peril we face today, Revelle wrote that "Human beings are now carrying out a large-scale geophysical experiment."[12] Based on their findings, Revelle and Suess warned other scientists that it was critical to determine the exact levels of CO_2 in the atmosphere and find out if those levels were changing—specifically rising.[13]

Finally, in 1958, American climate scientist Charles David Keeling began a project that continues to this day. Partially funded by Revelle and Suess, he started taking systematic measurements of atmospheric CO_2 at Mauna Loa Observatory on the big island of Hawaii and in Antarctica—two places situated far from any localized "noise," or sources of extraneous, irrelevant data—that could affect the scientific clarity of his results.[14]

Through this monitoring, Keeling produced concentration curves that provided the first unequivocal proof that CO_2 concentrations in the atmosphere were rising on a long-term basis.[15] Today, the "Keeling Curve" remains one of the major icons of evidence that CO_2 is accumulating in our air.[16]

After all was said and done, the media and many scientists mostly ignored the data collected by Plass, Revelle, Suess, and Keeling.[17] They instead chose to believe that Earth was actually cooling instead of warming. What a pity, considering that if we had started trying to solve the climate-change crisis when this information first came to light we would be forty years into a solution.[18]

When the rebellious decade of the sixties began, politicos started at least paying some attention to the greenhouse gas (GHG) issue and the potential long-term effects of too much atmospheric CO_2. During a 1965 speech to Congress, President Lyndon Baines Johnson noted that the current generation of Americans "has altered the composition of the atmosphere on a global scale through . . . a steady increase in carbon dioxide from the burning of fossil fuels."[19] Then, in 1969, New York senator Daniel Patrick Moynihan warned of a dangerous sea-level rise of ten feet or more. "Goodbye New York," he said. "And Goodbye Washington, DC."[20]

It has now been more than 120 years since Arrhenius first revealed that the activities of humans were changing Earth's climate. And it has been more than 50 years since politicians on Capitol Hill first started alerting the public about the potential hazards of climate change and the associated global warming. This means that the original scientific acknowledgment of climate change is now older than the atom bomb, the discovery of penicillin, and the recognition of DNA's role in genetic inheritance. It's also older than transatlantic jet flights, digital computers, and moon rockets.[21]

We believe this brief but concise summary proves—beyond a shadow of a doubt—that climate change is real and that the warming trend we are now experiencing is strongly associated with the activities of humans, namely the burning of fossil fuels. To claim that anything else is true is to attempt to rewrite history and ignore scientific fact.

From the 1960s on, the history of climate change is tightly enmeshed within the broader-based overall history of the American environmental movement. Since most of our current climate-change heroes arrived on the scene after the 1960s, we refer you to part 3 of this book for a continued discussion of the climate-change timeline as it's embedded within the overall American environmental movement timeline. In addition, part 4 provides a detailed look at each of our individual climate-change heroes, including how they are helping to solve our ongoing climate-change/global-warming crisis.

Chapter 3

HOW DOES CLIMATE CHANGE OCCUR?

As we've just discussed, scientists have now proven that greenhouse gases (GHG) in Earth's atmosphere—particularly carbon dioxide (CO_2)—trap solar radiation from the sun and form a blanket of insulating warmth around the planet. Every day, more and more CO_2—about 2.4 million pounds a second—is being released into our atmosphere.[1] Every time we burn fossil fuels—coal, oil, and natural gas—to power our modern lifestyles, we are also dumping more polluting GHG into the air we breathe. Once CO_2 and the other GHG—methane, nitrous oxide, and water vapor—are released, they then capture and retain the sun's radiative energy near Earth's surface, keeping it from escaping back into space.

So now let's take a more detailed look at how this radiation trapping works: Concentrations of CO_2 are measured in parts per million (ppm—the total molecules of CO_2 per million molecules of dry air), while those of two other primary greenhouse gases—methane and nitrous oxide—are measured in parts per billion (ppb). Carbon dioxide, nitrous oxide, and methane are all trace constituents of Earth's atmosphere. Taken together with water vapor—actually another greenhouse gas—these four components account for less than 1 percent of the total volume of our atmosphere.[2]

Nitrogen and oxygen—which account for the other 99 percent of the volume of Earth's atmosphere—are essentially transparent to infrared radiation. So when the energy of visible sunlight (shortwave) radiation bounces off the ground and reflects back as infrared (longwave) radiation, the infrared energy isn't affected by the 99 percent component comprised of nitrogen and oxygen. But it is absorbed by the 1 percent component of greenhouse gases.[3] Accordingly, some of the infrared energy that would otherwise directly escape into space is reemitted back to Earth's surface.

Without this natural greenhouse effect—primarily being carried out by CO_2 and water vapor—Earth's mean surface temperature would be a freezing minus 18 C (minus 1 F), instead of the habitable 15 C (59 F) that we currently enjoy.[4]

So, based on this atmospheric process, the greenhouse gas warming effect is a good thing—up to a certain point—because it keeps our planet from becoming a mass of rotating globular ice. But the point at which atmospheric CO_2 is a good thing has now been seriously surpassed. Earth's atmosphere is currently being overloaded with CO_2—meaning it's now becoming *too* warm—and, as we've previously mentioned, it's getting measurably worse with each passing day, month, and year.

Chapter 4

WHAT IS THE PROOF FOR CLIMATE CHANGE?

Levels of CO_2 in Earth's atmosphere have been continuously monitored since David Keeling began taking his readings on Mauna Loa in 1958—almost sixty years ago. Since that time, CO_2 concentrations have increased from an average of 315 ppm to more than 400 ppm today. These data show that CO_2 levels have been steadily increasing during this time and our planet has been warming in correlation with these CO_2 increases.

Over the past 130 years, the average global temperature has increased almost 1.0 C (1.8 F). More than half of this increase has occurred during the past thirty-five years. In fact, the sixteen warmest years on record have all occurred since 1998, and every one of the past thirty-five years has been warmer than the twentieth-century average.[1]

According to the Union of Concerned Scientists, "Scientists can now conclusively identify that human activity is responsible for the observed increases in CO_2. The carbon dioxide emitted by burning coal, natural gas, and oil has a unique 'fingerprint'—and the additional CO_2 in the atmosphere bears that signature."[2] The particular ratio of carbon isotopes in the atmosphere produces this fingerprint. These isotopes correlate with the type of carbon that is produced when fossil plant materials are burned.[3] In other words, burning the biomass of fossil fuels—which are derived from fossil plant materials—is primarily what is causing the increases in CO_2 in Earth's atmosphere.[4]

Minute changes in the atmospheric concentration of oxygen also verify that the added CO_2 is derived from burning plant materials. Furthermore, the concentrations of CO_2 in Earth's oceans have increased right along with the atmospheric concentrations, which proves that atmospheric CO_2 increases cannot be a result of releases of CO_2 from Earth's oceans. In fact, all of this evidence—taken collectively—clearly shows that increases in atmospheric concentrations of CO_2 are human induced and primarily associated with fossil-fuel burning.[5]

As we discussed in the introduction and as is described on the British Antarctic Survey's website, "Antarctic ice cores show us that the concentration of CO_2 was stable over the last millennium until the early nineteenth century. It then started to rise, and its concentration is now nearly 40 percent higher than it was before the Industrial Revolution. Measurements from older ice cores confirm that both the magnitude and rate of the recent increase are almost certainly unprecedented over the last 800,000 years."[6]

So, not only are the most recent measurements of CO_2 in our atmosphere the highest ever recorded in at least the last 800,000 years, they are also in excess of the 350 ppm "cautionary threshold" set by NASA's James Hansen and other climatologists in 2008.[7] In summary, the world is already 1.0 C (1.8 F) warmer than it was in preindustrial times—around 1776, when the United States was founded as a nation. At first blush, this air temperature increase may seem inconsequential over such a long time period—240 years prior to the publication of this book. But the truth is that this global temperature increase is already having devastating effects on the livelihoods of millions of people around the world.

If we allow global temperatures to rise another 3.0 C (5.4 F)—which could easily happen by the end of this century, if not much sooner—the overall quality of life for humans on Earth will be *unequivocally and dramatically altered for the worse.* In fact, right now in 2016, scientists all over the world are issuing dire warnings that just a 2.0 C (3.6 F) overall temperature increase may be the point of no return from which many human civilizations on Earth may not be able to recover.

Chapter 5

WHAT ARE THE CURRENT IMPACTS AND WHERE ARE THEY OCCURRING?

So, to reiterate, the greatest falsehood about climate change in most people's minds is that it is somehow a long-term thing. The general attitude is, "It's a long way off, so why should I worry about it now." In fact, with most adults—especially those of us in the baby boomer generation—it's "I'll be dead and gone before anything really happens, so why should I be concerned?"

Well, we are here to tell you folks that absolutely nothing could be further from the truth. We are seeing the effects of climate change *right here* and *right now*, and not just in one or two areas but in just about everything you can think of.

MELTING ICE SHEETS

If you haven't seen the phenomenally frightening, yet hauntingly beautiful, National Geographic documentary *Chasing Ice*, find it and watch it.[1] There is no way you can view this meticulously crafted production and not be awestruck. Watching sections of glacier—some as large as the island of Manhattan—give way and calve off into the North Atlantic Ocean will have you squirming in your seat in both amazement and agony.

The unprecedented warming of Earth's surface is causing melting and recession of the polar ice caps, the Greenland and West Antarctica ice sheets, and glaciers all over the world—again faster than anything previously measured in world history.[2] Data from NASA's (the National Aeronautics and Space Administration) Gravity Recovery and Climate

Experiment show that the "Greenland ice sheet lost 150 to 250 cubic kilometers (36 to 60 cubic miles) of ice per year between 2002 and 2006, while the Antarctic ice sheet lost about 152 cubic kilometers (36 cubic miles) of ice between 2002 and 2005."[3] Plus glaciers in all of Earth's major mountain ranges—including the Himalayas, the Andes, the Alps, the Rockies, and the Brooks Range—are all retreating more each year."[4]

Also, since the melting glaciers and ice sheets are not being replenished at their upper end, they are shrinking in overall size. This means more land—which was previously encased and inaccessible under the immense, flowing ice masses—is now open and available to increased oil and gas exploration. As this freshly exposed land is leased and developed by oil companies, a vicious positive feedback cycle is established. When newly extracted volumes of oil and gas are refined and burned, they will emit more CO_2 into the atmosphere. This will cause additional warming of the Earth and even faster melting of the glaciers and ice sheets. The process just continues to feed off of itself.

DISAPPEARING GLOBAL SEA ICE

Covering an average of 9.6 million square miles—2.5 times the overall land area of our fifty states—global sea ice is also declining at an alarming rate. According to the National Snow and Ice Center, measurements of the extent of Arctic sea ice taken in the month of September—when the ice reaches its annual minimum—have declined by more than 30 percent since 1979. Plus, these same surveys also show that the sea ice is much younger and thinner—meaning that it is comprised of less multiyear ice—than it used to be.[5]

By reflecting sunlight back into space, floating sea ice plays an essential role in regulating Earth's climate. Icebergs also provide vital habitat for maintaining the life cycles of polar species, including walruses, seals, and—as we have all heard about in great detail recently—polar bears.[6] Some biologists are even predicting that, due to this extensive loss of their critical sea ice habitat, polar bears may go extinct within just the next few decades.[7]

SEA-LEVEL RISE AND GLOBAL FLOODING

So what does this mean, exactly? These ice masses are just sitting there way out in the middle of nowhere, with no human populations or human-made structures anywhere in sight. So what if the ice keeps falling off?—It's not hurting anything or killing anybody. Plus, it's just landing in the open ocean, so it's not causing any flooding or other catastrophes—is it?

Unfortunately, the answer to this question is a scientifically documented YES. The chunks of ice that are now falling off into the sea—like ice cubes steadily flowing from an out-of-control refrigerator icemaker—are causing sea-level rise and flooding and are endangering human populations all over the world.

Here's why: When the intact glaciers are perched over landmasses, they have no effect on sea level. The astronomical volumes of water they entomb are frozen in time, suspended in space, and restricted from entering nature's hydrologic cycle—water cycle—of gas (evaporation) to ice (condensation) to water (rain and snow) and back to gas (evaporation) again. But as soon as this ancient glacial ice falls off into the ocean, it once again—maybe for the first time in several millennia—becomes part of Earth's hydrologic cycle and causes the levels of our oceans to rise.

Of course, a few chunks of ice calving off the face of a glacier aren't going to have any measurable effect on the ocean's height. Not even if the ice continues to fall off all summer long. That's just normal and routine behavior for a glacier. Right? Lose some ice during the heat of the summer and then rebuild a higher and deeper ice mass when the cold and snow comes during the rest of the year.

But "normal" and "routine" are not words that can describe what's going on with glaciers today. As so vividly demonstrated in the *Chasing Ice* documentary, because of the warming climate, ice sheets and glaciers are melting at a more rapid rate than they are being replenished.[8] This results in worldwide sea levels rising steadily each year.

This is indeed what is happening—as documented at many locations throughout the world. Sea-level rise has been happening even faster than scientists anticipated just a few years ago. If recent projections are accurate, a 1.1 C to 1.7 C (2.0 F to 3.1 F) warming could cause three feet or more of global sea-level rise by 2100. That's enough of an increase to displace 56 million people in eighty-four developing countries around the world.[9] Plus, picture this extreme scenario: a complete meltdown of the Greenland ice sheet would cause a twenty-three-foot increase in the level of the Atlantic Ocean, which would be enough to submerge the entire city of London.[10]

Right now, sea-level rise is threatening to swallow up many low-lying parts of the world—including portions of the US states of Alabama, Florida, Louisiana, and Mississippi.

State of Florida

I (Budd) am now living here in the capital city of the state that will be the US focal point for climate change and associated sea-level rise. As Laura Parker writes in *National Geographic*, "Florida is a good place to see the (hazards and) costs of climate change emerging into sharper view. Many coastal places are at risk, but Florida is one of the most vulnerable. (The state's) future will be defined by a noisy, contentious public debate over taxes, zoning, public works projects, and property rights—a debate forced by rising waters."[11]

The federal government's National Climate Assessment Team (NCAT) predicts that sea-level rise in Florida will be accompanied by extreme weather events—prolonged droughts, monsoon-like deluges, F5 tornadoes, and hurricanes with associated super storm surges, known locally as "king tides." Of course, the most disruptive effects of sea-level rise in Florida will occur along the state's 1,350 miles of coastline. More than half of the state's 825 miles of sandy beaches are already showing signs of severe erosion.[12]

With three-quarters of Florida's 18 million people living in coastal counties, local infrastructure is now valued at more than two trillion dollars. Much, if not all, of this will be wiped out by the projected impacts of sea-level rise. Plus, the nine million people living in Florida's coastal plains will all need to be relocated. As you can imagine, the costs associated with climate change–induced sea-level rise here in Florida are unfathomable!

With all these dire predictions, how can we possibly hope to stem the onrushing tide of climate change and sea-level rise? Civic leaders in South Florida—such as Broward County's Kristin Jacobs and Miami-Dade's Philip Blumberg—have taken it upon themselves to map out the future with their own initiatives. Their efforts receive little help from the state legislature—or from 2016 presidential candidate Senator Marco Rubio—or current Republican Governor Rick Scott, whose only responses to questions about climate change are to repeatedly profess that "I am not a scientist." Scott even instructed his staff in the Department of Environmental Protection to "cease and desist" using the terms climate change, global warming, and sea-level rise, under the threat of immediate dismissal.[13]

Working through civic leaders like Jacobs and Blumberg, four South Florida counties—Monroe, Miami-Dade, Broward, and Palm Beach—have already drafted a general to-do list that would "reengineer" the region, step-by-step, through 2060. This includes producing land-use plans that call for new massive structural fortifications—fortifying seawalls and installing heavy-duty pumping stations—along the most vulnerable portions of their municipal coastlines. The more daunting projects, involving moving utilities away from the coasts and protecting high-value real estate—universities, hospitals, airports, and the tourist areas that drive Florida's economy—will come next. "We will do what we always have done," says Joe Fleming, a Miami land-use attorney. "We will dredge and prop everything up."[14]

In reality, it will likely take technology that has not yet been imagined to overcome the challenges posed by South Florida's unusual geology, which consists of limestone bedrock that is honeycombed like a beehive. When it is mined, limestone consolidates into fill that is perfectly fine for creating "high ground" for the construction of roads, shopping malls, and community centers. But, in its natural state, limestone is a whole different ballgame, functioning like a piece of cork that has holes poked all through it. Limestone bedrock can't hold water, and it can't be sealed up. So while seawalls in the city of Miami Beach can be raised to restrict some storm surges, they won't stop water that bubbles up from beneath the ground.[15]

According to city engineer Bruce Mowry, "Miami Beach will never not exist—it will just exist in a different way. We may have floating residential areas. We could have elevated roads built up on pilings. We could convert a transportation corridor to water. People ask me, 'Bruce, can this be done?' I say, 'It can be done, but can you afford it?'"[16]

In summary, no matter how our state attempts to counteract predicted climate-change impacts, Florida will have a very different topography and population distribution by the year 2050. On second thought, maybe things won't be so bad after all. By then our home here in Tallahassee—currently located about twenty-five miles away from the open waters of the Gulf of Mexico—will be beachfront property!

Mississippi River Delta

I (Budd) have a personal connection with the Mississippi River Delta. New Orleans is my hometown. I was born there—although I don't remember anything about it from my childhood. In a situation that is quite apropos to climate change, we were flooded out of our second-story flat by a hurricane in 1948, when I was only one and a half years old.

One thing that has always stayed with me, however, is the fact that most of the city of New Orleans was built—and still remains—below sea level. Hurricane Katrina made this certitude abundantly clear to the rest of the world during her devastating rampage through the area in 2005. Unfortunately, this ill-advised city design siting and construction decision does not bode well for the future of the Crescent City.

Right now, the Mississippi River Delta is losing the equivalent of one hundred yards of land every hour. Go to the website Restore the Mississippi River Delta and you will see a continuous numerical counter positioned in the upper left hand corner. You will notice that every second this counter adds about five tons of "uncaptured sediment since January 1, 2014" to the total, which—as of August 4, 2016—stands at more than 261 million tons of eroded land.[17]

Unfortunately—as the rolling website counter shows—the Mississippi River Delta has been losing wetlands at an alarming rate. This loss is the direct result of unsustainable environmental management practices.[18] Every hour, an area of land the size of a football field turns into open water. Because of this, Louisiana's fragile coastal wetland ecosystems—some of the largest and most productive in the world—are facing total collapse.[19]

While this piecemeal process may go unnoticed from day to day, natural disasters like Hurricanes Katrina and Rita make the impact evident.[20] Intact coastal wetlands could have provided extensive protection against the force of these storms. According to Keenan Orfalea, blogger on the Mississippi River Delta website, "Coastal wetlands are the first line of defense against climate change impacts such as storm surge."[21] They are naturally designed to take the brunt of the hit and buffer abutting uplands from damage to both natural and human-made features. But the degraded coastal wetland systems were quickly overwhelmed by the sheer intensity of these storms. Then—in the case of Katrina, as we discuss later in this chapter—portions of the poorly engineered and maintained levees also did not hold up, allowing large sections of the city to flood and become uninhabitable long after the storm had passed.

Today, the land is still sinking rapidly throughout the Mississippi River Delta. Heavy sediment loads cannot rebuild soil fast enough to overcome the sea-level-rise-related drowning of vegetated wetlands. Accelerating sea-level rise also fragments the critical coastal habitat provided by barrier islands, while endangering human facilities, including homes, businesses, and associated infrastructure. Climate scientists also expect projected increases in extreme weather to exert a dramatic impact on US energy production and supply. Sea-level rise, high winds, and storm surge threaten several thousand offshore drilling platforms, dozens of refineries, and thousands of miles of pipelines.[22]

Sea-level rise will also cause more frequent flooding of, and severe impacts to, transportation corridors—highways, railroads, and airports. As the Union of Concerned Scientists reports: "The state of Louisiana has already spent tens of millions of dollars to remove storm-related debris and repair submerged roads. Now, in response to rising risks of storm surge and flooding—related in part to climate change—the state is elevating Highway 1 above the 500-year flood level. Highway 1 is the only road that connects Port Fourchon, which supports 75 percent of deepwater oil and gas production in the Gulf of Mexico, with the rest of our nation."[23]

"The sea is rising and the land is sinking," Louisiana state climatologist Barry Keim told *USA Today*'s Dan Vergano. "The two together mean that wetlands are disappearing here at unprecedented rates."[24] Coupled with the threat of more powerful hurricanes that are given a climate-change assist, Keim told Vergano that "you have to worry about the past repeating itself here."[25]

"Louisiana is in many ways, one of the best examples of starting to see some of the near-term implications of climate change,"[26] environmental policy expert Jordan Fischbach said. Fischbach was part of the team from Pardee RAND Graduate School's Pittsburgh office that created resources to help Louisiana determine their focus for coastal restoration projects. He went on to tell Vergano, "In some ways, I feel like it is the canary in the coal mine because they are seeing effects that [are changing] people's day-to-day lives."[27]

Venice, Italy

I (Budd) had the good fortune to visit Venice, Italy, several times. Not only is Venice one of the world's most entrancingly beautiful cities, but it also is unrivaled for its setting and urban design. Strolling along the narrow, winding city streets is like suddenly being transported to the middle of a Giorgione landscape painting. The unparalleled sights—the gondoliers poling their ornate craft along the shimmering canals, the rippling reflections of multihued flags and flower baskets, the armadas of boats of every size and color—are at once both iconic and unforgettable.

Sadly, even though Venice has been around for a very long time—founded in 421 CE—it would never be built in today's world. That's because there is no contiguous landmass under the city. Venice is essentially a collage of grand and sacred structures that are floating on top of an archipelago of 117 islands scattered throughout a shallow lagoon of the Adriatic Sea.

As you can imagine, sea-level rise has always been a particular concern for this flood-prone city. Now, on top of the rising-sea-level problem, comes the realization that Venice is still sinking—a dilemma that was thought to have been resolved in the 1970s by ceasing groundwater withdrawals for factory use.

Enter the city's expensive and oft-delayed system of underwater MOSE barriers—a nod to Moses and his parting of the Red Sea.[28] Blasting the city's engineering budget at a total cost—to date—of more than $6.0 billion (US), MOSE consists of a series of steel gates installed at the three inlets separating the Adriatic Sea from the lagoon surrounding Venice. In theory, these barriers would be raised in the event of flooding over 3.6 feet, higher than the 2.6-foot level that often floods the famed St. Mark's Square.[29]

Many Venetians remain skeptical of the MOSE project due to the high costs and concerns over environmental risks. When the flood barriers are raised, they will trap the considerable pollution and untreated sewage—Venice has no modern sewage treatment system—within the confines of the city's lagoon and famous canals.[30] While the gates are up, the contamination will not be able to dissipate and dilute itself as it normally does when it drains out into the Adriatic.

Whether or not this rather dramatic engineering concept works out, we have our concerns that it may not be able to save this Italian jewel of the sea for the long term. Something tells us that if we don't thwart the global rising sea levels as soon as possible, this entire colossus of artistic and cultural glories may slip beneath the waters of the Adriatic, never to be seen again. And that would be a shame, indeed, especially for those—like me (Mariah)—who have never had the opportunity to experience Venice's visual delights.

Island Nations of the South Pacific

No low-lying region in the world is immune to the ongoing effects of climate change today. Island groups throughout our oceans are fighting against continual inundation, wondering what the near-term future holds for their homelands—if any land will remain for them to call home.

Climate change poses particular hazards for island groups located in the Pacific Ocean. The Marshall Islands, the Kiribati Islands, the Maldives Islands, the Galapagos Islands, the Carteret Islands of Papua

New Guinea, the Taro Islands of Fiji, the Mariana Islands, and American Samoa—the list goes on and on. According to US Fish and Wildlife Service, all are facing immediate risks from "rising temperatures, sea-level rise, contamination of freshwater resources with saltwater, coastal erosion, an increase in extreme weather events, coral reef bleaching, and ocean acidification."[31]

As described and illustrated in a March 11, 2015, photo display on the *Guardian*'s website, rising sea levels have made every high tide a perilous incident on the Marshall Islands, which are a string of more than one thousand low-lying islands and coral atolls in the North Pacific Ocean.[32] This island group provides a tropical paradise of land-based living space for tens of thousands of native people.

During the past few years, many homes in the Marshallese island capital of Majuro have been wiped out by climate change–driven sea-level rise. When storm events are coupled with abnormally high "king tides," the ensuing deluge can be quite devastating. According to scientists studying the Marshall Islands, every 1.2 inches of sea-level rise causes a storm inundation to reach an additional hundred feet further inland.[33]

These storm surges sometimes lead to flooding that sends waves crashing through the native villages on Majuro and spilling across all the way into the lagoon on the island chain's interior. This destroys seawalls, desecrates graves in cemeteries, poisons drinking water wells, wipes out crops, washes away beaches, topples palm trees, and spreads disease.[34]

In a gut-wrenching video released on September 18, 2014, Marshall Islands' president Christopher Jorebon Loeak made this appeal to the world:

> Out here in the middle of the Pacific Ocean, climate change has arrived. In the last year alone, my country has suffered through unprecedented droughts in the north, and the biggest ever king tides in the south; and we have watched the most devastating typhoons in history leave a trail of death and destruction across the region.
>
> Lying just 6.6 feet (2.0 meters) above sea level, my atoll nation stands at the frontline in the battle against climate change. The beaches of Buoj where I used to fish as a boy are already under water, and the fresh water we need to grow our food gets saltier every day. As scientists had predicted, some of our islands have already completely

> disappeared, gone forever under the ever-rising waves. For the Marshall Islands and our friends in the Pacific, this is already a full-blown climate emergency.
>
> Some tell us that we should begin planning to leave. But how can we? And why should we? These islands are our home. They hold our history, our heritage, and our hopes for the future. Are the world's polluters asking us to give up our language, our culture, and our national identity? We are not prepared to do that—we will stay and fight. If the water comes, it comes.[35]

An equally sad scenario is playing out in the other Pacific island chains. Primarily composed of coral reefs and sand, the Republic of the Maldives is also the flattest country on Earth. With these topographical characteristics, the Maldives—as you might expect—are more vulnerable to climate change than any other place on the globe. In fact, many of the white sand beaches of this 1,190-island tropical paradise archipelago—that annually attract more than 600,000 tourists—face the very real possibility of soon disappearing beneath the waves, never to be seen again.[36]

On October 17, 2009, former Maldives president Mohamed Nasheed and eleven of his cabinet members signed a document calling for global cuts in carbon emissions.[37] This was—at the time—an unprecedented action for a small developing country, but what made it even more so was where this meeting took place.

To send a message to the world about climate change, President Nasheed held an underwater cabinet meeting.[38] It lasted for about thirty minutes and was held twenty feet down in a blue-green lagoon, on a small island used for military training. The cabinet members communicated using hand signals and whiteboards.

Although the highest upland area in the Maldives is sixteen feet above sea level, 75 percent of the landmass is only three feet above high-tide levels. About half of the human settlement's residential structures and approximately 70 percent of the critical infrastructure on the island, such as airports, power plants, landfills, and hospitals, lie within four hundred yards of the coastline. As a result, the predicted two feet of sea-level rise will massively affect the islands' abilities to function.[39]

Ganges-Brahmaputra-Meghna Delta in Bangladesh

More than 100 million people live in the Ganges-Brahmaputra Delta in Bangladesh and India. Accelerated sea-level rise caused by global warming is putting added stress on land, water, and food in this highly at-risk region.[40] Already dire, this situation has the potential to lead to mass slaughter of refugees and a border war between Bangladesh and the Indian state of Assam.

This potentially harrowing situation is featured in an episode of Showtime's *Years of Living Dangerously*.[41] Screen and stage actor Michael C. Hall, acting as Showtime's correspondent, travels to Bangladesh to examine firsthand the effects of climate change–induced sea-level rise and associated flooding on the nation's heavily populated southern delta region. Formed by a delta plain at the confluence of three of the world's mightiest rivers—the Ganges (Padma), the Brahmaputra (Jamuna), and the Meghna—Bangladesh's alluvial soil is highly fertile but extremely vulnerable to flood and drought. What Hall finds there is not a pleasant sight.[42]

Whole villages have been either inundated by flooding or wiped out by typhoons, displacing millions of people who no longer have a home. Without another option, the displaced families pack up what little they have left and move north into the teeming city of Dhaka—already one of the most densely populated urban areas on Earth.[43]

The country's capital, Dhaka, has experienced a population explosion of heretofore unprecedented proportions. Migrants arrive from other areas of Bangladesh at a rate of 400,000 per year—one of the fastest growth rates of any city in Asia. Within a few short years, the capital multiplied from four million to fifteen million people.[44]

Reaching the city's limits, the displaced families shoehorn themselves into the squalid shantytowns of the city's interior. As they walk along through the steaming, narrow paths of a plywood- and-tin-encased endless sea of humanity, Dr. Atiq Rahman, executive director of the Bangladesh Centre for Asian Studies, tells Hall that a flimsy shelter with a single functioning spigot and a latrine trench is considered luxury living in Dhaka.[45] It is at once a riveting, sickening, and maddening scene for well-off Americans to see and know that millions of people are trapped in a world like this. How could such a thing happen, we ask ourselves, in a supposedly "civilized world"?

So where will the farming families that are being displaced by sea-level rise go in the future? The most obvious answer—at least from a proximity perspective—is to move westward toward Bhomra, where there is a border crossing into India. So to check out this possibility, Hall and Dhaka journalist Morshed Ali Khan travel to the Bangladesh-India border. What they find there is even more shocking than the shantytowns of Dhaka. India has posted legions of sequestered sentries, who have orders to shoot on sight any "unknown people" who attempt to cross their border illegally.[46]

To show how dire this situation really is, Hall and Ali Khan get permission to walk into the no man's land between the two border crossings. As they walk briskly and cautiously along, their sense of fear is palpable. They talk about how many hidden rifles of Indian sharpshooters may be focused on them at that very moment. Becoming increasingly afraid, they turn around and go back before they reach the Indian border.[47] As you watch the piece, you can sense the terror exhibited in the narrators' movements.

The point of this episode is frighteningly clear. As the sea continues to rise, millions of Bangladeshis are going to be displaced, without any place to go within their own country. Desperate to find a home for their families, they are going to flow toward the border with India. India's prime minister, Narendra Modi, has publicly voiced his acceptance of migrants from Bangladesh—but only if they are Hindu. India's leftist politicians express distress over this tendentious policy, but it is not likely to change. The stipulation is rooted in a long-standing Indian discrimination against Muslims.[48] And Bangladesh's population is more than 90 percent Muslim.

So what is going to happen when thousands of Bangladeshi Muslims descend on India looking for a fresh start? This kind of pressure cooker environment is primed for intensive upheaval. There is a direct relationship between resource scarcity and conflict, which history has shown is followed by an explosion of violence.[49] Will the result of this tension at India's borders be a massive slaughter of men, women, and children at the hands of the border guards and backup military troops? Or will violent fighting bring tremendous losses of life on both sides of the border? It's all too disturbing to even contemplate.

INCREASED STORM INTENSITY

It seems pretty straightforward—the higher sea levels rise, the stronger the storm surges associated with extreme weather events like hurricanes and typhoons will be. During the past ten years, we have seen direct evidence of this happening both here in the United States and around the world. Whether we are talking about the loss of life or loss of structures—or in most cases both—it is surges of water, not sustained winds or gusts, that cause the most damage during severe coastal storms.

2013 Superstorm Sandy

The so-called "Superstorm Sandy" that wrecked the northeastern coast of the United States—making landfall at eight p.m. on October 29, 2013—was the second costliest storm (an estimated $75 billion in damages) ever recorded in US history.[50] Sandy was also the largest Atlantic hurricane on record, with winds spanning 1,100 miles.[51] The storm killed an estimated 230 people in eight countries, including more than 100 in the United States.

So what caused this enormous amount of damage to homes, businesses, and other facilities? First, the storm hit on the day of the maximum monthly high tide (full moon), which was coupled with the Atlantic Ocean's surface being a foot higher than normal because of sea-level rise associated with climate change. This produced a record storm surge—13.88 feet above the average of the daily lowest tide of the month, as measured at the Battery in Lower Manhattan. Harvard geologist Daniel P. Schrag called Hurricane Sandy's thirteen-foot storm surge an example of what will be, by midcentury, the "new norm on the Eastern seaboard."[52]

Next, the warming atmosphere resulted in abnormally high offshore sea-surface temperatures for late October along the East Coast of the United States—more than 3.0 C (5.4 F) above normal. As we have previously discussed, the capacity of the atmosphere to hold water increases in lockstep with rising temperatures. This leads to higher rainfall and stronger storms.

Finally, as they move north, Atlantic hurricanes typically are forced east and out to sea by the Prevailing Westerlies. In Sandy's case, a ridge of high pressure over Greenland blocked the Prevailing Westerlies from pushing the hurricane out to sea. This resulted in a negative North Atlantic Oscillation, which formed a kink in the jet stream and caused it to double back on itself off the East Coast of the United States.[53] Sandy was caught up in the jet stream's northwesterly flow. The blocking pattern over Greenland also stalled an arctic front that combined with Sandy. According to Mark Fischetti of *Scientific American*, the jet stream's unusual shape was caused by the melting of sea ice.[54]

2005 Hurricane Katrina

Hurricane Katrina was the costliest natural disaster in US history, as well as one of the five deadliest hurricanes. The storm is currently ranked as the third most intense US land-falling tropical cyclone, behind only the 1935 Labor Day Hurricane and 1969's Hurricane Camille. Overall, at least 1,833 people died in Katrina and subsequent flooding. This made Katrina the deadliest US hurricane since the human-made travesty called the Lake Okeechobee Hurricane in 1928.[55] Katrina's total property damage was estimated at $108 billion, more than four times the damage wrought by Hurricane Andrew, which roared across the southern Florida peninsula in 1992.[56]

Katrina caused severe destruction along the Gulf Coast from central Florida to Texas, much of it due to the extreme storm surge. The most significant number of deaths occurred in the city of New Orleans. There, most of the city flooded when the levee system—that was built to protect this mostly-below-sea-level urban area from just such an occurrence—failed catastrophically, in many cases just hours after the storm had moved through the city proper into area further inland.

Overall, 80 percent of the city of New Orleans flooded, with some areas submerged in over ten feet of water.[57] However, the worst property damage occurred in coastal areas, such as the Mississippi beachfront towns where over 90 percent of the landscapes were inundated. Boats and casino barges rammed buildings, pushing cars and houses inland. In some cases, floodwaters extended as far as twelve miles inland from the gulf beaches.[58]

While it's difficult to say for certain that climate change contributed directly to Hurricane Katrina's destructive and deadly force, one thing is certain: the warming atmosphere is going to increase the future severity of oceanic cyclones—called "hurricanes" when they form in the Atlantic Ocean and Gulf of Mexico and "typhoons" when they occur in the Pacific Ocean. Warmer air temperatures mean warmer water temperatures, and warm water is what fuels the intensity of cyclonic events such as Sandy and Katrina.

As reported by Daniel Stone, writing for the *National Geographic News* on March 19, 2013, "Using modeling data focused on the conditions in which hurricanes form, a group of international researchers at Beijing Normal University found that for every 1.0 C (1.8 F) rise of Earth's temperature, the number of hurricanes in the Atlantic that are as strong as or stronger than Hurricane Katrina will increase twofold to sevenfold."[59] The study concluded that the global warming that occurred during the twentieth century resulted in a doubling of the number of debilitating storms such as Katrina. But if the earth continues to warm throughout the twenty-first century, the frequency of Katrina-like storms could increase by as much as 700 percent. This means that the US Atlantic coastline could be impacted by a Category 5 hurricane nearly every year![60]

2013 Typhoon Haiyan

Typhoon Haiyan was one of the strongest tropical cyclones ever recorded. Blasting ashore on November 8, 2013, the monster storm devoured entire sections of Southeast Asia, particularly in the Philippines.[61] At least 6,300 people died in the Philippines alone.[62]

Writing for the *Guardian*, John Vidal and Damian Carrington reported that:

Just as the world was beginning to take in the almost unimaginable devastation wrought by Typhoon Haiyan, a young Filipino diplomat, Naderev Sano, was getting ready to lead his country's negotiations in the UN climate talks in Warsaw, Poland. Yeb, as he is known, is a scientist and head of his country's national climate commission and had flown out of Manila just hours before the vastness of Haiyan had become apparent.

By Monday morning, Sano knew that the Philippines had been struck by possibly the strongest storm ever measured, killing many thousands of people and making millions homeless. He took the floor and, with some trepidation in front of the delegates of 190 countries, gave an extraordinary, passionate speech in which he clearly linked super typhoon Haiyan to manmade climate change and urged the world to wake up to the reality of what he said was happening from Latin America to Southeast Asia and the US. He lambasted the rich countries, and dared climate-change deniers to go to his country to see for themselves what was happening.

When he sat down, sobbing, he was given a standing ovation.[63]

2015 Snowmageddon in Boston

Anyone living in Boston during the "winter of their discontent," 2014–2015, will likely tell you that they had never experienced anything like this before. This was especially true during the month of February, when the city and its suburbs were hit with almost sixty-five total inches of snow, obliterating the previous monthly record by more than twenty inches. So how can Earth be warming, if winters are unlike anything anyone has ever seen before?

Here's what Penn State climate researcher Michael Mann had to say about the 2014–15 winter in New England in an article for the *Washington Post*: "Sea surface temperatures off the coast of New England . . . [were] at record levels, 11.5C (21F) warmer than normal in some locations. . . . There [was] a direct relationship between the surface warmth of the ocean and the amount of moisture in the air. What that [meant was] that this storm [was] feeding off these very warm seas, producing very large amounts of snow as spiraling winds of the storm squeeze[d] that moisture out of the air, cool[ed] it, and deposit[ed] it as snow inland. Warmer oceans also increase the temperature contrasts that winter storms encounter when they hit the East Coast—and this ups their strength."[64]

In the same article, Kevin Trenberth, a climate expert at the National Center for Atmospheric Research (NCAR), further clarified the issue: "Heavy snows mean the temperature is just below freezing, any cooler and the snow amount would be a lot less. . . . Warmer waters off the coast help elevate winter temperatures and contribute to the greater snow amounts. This is how global warming plays a role."[65]

Worldwide Danger

While the US media certainly adheres to the unfortunate journalistic philosophy of "if it bleeds, it leads," they often tend to be isolationists when it comes to even mega-disasters in the world's developing nations. For example, we would bet that not even 1 percent of the US population was aware of the devastation and death caused by Typhoon Haiyan in the Philippines or Cyclone Nargis, which killed at least 85,000 people in Myanmar. Imagine if either of these events had happened in the United States. The media would have been all over it for months. But countries like the Philippines and Myanmar sometimes merit barely a one or maybe two-day blip on a US network's radar screen.

We bring this up because climate change is causing increases in extreme weather events all over the world, not just here at home. To gain a full understanding of the magnitude of these impacts, we have to pay attention to what is happening worldwide, especially in developing countries, where the death and destruction may be magnified due to a combination of primitive building construction and unsophisticated advance-warning systems.

INCREASED WILDFIRES AND OTHER FOREST IMPACTS

In another episode of *Years of Living Dangerously*, former California governor Arnold Schwarzenegger is embedded with a team of Hotshots—

an expertly trained firefighting crew—battling a wildfire that is threatening a small Montana town. While he is listening in on the team's plan of attack for the day, Schwarzenegger comments, "There is no longer a wildfire season in the American West, wildfires now happen all year round. As governor I have seen the tremendous changes over the last few years; the amount of land that we have lost, the trees that we have lost, the homes that we have lost, lives that have been lost, and it is due to a large extent to global warming."[66]

The fact that western wildfires now can—and, in fact, do—happen during any month of the year is clearly tied to climate change. The warming atmosphere is drying out the land and causing massive droughts all over the western United States. During 2015, record high daily, monthly, and annual temperatures were recorded as far north as Boise, Idaho, and Billings, Montana.[67] Also, record low snow packs were measured in nine western states, including throughout the High Sierra and Cascade Mountain Ranges.[68] All these factors create prime kindling for lightning strikes to ignite tinder-dry understories into canopy-high blazing infernos.

Loss of Permafrost

While a scorched landscape exhibits a myriad of problems intrinsic to fire damage—loss of homes, loss of land, loss of timber, and so on—it may also cause a loss of permafrost.[69] Permafrost—or permanently frozen soil—contains 1,700 gigatons of carbon—over twice what is currently in the atmosphere.[70] An accelerated loss of that permafrost may significantly amplify global warming, create radical changes in ecosystem structure, and cause infrastructure damages.[71] All three of these consequences from melting permafrost are troubling, but the first point—that lost permafrost equals amplified climate change—gives us the greatest pause. If a warming planet releases the massive amount of carbon that is currently safely ensconced in the permafrost, the release of this carbon will exacerbate climate change.[72] This positive feedback loop—thawing permafrost releases carbon, which causes the global temperature to rise, which thaws additional permafrost, and so on—is a process called the permafrost carbon feedback.[73] To be blunt, if the permafrost does begin to thaw and release carbon due to this positive feedback, we will spiral toward serious, inexorable climate change. The permafrost carbon feedback will be one of several irreversible processes—along with dry-season rainfall reductions in several regions, and sea-level rise[74]—to which we cannot apply the brakes.

This feedback loop could cause global warming to occur 20 to 30 percent faster than fossil fuels alone, yet it has not been accounted for in the climate modeling from the IPCC's Fifth Assessment report.[75] The report recognizes permafrost carbon feedback, but the IPCC lacks confidence in the veracity of the science of this phenomenon—the field of study is just too new and the outcomes are not well construed.[76]

While it is well understood that fire clears the thick layer of "duff"—the blanket of twigs, needles, moss, and leaf litter that insulates the forest floor, keeping the permafrost layer cold[77]—it is less clear how this will affect carbon sequestration and the permafrost. A fire will burn through the duff, creating a blackened landscape and leaving the soil bare and exposed to the sun's warming rays; black ground absorbs more heat than the duff, and the duff insulates the ground, so the permafrost will be compromised, releasing carbon. Additionally, there will be an immediate release of carbon from the duff when it burns.[78] Ecologist Jon O'Donnell told Sarah Zhang that experts don't know what might happen if all that carbon gets released. "It could be fine," Zhang wrote, after her conversation with O'Donnell. "The trees that spring up immediately after a fire—hardwoods like aspen—might sequester more carbon than the black spruces that they replace. . . . But if that doesn't happen and the permafrost becomes less of a carbon sink, more carbon in the atmosphere could mean more extreme weather events, like the dryness enabling the current fires."[79]

As Joseph Romm describes it in his book, *Climate Change: What Everyone Needs to Know*, "The Arctic acts like a freezer—a very large carbon freezer—and the decomposition rate is very low."[80] The insulation that the duff provides for the permafrost is very important in maintaining this low rate of decomposition. Burning the duff is like ripping the blanket off of the permafrost[81], or as Romm describes it, "We are in the process of leaving that freezer door wide open. The tundra is being transformed from a long-term carbon locker to a short-term carbon unlocker."[82] Due to this, there is a large faction of scientists, Joseph Romm included, who believe the IPCC projection of future warming represents a gross underestimation of impending climate impacts.[83]

Proliferation of Bark Beetles

Another facet of the ongoing desiccation (drying out) of western forests is the proliferation of bark beetles—an insect no bigger than a grain of rice. The longer summers enable these destructive pests to reproduce up to twice each year and kill trees with their toxic secretions.

For centuries, scientists who manage the world's forests have been fighting prolonged battles with the devastation wreaked by bark beetles. But the scale of the outbreaks of these tiny terrors in North America has been unprecedented in recent decades. Since the 1990s, almost 30 billion conifers—from Alaska to Mexico—have been lost to enormous bark beetle swarms.[84]

Just by itself, the mountain pine beetle has killed enough trees to cover the entire state of Washington—a mind-blowing total of more than 44,800,000 acres (70,000 square miles). A pair of recent studies clearly demonstrate how climate change is contributing to this seemingly unstoppable beetle assault and how much worse things are likely to get as our Earth continues to warm.[85]

Published in the journal *Ecology*, the first study shows how severe drought can produce population explosions of these insatiable wood chompers and how the rapacious tree-killing is likely to continue even if the drought subsides.[86] The second and more obvious effect of climate change on bark beetles is based on a study published in *Proceedings of the National Academy of Sciences*. These findings prove that as our world continues to warm, the beetles will wreak their havoc higher and higher up our mountainsides, killing vast numbers of trees that have limited resistance to their infestations.[87]

DROUGHTS AND FAMINE

As you might expect, warming atmospheric temperatures associated with climate change are creating life-sapping drought conditions all over the world. This is especially true on the west coast of the United States, where years-long drought now covers 97 percent of the state of California. In 2015, nearly half of the state fell into the worst category of "exceptional drought"—this number has since been downgraded to 21 percent.[88] Although severe civil unrest is unlikely in California, tension is indeed growing between political leaders in northern and southern parts of the state over who gets the increasingly scarce water from rivers, underground aquifers, and snow melt, all of which are declining.

In fact, climate change is already inflicting its triple whammy of misery—no fertile ground, no water, and, eventually, no food—on many other parts of the world as well. According to Terrell Johnson, reporting on the Weather Channel's website, "Brazil, North Korea, Puerto Rico and South Africa all are in the grip of their worst drought in years or even decades, situations that threaten potentially dire consequences for millions of their citizens."[89]

Here in the United States, we are fortunate to be able to cover our bases—at least for now—and still feed ourselves, albeit sometimes at higher grocery store prices than we prefer to pay. Many developing countries are not so lucky. When dramatic drops in production—especially grain crops—occur in major agricultural regions, it severely impacts the countries involved and their human populations. This—in turn—leads to widespread famine, severe hunger, and loss of life, especially for infants and young children. As you will read next, prolonged drought may also breed civil unrest and—in extreme cases—revolution.

CIVIL WARS AND REVOLUTIONS

As we have discussed, climate change can make storms stronger and droughts longer. But can it cause war? According to several sources, the answer to this question is, unfortunately, an emphatic "yes."

In yet another episode of the Showtime series *Years of Living Dangerously*, *New York Times* columnist Thomas Friedman travels to the Middle East.[90] His mission is to examine what part climate change played in the ongoing civil war that has been ripping Syria apart. This war has resulted in thousands of deaths at the hand of the ruling president Bashar al-Assad's government and sent millions of refugee families fleeing toward the country's borders with Lebanon, Jordan, Iraq, and Turkey.

Friedman reports that climate change is indeed a contributing factor to this horrible humanitarian disaster. When the drought in Syria became so severe that young men could no longer feed their families, they demanded relief from the Assad government. When none came, they decided to take matters into their own hands and began organizing a movement to overthrow the dictatorial regime.[91]

As Friedman reported in a 2013 op-ed piece in the *New York Times*:

> Between 2006 and 2011, some 60 percent of Syria's land mass was ravaged by drought and . . . it wiped out the livelihoods of 800,000 Syrian farmers and herders. . . . "Half the population in Syria between the Tigris and Euphrates Rivers left the land" for urban areas during the last decade said [Syrian economist Samir] Aita. And with Assad doing nothing to help the drought refugees, a lot of very simple farmers and their kids got politicized. "State and government was invented in this part of the world, in ancient Mesopotamia, precisely to manage irrigation and crop growing," said Aita, "and Assad failed in that basic task."[92]

Of course, Assad wasn't about to give in or relinquish any power, so he came out swinging with everything he had. Unfortunately this included biological weapons—such as deadly Sarin nerve gas—that sickened and wiped out thousands of innocent Syrian men, women, and children.[93]

As author Mark Fischetti explains in *Scientific American*, a study in the *Proceedings of the National Academy of Sciences* supports Friedman's findings: "Drought in Syria, exacerbated to record levels by global warming, pushed social unrest in that nation across a line into an open uprising in 2011. . . . Drying and drought in Syria from 2006 to 2011—the worst on record there—destroyed agriculture, causing many farm families to migrate to cities."

Richard Seager, a coauthor of the study and a climate scientist at Columbia University's Lamont-Doherty Earth Observatory, told Fischetti that the cities were overflowing with both these Syrians and refugees fleeing the war in Iraq, creating serious social stress.[94]

According to Seager's findings, the climate all throughout the Middle East is becoming drier and hotter because of climate change. Water resources will become more limited, making agriculture more difficult, and the risk of conflict will rise due to these stresses.[95]

Research from around the world suggests that climate change–related rising temperatures and drought will continue to increase the potential for violent conflicts. Cases in point include the growing likelihood of civil war in sub-Saharan Africa and the 2011 revolution in Egypt.[96]

INTERNATIONAL SECURITY THREATS

The world's Public Enemy Number One is now the Islamic State of Iraq and Syria, otherwise known as "ISIS," or the "Islamic State." They have taken over the terrorist reins from al-Qaeda because they are much better organized and skilled at recruiting. In fact, word has it that they are very much set up like a high-end profitable corporation, with professional money managers, fundraisers, marketers, and recruiting specialists. They have their tentacles spread out all over the world—even including cells in the United States—using the Internet to tease and tempt disaffected youth to come and join their cause in destroying the "evil capitalistic infidels."

So exactly how did ISIS become so all-powerful so quickly in the disturbing dregs of organized terrorism? Of course, many factors have contributed to the "success" of ISIS, and one of these factors is certainly climate change. As economies wither under increasing droughts and famines in developing countries, people—especially young men—become desperate in seeking ways just to survive.

Then along comes ISIS, offering them a place to become professionally trained, receive regular paychecks, and serve "in the name of Allah" for a noble cause. Moreover, they are told that there can be no more honorable death than to give their lives fighting against infidel capitalists.

The United Nations Intergovernmental Panel on Climate Change (IPCC) has predicted that human influences may make drought in the Middle East progressively worse. This means the world's hotbed of anti-capitalist sentiment could soon become the breeding ground for more ISIS-type terrorist organizations.

Richard Seager's study confirmed that finding. "Climate change is very much a cause of concern for national, regional and international security," Seager told *Scientific American*.[97] In a speech on October 13, 2014, the US Department of Defense Secretary Chuck Hagel noted that "climate change is a trend that will affect national security."[98] Hagel deemed climate change a "threat multiplier," citing the intensification of challenges such as global instability, poverty, and conflict that would lead to disputes and potentially terrorism, pandemic disease, and shortages of basic necessities.[99]

ECOLOGICAL IMPACTS

Inundation and Loss of Coastal Wetlands

As we have previously discussed, we fully recognize the tumultuous and overwhelming impacts that climate change will have on human populations. If we maintain our current course over time, hundreds of millions of people will be ripped from their homelands and relocated—if that will even be a possible option.

As we are both biologists, we are also extremely concerned about the brutal impacts that unmitigated climate change is having—and will continue to have—on our coastal resources, oceans, and seas.

The incredibly varied ecology of our nation features some 88,000 miles of tidal shorelines that are embedded with vast areas of brackish wetlands.[100] Our major rivers flow into the oceans and Gulf through more than one hundred different estuarine systems. Serving as transition zones between freshwater and saltwater, these coastal habitats are among the most productive ecosystems on the face of the Earth.[101]

Unfortunately, within just a few decades, rising sea levels will likely obliterate the majority of our coastal salt marshes and seagrass beds. These areas are ecologically known as the primary "nursery areas" for the majority of life in our oceans and seas—including commercially valuable fishery and shellfish stocks. Rising oceanic tides will turn these vegetated natural resource areas into permanent open water.

I (Budd) can tell you from twenty-five years of experience as a professional wetlands scientist, changing vegetated wetlands into open water is considered as much of an impact as actually filling these same wetlands. When this transformation is made, many of the vital functions of the vegetated wetlands—wildlife habitat, pollutant cleansing, groundwater recharge, flood control, storm surge protection—are lost forever.

The state of Louisiana is home to about 40 percent of our nation's coastal wetlands. In the July–August 2015 issue of *Audubon* magazine, David Gessner writes, "There are times when you can't be sure of much in the Gulf [of Mexico], but amid all the confusion, one thing is undeniable: Habitat is going away, and it is going away fast, the land sinking and sea rising like nowhere else on earth, to the point where organizations working in the region report that the River Delta loses about 'a football field' of coastal wildlife habitat every hour. We are not talking about geological time here but about whole marshes and small islands that have disappeared since I last visited."[102]

Our nation's largest estuary—the Chesapeake Bay—provides habitat for more than 3,600 species of plants, fish, birds, and other animals.[103] However, if global warming proceeds unabated, rising sea levels will continue to reshape the region's coastal landscape significantly, wiping out thousands of acres of bordering wetlands and their critical nursery and rearing areas for oceanic fish, crustaceans, and mollusks.

The Delaware Bay is our second-largest estuarine system. University of Maryland scientists have already proven that this system is losing wetlands to sea-level rise. In fact, their research shows that many of the Delaware Bay's coastal marshes could be permanently inundated by the end of the twenty-first century.[104]

In the coastal areas of Florida's Everglades National Park, mere inches of sea-level rise can result in miles of habitat change. Below sea level at the coast, Everglades' wetlands rise a miniscule two inches per mile as you move north up the state's peninsula. This means that every two-inch increase in sea level will cause the loss—through saltwater encroachment—of a mile-wide swath of freshwater Everglades' wetlands.[105]

This initial saltwater intrusion starts a domino effect, essentially shifting Florida's remarkable coastal ecology northward. Freshwater areas further upstream become more vulnerable to saltwater intrusion from storm surges. These storm surges transport mangrove seeds that then become established and begin the process of permanent habitat alteration.[106] As sea-level rise continues over the years, the salt-intolerant sawgrass marshes—Marjory Stoneman Douglas's "rivers of grass" for which the Everglades are so well-known—gradually die off and give way to more saline and mangrove-dominated habitats.[107]

Ocean Acidification and Loss of Coral Reefs

When I (Budd) was a young boy back in the 1950s, the general, misguided opinion was that our oceans couldn't ever be polluted because "they are so vast." Accordingly, it was considered okay to dump whatever you wanted into the ocean because no one would ever see any measurable consequences of your actions. Besides, it was far better to get rid of unwanted stuff offshore than to bury it somewhere in our relatively limited land masses, where it could come back to haunt us.

My (Budd's) dad, who was an "environmental activist" before anyone had ever heard of the term, insisted that the philosophy that the ocean could contain all our waste was ludicrous. He loved saltwater and always maintained that our oceans and seas had to be protected at all costs.

If he were still alive today, I'm sure he would be thinking—if he didn't just say it outright—"See . . . I told you so!" Our oceans—as we currently know and love them—are in grave peril. And the really sad part is that very few people even realize what is happening.

According to the Smithsonian's Ocean Portal Team, "Ocean acidification is sometimes called 'climate change's equally evil twin,' and for good reason: it's a significant and harmful consequence of excess carbon dioxide in the atmosphere that we don't see or feel because its effects are happening underwater. At least one-quarter of the carbon dioxide (CO_2)

released by burning coal, oil, and gas doesn't stay in the air, but instead dissolves into the ocean."[108]

The result is a harrowing scenario that's rapidly making seawater much more acidic. This bodes potential disaster for all the creatures dwelling in our oceans—from tiny plankton and copepods, up to oyster beds and massive coral reefs, then ending near the top of the food chain with salmon, sea otters, seals, sharks, and whales. The final and ultimate impact will—of course—be on the people who depend on the oceans for their way of life and food sources.[109]

Ocean acidification is really just simple chemistry. When CO_2 dissolves in seawater, two important things happen. First, the pH of seawater drops, making it more acidic. Then, the increasingly acidic seawater binds up carbonate ions, making them less abundant. This, in turn, causes the populations of many sea-dwelling, shelled organisms—including corals, oysters, and mussels—to rapidly decline since they need carbonate ions to build their shells.[110]

Our oceans have absorbed approximately 525 billion tons of CO_2 from the atmosphere since the Industrial Revolution started. The current rate of oceanic absorption is an estimated 22 million tons per day.[111] At first, scientists thought this might be beneficial since it leaves less CO_2 in the air to warm the planet. But in the past decade—just as my (Budd's) dad could have told them—they've realized that this slowed warming has come at the great cost of changing the chemistry of our oceans. Ocean water has become 30 percent more acidic just in the past two hundred years. This represents a faster change in oceanic chemistry than anything that has occurred in the past 50 million years.[112]

As reported by the Smithsonian's Ocean Portal Team, "Scientists formerly didn't worry about this [CO_2 absorption] process because they always assumed that rivers carried enough dissolved chemicals from rocks to the ocean to keep the ocean's pH stable. (Scientists call this stabilizing effect 'buffering.') But so much carbon dioxide is dissolving into the ocean so quickly that this natural buffering hasn't been able to keep up, resulting in relatively rapidly dropping pH in surface waters."[113]

The Ocean Portal Team continues, with this rather ominous assessment, "As those surface layers gradually mix into deep water, the entire ocean is affected. Such a relatively quick change in ocean chemistry doesn't give marine life, which evolved over millions of years in an ocean with a generally stable pH, much time to adapt."[114]

The Center for Biological Diversity provides this alarming summary: "Two critically important planetary ecosystems, coral reefs and the polar regions, are on the front lines of the acidification crisis. Coral reefs critical to the protection of coastlines across tropical and subtropical parts of the world will disappear as the rate of erosion exceeds the rate at which corals can rebuild—with staggering repercussions for related ecosystems like mangrove and seagrass."[115]

Our north and south polar regions may be hit even harder. It is here that vast quantities of marine plankton—which form the base of the oceanic food chain—may be lost. As an example, pteropods (tiny sea snails that are also known as "sea butterflies") are a critical staple in the diet of commercially important fisheries—such as salmon, mackerel, herring, and cod—as well as baleen whales.[116] But marine plankton numbers are rapidly declining because the increased acidification is also dissolving their shells. The Center for Biological Diversity again offers this dire assessment: "It's clear that this crisis may spin out of control, with devastating effects on vast numbers of species, from small shell-building oysters and reef fish to crabs, whales, and sea otters."[117]

While it's difficult to say exactly what the ever-increasing dumping of CO_2 will ultimately do to our oceans, one thing is for certain: if we don't take dramatic actions soon—within the lifetimes of our children and grandchildren, or even sooner—our oceans and seas will be vastly different places from what we all know today. And Budd's dad would definitely not be happy!

Saltwater Encroachment into Freshwater Systems

What do you think is the most valuable natural resource on Earth? Gold? Diamonds? Oil? Our bet is if you ask anyone who has ever been really thirsty they will not hesitate with their answer of clean, cold fresh water. Of all the immense volume of water here on Earth, only 2.5 percent of it is fresh and drinkable by humans. Without access to clean, potable water, we would all soon die. Furthermore, if we do not have water, we also will not have food, because everything on Earth—from vegetables and fruits to chickens and beef cattle—requires water to grow.

Something else to think about regarding impacts from climate change: when sea levels rise, what is going to happen to our freshwater drinking and irrigation supplies in coastal areas, where the bulk of Earth's human population lives—40 percent within sixty-two miles (one hundred kilometers)?[118]

With regard to human uses of freshwater, sea-level rise often produces several calamitous results. Saltwater intrusion can contaminate groundwater drinking supplies, destroy irrigation sources, and inundate agricultural fields. Because of their gently sloping topography, low-lying coastal areas are particularly vulnerable to freshwater contamination by saltwater intrusion.[119]

Potential contamination of freshwater aquifers by intruding saltwater has always been of critical concern to hydrogeologists and wetland scientists. If the saltwater contamination is severe, the aquifer may no longer have any potential for human use. Even when it's determined that an aquifer can be saved, highly sophisticated and expensive technology must be employed to treat the aquifer water and make it potable again.

Instead of talking generalities here, let's consider what impacts sea-level rise could have on freshwater aquifers in South Florida. As we have already discussed, most of Florida is underlain by limestone rock that's honeycombed like a beehive, known as Karst topography. So no matter what we do to try to stop saltwater intrusion associated with sea-level rise and storm surges, the freshwater aquifers in Florida's coastal areas are going to be contaminated.

If the seawater storm surges are not contained on the surface, then they will flow into storm-water management systems and eventually infiltrate into the underlying groundwater. If we structurally engineer our coastline counties to hold back the storm surges, then the intruding saltwater will just move under the mainland's porous subsurface, eventually finding pathways through the limestone and bubbling up into and contaminating the aquifers from underneath.

In fact, this process is currently happening. According to December

2010 report titled "Climate Change and Sea Level Rise in Florida," prepared by the Florida Oceans and Coastal Council of Tallahassee, Florida, "Florida's Biscayne Aquifer, the principal water supply to southeastern Florida and the Florida Keys, is recharged by rainfall and the freshwater Everglades. Surficial coastal aquifers are already experiencing saltwater intrusion. Rising sea level will increase the hydraulic backpressure [causing a reversal of normal flow] on coastal aquifers, reduce groundwater flow toward the ocean, and cause the saltwater front to move inland, thus threatening to contaminate water-supply wells in coastal areas with seawater."[120]

Of course, saltwater encroachment can also contaminate valuable surface resources. In some cases, it's a matter of salt water from sea-level rise dramatically altering the salt-fresh water mix in brackish estuaries and causing dramatic shifts in the local fishery populations.

Here's an example: Living here in Tallahassee, I (Budd) know the Apalachicola River estuary and environs very well. My wife and I regularly take day trip excursions to beautiful St. George Island State Park—annually listed among the top ten beaches in the United States. After having a picnic and then walking the pristine white "sugar sand" beach, we typically head into the town of Apalachicola for an early dinner. A dozen oysters on the half shell—which must come straight out of neighboring Apalachicola River estuary—are always first on our list of palatable delights. So imagine our surprise when we watched episode four of *Years of Living Dangerously* (*YLD*) that featured Apalachicola Bay and the fact that the area's "world-famous oysters" had all but disappeared.[121]

For decades, the Apalachicola River's estuary in Florida's Panhandle had a reputation for producing some of the tastiest oysters on the planet. As author Dennis Pillion describes it: "In 2012, Apalachicola Bay oyster fishermen harvested more than 3 million pounds of oyster meat, roughly 92 percent of the Florida oyster harvest and 10 percent of the harvest nationwide." Then something very dramatic happened. Just a year later in 2013, the oyster harvest dropped to around 1 million pounds and NOAA declared the Apalachicola Bay "fishery disaster."[122]

This point was vividly driven home by a multigenerational oysterman who took the actor/correspondent Ian Somerhalder and the *YLD* crew out to what had been his most productive oyster reef. For decades, this oyster bar had been providing enough oysters to support his family. But when he dropped his dredge into the water to demonstrate his technique, he came up with essentially nothing—just a few lone oysters in the bottom of the netting instead of the hundreds he had been used to getting with each drag. He then lamented that this was the way it had been for a while, saying that he considered the oysters reefs all over Apalachicola Bay to be dead and gone forever.[123]

Shannon Hartsfield, local president of the Franklin County Seafood Workers Association, recognizes that shutting down the Apalachicola oyster harvest may be the only hope for the oyster population to recover.[124] The community depends on the jobs, which will disappear with the oysters, and the ecosystem depends on the filtering services that the mollusks provide the bay.

There are two primary reasons that the oysters have disappeared from Apalachicola Bay, and they are both associated with ongoing climate change. The first is the drought-related overconsumption of water in the upper reaches of the Apalachicola River by the state of Georgia. (Note: This situation is an ongoing controversy that has yet to be resolved between the states of Florida and Georgia.) The increasing upstream (in Georgia) use of the fresh water in the river means that less fresh water is coming into the bay. The second reason is that sea-level rise is increasingly pushing salt water further up into the bay. Taken together, both of these factors are increasing the levels of salt water in Apalachicola Bay past the salinity tolerance limits for the production and growth of oysters.

FOSSIL-FUEL EXTRACTION ACCIDENTS

From the perspective of climate change leading to more frequent and devastating disasters associated with fossil fuels, we ask you to consider the following cycle of events: Our primary dependence on fossil fuels is causing us to look for, develop, and extract oil and gas from more and more places that are less and less safe. When a problem arises at one of these unsafe locations—for example, the deep water of the Gulf of Mexico—we may not have the technology in place to handle it. When we don't have the technology in place to handle even routine problems, people can die, the economy can be devastated, and millions of plants and animals may be wiped out.

The infrastructure that extracts and delivers fossil fuels can breakdown, often causing much damage. According to Benjamin K. Sovacool, director of the Danish Center for Energy Technology: "279 major energy accidents occurred between 1907 and 2007—causing 182,156 deaths with $41 billion in property damages."[125]

2010 BP Oil Platform Explosion and Oil Gusher

By now, almost everyone knows the horrific backstory for the *Deepwater Horizon* oil rig explosion and oil spill. An accidental explosion on April 20, 2010, destroyed the oil platform and killed eleven workers. Since we talk at length about this calamitous event as part of the US Environmental Timeline in chapter 21 of part 3, we won't repeat the details here. Suffice to say, that while the accident itself might be understandable, what happened afterward—allowing uncontrolled and untreated crude to flow for months—was an unforgivable atrocity of corporate and bureaucratic chaos.

2010 Massey Energy Coal Mine Explosion

On the afternoon of April 5, 2010, high methane levels in the Upper Big Branch Coal Mine caused a massive explosion taking the lives of twenty-nine of the thirty-one employees working in the mine's depths.[126] Located in the town of Montcoal in Raleigh County, West Virginia, this mine was operated by Performance Coal Company, a subsidiary of Massey Energy. The accident was the worst in the United States since 1970, when thirty-eight miners were killed at Finley Coal Company's No.15 and No.16 mines in Hyden, Kentucky.[127]

A state-funded independent investigation found Massey Energy was directly responsible for the blast, owing to flagrant safety violations.

After the incident, the federal Mining Health and Safety Administration (MHSA) issued 369 citations for safety violations, resulting in $10.8 million dollars in penalties—an exceptionally paltry price to pay for twenty-nine lives.[128]

2013 Lac Megantic Rail Disaster

Patrons at the Musi-Café in the quaint Canadian village of Lac-Mégantic in the Eastern Townships of the Canadian province of Quebec were partying at just past one in the morning on July 6, 2013, when their revelry suddenly turned to horrific mayhem and death. Once again, the extreme dangers associated with the fossil-fuel industry—in this case, the transport of crude oil—were blatant.

The Lac-Mégantic Rail Disaster occurred when an unattended seventy-four-car freight train carrying crude oil rolled downhill and derailed. The resulting fireball, created by the explosion of multiple tank cars, killed forty-two people and destroyed more than thirty buildings in the town's center—roughly half of the historic downtown area.[129]

2015 Santa Barbara Oil Spill

Despite what the fossil-fuel companies will admit to, oil spills and leaks occur all the time in the industry. Another oil spill occurred in Santa Barbara, California, on May 19, 2015. An earlier incident had transpired there in 1969—at the time considered the worst fossil-fuel disaster in US history. This latest spill was much smaller, with a ruptured pipeline owned by Plains All American Pipeline Company sending "only" 21,000 gallons (500 barrels) of unrefined crude into the blue waters of the Pacific Ocean.[130]

However, plenty of people living in the area will tell you that this spill was worse than the 1969 debacle, which unleashed more than three million gallons of oil into the open ocean.[131] Because this leak happened on land, it also fouled the near-shore beaches of the splendidly scenic and mostly undeveloped Gaviota Coastline.

Alberta Tar Sands: A Pending Disaster

While it's not our intent to further berate the fossil-fuel companies and the US government, it is our intent to emphasize that coal, oil, and gas extraction and production have always been—and *will always continue to be*—extremely dangerous activities.

Every minute of every day, extraction and production workers are handling volatile materials that could blow up in their faces. From coal mine explosions in the Appalachian Mountains, to oil spills on the California coast, train wrecks in the Canadian outback, and tap-water infernos in Pennsylvania, another fossil-fuel disaster is just an errant spark away from happening.

That's why the process of extracting, transporting, and then refining the Alberta Tar Sands—also known as Athabasca Oil Sands, located in northeastern Alberta, Canada—is an accident waiting to happen. Collectively, the tar sand deposits contain about 1.7 trillion barrels of bitumen, which is comparable in volume to the world's total proven reserves of conventional-grade petroleum. That's said to be enough crude oil to supply North America's energy needs for the next hundred years.[132] Is it worth putting human lives at risk, both directly due to the hazards inherent to the process, and indirectly via the contribution of emissions to the climate-change problem?

The answer is an emphatic NO! Tar sands bitumen is not the flowing black "liquid gold" that many people think of when they think of oil. In fact, in many circles, tar sands are known to be the dirtiest and most dangerous fossil fuels ever extracted on Earth.[133] Due to their sheer volume and processing difficulties, the Alberta Tar Sands pose the greatest current fossil-fuel threat to the North American mainland. Since the Keystone XL pipeline project—proposed to transport this crude to Gulf Coast refineries—is such a critical event in the history of the US environmental movement, we refer you to chapter 21 for details.

Chapter 6

WHAT CAN WE DO TO COUNTERACT CLIMATE CHANGE?

Our ultimate goal in writing this book is to provide an array of solutions for stopping climate change. In parts two and three, we first take a look back at our nation's past environmental heroes—visionaries like John James Audubon, George Perkins Marsh, Harriett Lawrence Hemenway, John Muir, Robert Marshall, Roger Tory Peterson, Ding Darling, Marjory Stoneman Douglas, Aldo Leopold, David Brower, Rachel Carson, and Dana Meadows. We discuss the critical and often dramatic contributions each has made to the history of the US environmental movement. This includes analyzing the ways they accomplished what they did and how their practices could be applied to solving our current climate-change crisis.

Next, in part 4, we present the scientists, activists, and celebrities who are taking the lead in working on climate change—including people like Bill McKibben, Naomi Klein, James Hansen, Al Gore, Gus Speth, Thomas Lovejoy, Paul Hawken, Janine Benyus, Naomi Oreskes, Mark Ruffalo, and Leonardo DiCaprio. In many cases, we have interviewed these current climate-change heroes so that we can capture—first-hand—their thoughts and ideas on providing solutions.

Finally, in part 5, we present a carefully crafted list of the climate-change solutions that have emerged from our research and analysis in all of our book's previous sections. We fully believe that the synopsis of solutions that we provide in part 5 represents the best and brightest thoughts on how we can deal with this looming crisis and allow people to continue living sustainably on Earth without dramatically altering their existing quality of life. In fact, we are convinced that these solutions—if implemented quickly enough—will actually improve the sustainability and longevity of life for Earth's human populations. We truly hope you agree with our findings.

Part Two

THE US ENVIRONMENTAL MOVEMENT

A Story for the Ages: The Beginning through 1968

Chapter 7

IN THE BEGINNING

An American Melodrama (~12,000 BCE to 1799)

THE BERING LAND BRIDGE AND NATIVE AMERICANS

So, let's start at the beginning—at least in the United States. About 12,000 years ago, during the last ice age, much of the earth's water supply was locked in glacial ice. As sea levels gradually dropped worldwide, a land bridge emerged from the sea and connected eastern Asia with North America.[1] The first residents of what was to become North and South America lived on this land bridge, called Beringia, until sea-level rise forced them to migrate.[2] Today the land on which these first peoples made this passage lies deep beneath the Bering and Chukchi Seas.[3]

By most accounts, these original tribes had a deep and abiding respect for the natural world. They treated animals as equals. They hunted animals, of course, but many tribes believed in praying for permission from the animals' spirits before starting the hunt. They lived communally, with no property ownership. The thought of owning and selling land repulsed them.[4]

They hunted, fished, and raised crops out of necessity, to put food on their tables. They believed that the Great Spirit who created the Earth—or some version thereof—watched over them and cared for their needs. Most also believed they had a duty to protect and manage the bounty of the land provided to them.[5] These Native American tribes were then—in essence—the original heroes of the American environmental movement.

ARRIVAL OF THE EUROPEAN COLONISTS

But where you have heroes, you inevitably have villains. And the villains arrived in the form of white-skinned Europeans seeking newfound wealth and health in a wondrous land they called "The New World."

At first, the Europeans appeared to be awe-struck adventurers—landing their tiny sailing ships in unwelcoming outposts like St. Augustine (Florida, 1565), Jamestown (Virginia, 1607), and Plymouth Rock (Massachusetts, 1620). For years they suffered, trying desperately to survive in the face of famine, disease, bad weather, and mental depression. They lost many of their original numbers but staunchly persevered, learning how to live off the land and start their lives anew.

Often owing to the friendship and accommodating favors of the local Native American tribes, the European colonists were eventually able to start their roughhewn farms and villages. The native people taught the newcomers much about living off the land and making the best of what was available to them. They taught them to plant and grow crops, make bows and arrows, and hunt local game. They taught them how to make nets and capture fish in the streams and rivers. They taught them how to prepare for inclement weather, and which native medicinal plants would help their sick and elderly. In fact—without the wisdom and guidance of American Indians—the thirteen "original colonies" of the fledgling United States of America might never have seen the dawning of a new day.

It is impossible to know what our country's original citizens might have done had they known what the colonists had in store for them. How could they have anticipated that their lands and cultures would be progressively stripped away by descendants of the same people they had befriended and rescued? How they would unjustifiably become the "hated heathen"—enemies to be first battled and then banished to live in misery on remote reservations where none of their people had ever before set foot.

By understanding our link to the past, we can better see how those experiences shaped the United States and perhaps find the insight to act more wisely.

—Benjamin Kline, *First Along the River*[6]

THE AGE OF ENLIGHTENMENT AND HUMANITY'S DOMINION OVER NATURE

What the American Indians also did not understand—at first, at least—was that these newcomers from far away came with an attitude about the environment that was vastly different from their own beliefs. As we previously described, most of the native tribes believed that humans were put on earth to live in harmony with the land. But the European settlers supported the religious and philosophical foundations from their old countries. Their elders taught them that nature's bounty existed solely to serve the needs of themselves and their families.

The Age of Enlightenment and the teachings of Christianity in their homelands taught the Europeans that humans had and should maintain dominion over the natural world. And—in this new land of unlimited abundance—that meant human expansion no matter the cost to the environment.

The colonists believed that nothing in nature should impede progress. In fact, continuing struggles with the environs were expected. And technological advancements were made specifically to assist in overcoming the environment and helping humanity achieve a higher standard of living.

In his treatise on nature and the American consciousness, Joshua Johns writes, "Two primary views of the wilderness were contested: the wilderness either contained savagery and temptation which threatened the authority of the community or it represented a new Garden which could flourish with the proper cultivation by the European settlers. Although these contrasting views of the wilderness shared the goal of establishing a civilization by removing the obstacles presented by the natural environment, the state of wilderness that originally characterized the young nation eventually became the source of national pride and identity for America."[7] In many early American colonies—especially in Pennsylvania and Virginia—the wilderness represented "the Garden" and was a place that had to be taken over and controlled by humans so that their communities could be safely established.

From the colonists' perspective, the land supplied the raw materials for building society, and nature was best used for this purpose. While these outlooks varied slightly from place to place, the end goal within the colonies was always the same—"destroy the savage wilderness and make it bloom with European civilization."[8]

Soon, the settlers' deep-seated and faith-based beliefs that nothing should stand in the way of progress led them to push further west into the wilderness that bordered their colonial villages and towns along the coasts. After all, they were sent here to explore and conquer and that—by God—was exactly what they intended to do!

As opposed to the Native Americans, the white settlers believed that individual land ownership was a good thing. As their colonies grew stronger and larger, they first created villages, then the villages became towns, and the towns grew into cities. All this growth, of course, had a pronounced effect on the local Indian tribes in the eastern United States. Through a series of wars during the 1600s and 1700s, these original Americans were continually pushed deeper and deeper away from their prime lands and into remote corners of the landscape—far away from the places they knew, loved, and called home.

Chapter 8

MANIFEST DESTINY

Explore and Conquer (1800–1849)

OUR NATION EXPANDS DRAMATICALLY

Most of the colonists felt uneasy about abandoning the rough-hewn lifestyles they had worked so diligently to carve out on their small plots of land and in their rugged communities. Additionally, the dangers and uncertainties of the untamed wilderness to the west gave pause to all but the hardiest among them. For several generations, white immigrants continued to arrive—in greater and greater numbers—and join the growing settlements, most of which were either close by the Atlantic Ocean or situated on one of its principal tributaries. Fighting their way through a litany of hardships while maintaining their religious beliefs, the colonists dutifully defended their places on the landscape, eventually evolving into the "Thirteen Original Colonies" that in 1776 became the United States of America.

While many of the Founding Fathers—most notably George Washington, Thomas Jefferson, and Benjamin Franklin—were resourceful thinkers and planners, few had ideas and thoughts that varied from the theory of "Progress at All Costs" for this burgeoning new nation. For the most part, an environmental ethic did not exist—at least not one strong enough to overcome the still-dominant theological beliefs based on humanity's right to dominion over nature. This innate belief system seemed especially true given the apparent limitless bounty of natural resources that expanded to fill the horizons and beyond in all directions.

In 1803, President Thomas Jefferson capitalized on the national ambition to expand when he purchased the territory of Louisiana from the

French government for $15 million. Occupying more than 820,000,000 square miles, the "Louisiana Purchase" stretched from the Mississippi River to the Rocky Mountains and from Canada to New Orleans, and it doubled the size of the United States.[1]

To Jefferson, westward expansion was the key to the nation's health. He believed that a republic depended on an independent, virtuous citizenry for its survival and that independence and virtue went hand in hand with land ownership, especially the ownership of small farms. "Those who labor in the earth," he wrote, "are the chosen people of God."[2] To provide enough land to sustain this ideal population of virtuous yeomen, the United States would have to continue to expand—and westward was now the only way to go.[3]

WHERE ARE THE ENVIRONMENTAL HEROES?

In the first fifty years after our grand nation was officially founded, in 1776, no past environmental heroes had yet emerged—although a few people with early conservation proclivities came close. In their 1824 *Farmer's Guide*,[4] Solomon and William Drown were the first settlers to advocate land conservation in North America. They encouraged the agrarian-based settlers—90 percent of the young country at the time—to practice contour plowing and crop rotation to control rampant soil erosion and preserve the fertility of the soils.[5]

Also in 1824, Jean Baptiste Joseph Fourier—a French physicist—became the first person to describe the "greenhouse effect" on Earth's weather. Unfortunately, Fourier had no way to test his hypothesis about the trapping of infrared radiation close to Earth's surface, and his ideas disappeared into scientific oblivion—for a while, at least.[6]

FINALLY—JOHN JAMES AUDUBON

Then finally—in 1827—our first legitimate past environmental hero came along when John James Audubon arrived on the scene, hawking his grandiose ideas for the world's first monumental natural history exposé. But that's jumping the gun a bit. Let's first go back and take a look at exactly who this Audubon fellow was and why he was trying so hard to sell his paintings of wild birds.

With his devilish good looks, flowing chestnut-colored locks, and stern visage, Audubon was an enigma in every sense of the word. Arguably the world's most prominent conservationist—at least in name alone—Audubon is usually depicted with a hunting rifle nestled in his arms.

Oddly, in the minds of many, Audubon could not write a decent sentence. He also could not draw very well, at least not when he tried sketching people. But there was one thing he was passionate about and, at this, he was very, very good. In fact, many—both then and now—consider him the best wildlife artist that ever lived. He could expertly craft meticulously detailed portraits of any wild bird. And his fondest dream was to show his magnificent avian portraits to the whole world.

Despite the eventual repute of his name in the United States, Audubon was not born as an American. He first saw the light of day on April 26, 1785, in Saint Domingue (now Haiti) as the illegitimate son of a French sea captain and his mistress.[7]

In 1803, Audubon's father sent his bright, brash teenager to the United States to escape conscription into Emperor Napoleon's army.[8] Despite many attempts, John James Audubon did not have business acumen. He tried to run a general store, a lead mine, a farm . . . all of these endeavors failed, to the point that he landed in jail.[9] He continually shirked his obligations as a businessman in favor of chasing birds in the woods. Audubon held a deep desire to paint every bird that nested in or visited the North American landscape. The fire to accomplish this burned so indomitably in his mind that he traveled throughout the eastern United States, observing and painting birds and trying every way he could think of, including painting portraits—his skills had improved by then—to raise the money to finance his ambitious goal.[10] His wife, Lucy Bakewell, worked as a tutor to support his travels during these years.[11]

Audubon even traveled to the United Kingdom and France in hopes of finding suitable financial backers. Finally, in 1827, while living in London, he connected with two talented engravers, Robert Havel Sr.

and his son, Robert Havel Jr. Soon after that, production of what is still considered by many to be the greatest natural history masterwork ever created, *The Birds of America*, was finally underway.[12]

Taking more than twelve years to complete, including several interim partial sets of plates, *The Birds of America* featured hand-colored, life-sized prints of every bird identified—at the time—on the North American continent. The publication featured 435 images presented in what was known as a "Double Elephant Folio" due to the enormous size of the paper (29 x 36 inches) required to reproduce Audubon's meticulous work.[13] The book even included six species that have gone extinct since its first publication—the Carolina parakeet, Labrador duck, great auk, Eskimo curlew, passenger pigeon, and pinnated grouse. Always a taker of copious notes, Audubon also possessed enough written material to produce a sequel tome entitled *Ornithological Biography*, which documented the life histories of birds and became a scientific treasure in its own right.[14]

Of course, Audubon's sumptuously detailed prints quickly created a sensation among nature lovers and early environmentalists all over the civilized world. And the lingering effect of his avian artistry remains unparalleled even in today's society. Millions of homes throughout the United States and around the world still have their living and dining rooms graced with boldly emblazoned, life-sized prints of birds bearing the signature of John James Audubon in the lower right corner.

Assembling *The Birds of America* required legendary strength and endurance, and Audubon epitomized the spirit of young America—when our nation's wilderness was still limitless and beguiling. Author and literary critic Lewis Mumford called Audubon "an archetypal American who astonishingly combined in equal measure the virtues of George Washington, Daniel Boone, and Benjamin Franklin" and "the nearest thing American art has had to a founding father."[15]

Audubon's influence in the field of ornithology has never been matched. The majority of his later ornithological works were elevated by his insistence on accuracy and details in his paintings. He also had a deep appreciation and concern for conservation. Many of his writings sounded the alarm about the destruction of birds and their habitats. It is fitting that today, through the National Audubon Society, we carry his name and legacy into the future of environmental conservation and natural resource protection.[16]

Although Audubon had no actual role in the organization that bears his name, there was a strong connection through his widow, Lucy Bakewell Audubon. She tutored George Bird Grinnell—one of the founders of the early Audubon Society in the late 1800s. Owing to Ms. Audubon's tutelage, Grinnell gained a deep appreciation for and thorough understanding of the sheer magnificence of Audubon's accomplishments. This led to his choosing this grand ornithologist's name as the inspiration for the organization's earliest work to protect birds and their habitats. Today, the name *Audubon* evokes everything about birds—including their habits, habitats, and conservation throughout the world.[17] In addition to the National Audubon Society, more than twenty-five separate entities in the United States—schools, roads, bridges, parks, sanctuaries, and so on—bear his name.[18]

So what can we take away from our first past environmental hero, John James Audubon, that will help us solve the climate-change crisis? First, to accomplish your goals, you may have to do some things you would prefer not to do. Many people know that Audubon was the foremost bird artist and one of the most ardent conservationists of his day. But what you might not know is that Audubon was also one of the most skilled bird hunters in the United States. Not because he enjoyed killing living things or needed trophies in his den but because—in the early to mid-1800s—it was the only way he could acquire the specimens he required for producing his art.

Next is the skill of mental dedication and endurance simply known as good old-fashioned stick-to-itiveness. While enduring the slings and arrows continually hurled at him by critics, Audubon steadfastly maintained his unwavering trek toward his flawless artistic and conservation goals.

Also, while diligently pursuing and producing his portfolio of avian artistry in our young nation's wild lands, Audubon was pounding the pavements in cities and towns throughout the United States and Europe, seeking the artistic and financial connections to make *The Birds of America* happen. Such extreme dedication to a cause is going to be required on the part of millions of people throughout the world to successfully make living with climate change a reality.

While thousands of Americans were being mesmerized by the vast diversity of avian life depicted in Audubon's *The Birds of America*, our nation's real treasure trove of natural resources was disappearing quickly, however. Westward expansion was eliminating wilderness and wantonly exploiting natural resources—often beyond recovery.

GO WEST, YOUNG MAN

Near the midpoint of the nineteenth century, the admonition of newspaper editor Horace Greeley to "Go West, Young Man" was starting to catch on with adventurous young men. It was suddenly considered de rigueur to buy a Conestoga wagon, load up the family's belongings, hitch up the team, and join one of the hordes of wagon trains now beginning to trammel their ways across the open prairies of the Midwestern United States. These hearty folks were looking to fulfill their destinies and—if everything worked out as they dreamed—end up with fabulous wealth and happiness beyond belief.

As coined by John L. O'Sullivan in 1845, Manifest Destiny was "a term for the attitude prevalent during the 19th century period of American expansion that the United States not only could, but was destined to, stretch from coast to coast. This attitude helped fuel western settlement [and] Native American removal."[19] While Manifest Destiny expressed typical imperialist thinking, imperialism developed a special meaning in the American context—that only European-Americans should decide the fate of the North American continent.[20] In other words, the opinions of everyone else, including Native Americans and non-European minorities, didn't matter.

The dominant Manifest Destiny belief postulated that young white men and their families had an obligation to spread their agrarian lifestyles beyond the now-conquered Eastern United States and into the western frontier. This belief—coupled with the still-prevalent belief that humans had a sworn duty to dominate and control nature—led to widespread destruction of the native grassland prairies that occupied most of the Midwest.

Dovetailed with the beliefs expressed in the idea of Manifest Destiny, the California Gold Rush (1848–1855) punctuated the coast-to-coast expansion of the United States landscape. Beginning on January 24, 1848, when James W. Marshall found gold at Sutter's Mill in Coloma, California,[21] the news of this treasure brought—mostly by sailing ships and covered wagons—some 300,000 gold-seekers (called forty-niners—as in, 1849) to the present state of California. While the majority of these nugget seekers were Americans, the Gold Rush achieved enough notoriety that droves of people from other parts of the world—Central America, South America, Europe, and even Asia and Australia—also showed up.[22]

Chapter 9

A MULTIPLICITY OF HEROES AND THE CIVIL WAR (1849–1869)

TRANSCENDENTALISM: RALPH WALDO EMERSON AND HENRY DAVID THOREAU

While the exploration or exploitation—take your pick—of the American West was just beginning to flourish, two more of our past environmental heroes—Ralph Waldo Emerson and Henry David Thoreau—were sitting, thinking, and writing in the newly-minted Commonwealth of Massachusetts. As the original transcendentalists, Emerson and Thoreau believed that there was much more to life than working feverishly and accruing wealth. Their thoughts and words were the first cries in the wilderness about living simply and compatibly with the natural world, and their words are still inspiring millions of people around the planet who want to make peace with—instead of continually exploiting—their environment.

Emerson, generally considered the "Father of Transcendentalism," and Thoreau's mentor, was born in 1803, fourteen years before Thoreau. His most famous work, *Nature*[1]—published in 1836—explained his belief that God was suffused throughout the natural world, and was not a separate, divine countenance living off in some heavenly sphere.[2] Meanwhile, Thoreau took the teachings he gleaned from Emerson and turned them into two books that ran completely counter to the religious and social forces that were then driving our nation's expansion. Today, Thoreau's works form a significant portion of the backbone of the US environmental movement.

In 1849, Thoreau published his essay *Civil Disobedience*,[3] which—while much less famous than his monumental work, *Walden*[4]—opened many people's eyes to the abject horrors perpetrated right in front of them. First and foremost among Thoreau's described atrocities was slavery—foreshadowing the tragic war that was only slightly more than a decade away from sending the United States spiraling into the depths of human chaos and pathos.[5]

With his face-framing beard and dark wavy hair, Thoreau could have easily passed for Abraham Lincoln's brother—appropriate considering they were both brandishing the same moral sword against the institution of slavery. They each, however, advocated different methods of dealing with this scourge on the American landscape. While Lincoln believed in achieving his desired results by operating within the law of the land, Thoreau insisted that the country should stand against slavery, even if that led to civil war and the destruction of the Union.

Ever since its publication, Thoreau's *Civil Disobedience* has inspired many leaders of protest movements around the world. Nicknamed the "Prophet of Passive Resistance" by some, Thoreau and his writings have provided supreme spiritual guidance for inspirational figures such as Mahatma Gandhi, John Fitzgerald Kennedy, and Martin Luther King Jr.[6]

Thoreau spent much of his life looking for the ultimate truth in the natural world. At night, he enjoyed hours of "looking through the stars to see if [he] could see God behind them."[7] His two older siblings—Helen and John Jr.—who were schoolteachers, paid Thoreau's tuition to attend Harvard, where he immersed himself in classic literature, philosophy, and languages.

After graduating in 1837, Thoreau returned to Concord, Massachusetts, and opened a school with his brother, John. While Thoreau enjoyed teaching, he always fancied himself as a writer and soon after he left Harvard began keeping a detailed personal journal. Henry's brother contracted tuberculosis in 1841, forcing the brothers to close their school; John died in Henry's arms from lockjaw one year later.[8] When the school closed, Henry realized he needed to find another way to make a living because writing was not paying the bills, so he turned to his family business—pencils.[9]

Inconsistent with most popular beliefs about his life, Thoreau was—at times—a successful businessperson. The Thoreau family's pencils were the first produced in the United States, and they equaled the worldwide standard—the German-made Faber pencils.[10] After his father's death in 1859, Thoreau took over as head of the family business and, characteristically, started recycling the company's scrap paper for lists, notes, and drafts of his natural history essays. He also maintained his own active and highly respected local practice as a self-taught land surveyor.[11]

But let's get back to the Thoreau story with which everyone is most familiar. In 1845, Thoreau built a small home for himself on Walden Pond in Concord, on property owned by Emerson. Thoreau desired a simpler type of life, so while at Walden, for two years and two months, he experimented with working as little as possible, rather than engaging in the standard pattern of six days on with one day off. He felt that this fresh approach helped him avoid the misery he saw around him, once famously writing, "The mass of men lead lives of quiet desperation."[12] To his critics, who were perhaps trying to counter this desperation in their lives, Thoreau wrote: "If a man does not keep pace with his companions, perhaps it is because he hears a different drummer. Let him step to the music which he hears, however measured or far away."[13] After two years and two months, Thoreau left Walden Pond and moved back into his parents' home and then into a house owned by Emerson, who was conducting a lecture tour in Europe. As Thoreau writes in *Walden*, "I left the woods for as good a reason as I went there. Perhaps it seemed to me that I had several more lives to live, and could not spare any more time for that one."[14]

Repeated questioning by the Concord townspeople about how he was living at Walden Pond inspired Thoreau to write his best-known collection of essays. Finally published in 1854—after initial public rejection and seven complete drafts—Thoreau's *Walden* emphasized living life in close harmony with the natural world. Since its publication, *Walden* has served as a source of supreme inspiration for countless naturalists, writers, and—in more recent decades—environmentalists.

Most important for the issue of climate change are Thoreau's dual beliefs that we can achieve significant changes in cultural and societal mores by passionate, passive resistance and sustainable living in harmony with the natural world. Sometimes it's not the earliest or the most aggressive bird that gets the most worms but the one that stays most focused on the long-term task of raising healthy chicks.

JOHN BURROUGHS: FATHER OF THE MODERN NATURE ESSAY

Around the same time as Thoreau was writing his influential works, another prominent American essayist was making a strong case for natural resource protection, establishing a reputation that demands his inclusion as one of our past environmental heroes. A gentle man who eschewed the limelight, John Burroughs—considered the "Father of the Modern Nature Essay"[15]—was constantly writing and working behind the scenes to protect the natural world he so ardently loved. In fact, Burroughs advocated for the protection of our natural resources in the 1850s—decades before there were any national parks or official conservation movements.

Burroughs was born in 1837 on his family's farm near Roxbury, New York. As a young boy, he developed a deep passion for the Catskills' woods and fields around him. He became a teacher when he was only seventeen—easily securing the good will of the pupils with a knack for imparting knowledge.[16] He saved his teaching wages, supplementing them with money earned working on a farm, to put himself through the Hedding Literary Institute at Ashland, New York, in the fall of his seventeenth year.[17]

Watching with chagrin as rapid westward expansion and industrialization systematically ate away the wilderness of his country, Burroughs decided to help save America's natural resources from disappearing forever. He used what he could do best—writing natural history essays—to help people visualize and understand the irreplaceable value of what they had. Then he taught them to feel passionately about protecting these resources. Very few people living today are aware of the tremendous influence Burroughs's nature essays had on the consciousness of the American public during the nineteenth century.

Burroughs was one of the most famous authors of his day. He had a knack for describing the natural world vividly and simply. His prose communicated the value of slowing down and taking the time to really observe and appreciate the great outdoors. His message resonated with all ages, but especially with children. No image of Burroughs fits his grandfatherly persona better than one of him sitting on a hillside, his long white beard flowing down while he tells a tale about his exploits in the natural world to a group of visibly enthralled youngsters. By encouraging his readers to understand and share a sense of their purpose and place in the landscape, Burroughs championed the importance of keenly observing and understanding what was happening in the natural world.

As is the case with most of our past environmental heroes who did not have the auspicious fortune of being born into wealthy families, Burroughs had to take other jobs to support his writing lifestyle. While he worked as a clerk for the US Treasury Department in Washington, DC, during the Civil War, he continued to pursue his interests in botany and ornithology. In Washington, he developed a fast friendship with the poet Walt Whitman, eighteen years his senior. Burroughs's first book, published in 1867, was entitled *Notes on Walt Whitman as Poet and Person*[18] and was partially written by Whitman.[19]

Summarizing his love affair with the birds in DC, Burroughs wrote *Wake-Robin* in 1871.[20] Although he enjoyed the city, he missed his boyhood Catskills and he returned to them in 1873 to build a house he named "Riverby" along the western shore of the Hudson River, about eighty miles north of New York City.[21] Then—about twenty years later—yearning for a more pristine writing environment, he built what he called "Slabsides," a rustic cabin located more than a mile into the woods from "Riverby."[22]

Slabsides was where Burroughs's profound and personal connections with the literary world took off. A parade of dignitaries—including Thomas Edison, Henry Ford, Harvey Firestone, and John Muir—regularly stopped by to visit and chat with him there.

Through the lasting friendships he built with his more prominent visitors, Burroughs began to have an important influence on the emerging preservationist movement. By capitalizing on his newfound conservation clout, he also inspired political leaders to work at protecting wild lands and wildlife. He continually encouraged his readers to get out and explore the natural world, telling them, "Each of you has the whole wealth of the universe at your very door."[23]

Over a period of sixty years, Burroughs wrote more than three hundred nature essays and articles, published in leading magazines, along with twenty-seven books. When Burroughs died in 1921, Clyde Fisher, then curator of visual instruction at the American Museum of Natural History, wrote in *Natural History Magazine*, "John Burroughs did perhaps more than anyone else to open our eyes to the beauty of nature."[24] In an email interview, Ginger Wadsworth, author of the children's book, *John Burroughs: The Sage of Slabsides*, wrote this description of Burroughs—which would have given him the ultimate pleasure in knowing that his life's goal had been accomplished: "His essays teach us to slow down and look around. They encouraged people of all ages to go out their backdoors and experience nature."[25]

More than anything else, John Burroughs had a remarkable predilection for moving people to action through his writing. For the writers among us climate-change activists, this is a critical skill for getting people involved with the solution process. Before you can convince people to act, you have to convince them to care, and that is exactly what world-class writings—such as those of John Burroughs—are carefully tailored to accomplish.

FREDERICK LAW OLMSTED: THE WORLD'S FIRST LANDSCAPE ARCHITECT

During the last few years before the American Civil War turned brothers against brothers—the darkest four years in our young nation's history—two other prominent citizens, Frederick Law Olmsted and Charles Darwin, left their distinctive marks on the US environmental movement. Even though they were both raised in wealthy, aristocratic families and looked remarkably alike, with their flowing white beards and bald pates, these two men made contributions that could not have been more different.

In 1857, in New York City, Frederick Law Olmsted was using his skills to help people live in harmony with the environment, by transforming New York's Central Park from a desolate brown dumping ground into the world's first showcase of urban green open space. As a child, Olmsted gained a deep and abiding respect for the natural world from both his father and his stepmother. This upbringing implanted within him the belief that access to the peace and solitude provided by open spaces and natural areas was one of the secrets to a happy life.

Born in Hartford, Connecticut, in 1822, Olmsted, as a young man, wasn't quite sure what he wanted to do with his life. He tried his hand at many professions—most notably as a journalist and book author—but none of those fulfilled him. Through all of these unsatisfactory career choices, Olmsted's thoughts kept returning to an idea he had been mulling over in the back of his mind ever since he was a child roaming the rolling open spaces of the Connecticut River Valley.[26]

Frederick believed that America's burgeoning cities should be more hospitable—making them enjoyable places to live instead of just urban commerce centers crammed with tall buildings and dense with gray pavement. As Olmsted saw it, the best way to improve the livability of a city was to create more open green space—places where residents could take a break from their workaday worlds and just sit, relax, dream sweet dreams, and enjoy themselves for an hour or so.[27]

What a novel idea for a profession, Olmsted thought—instead of designing buildings to shelter people's bodies from the outside world, design outdoor spaces that could expand people's minds to enjoy the intrinsic values of nature. Even better, he realized, would be creating a network of green spaces that tied urban cityscapes together and made it possible to walk for long, uninterrupted distances in a quiet environment. These thoughts later became the seed for Boston's famous "Emerald Necklace"—the first urban greenway system found anywhere in the world.

The more Olmsted thought about it, the more he realized he was on to something. In 1857, his big break happened when he was hired by the city of New York as superintendent for the reconstruction of Central Park. His work on Central Park's design set a standard of excellence that continues to influence landscape architecture in the United States.

At the end of his twenty-five-year career, Olmsted and his firm had designed more than five hundred projects throughout the United States—mostly of the urban-improvement variety. In addition to New York City's Central Park, Olmsted was the designer of the US Capitol Grounds, the Biltmore Estate property in Asheville, North Carolina, and the Stanford University Campus in Palo Alto, California. He also served as site planner for Chicago's 1893 World's Columbian Exposition. Since he emphasized emulating the scenic value of the natural world in his work, Olmsted also spent great deal of time helping humans experience nonurban parks. He was head of the first Yosemite National Park Commission and leader of the campaign to protect Niagara Falls.[28]

Olmsted was one of the first people to practice this field and is now widely considered to be the "Father of American Landscape Architecture."[29] His main goal—no matter what he was working on—was to improve the human experience. He wanted his parks to be available to all people, no matter their cultural status or lifestyle. Also, in one of the first official instances of social justice, Olmsted's antislavery letters were published individually, and then, in 1861, were collected into one book, entitled *The Cotton Kingdom*.[30]

Even though Olmstead died just after the turn of the twentieth century, the landscape architecture firm he founded successfully lived on until 1979—in the capable hands of his sons and their successors. Today, his home and office are owned and managed by the National Park Service as the Olmsted National Historic Site, located in Brookline, Massachusetts. Many of his conceptual drawings and detailed plans also can be found in the Library of Congress in Washington, DC.[31]

> ***What architect so noble . . . as he who, with far-reaching conception of beauty, in designing power, sketches the outlines, writes the colors, becomes the builder and directs the shadows of a picture so great that Nature shall be employed upon it for generations, before the work he arranged for her shall realize his intentions.***
>
> **—Frederick Law Olmsted**[32]

Olmsted dovetailed his passion for the natural world with his profound belief that everyone should have access to quiet green spaces for solitude and reflection, away from the din and clamor of our nation's expanding urban areas. He turned his quest and extraordinary vision for improving the human condition into a unique profession that endures today throughout the world. His dedication to realizing his childhood dream by inventing something the world had never before seen should certainly be something imitated by today's climate activists. Perhaps new, yet undreamed of, technology holds the key to designing a world future compatible with climate change.

CHARLES DARWIN AND HIS *ON THE ORIGIN OF SPECIES*

As we mentioned previously, Frederick Law Olmsted's influential look-alike was English naturalist Charles Darwin. Almost everyone who has studied science, and many of those in other fields of study, know that Darwin's theory of evolution became the foundation of modern evolutionary studies. What is not so well known is that the release of Darwin's monumental 1859 book *On the Origin of Species*,[33] caused a flurry of discussion, debate, criticism and support.[34] The best treatise on the matter comes from Asa Gray, writing for the July 1860 issue of the *Atlantic*. He describes humans' natural tendency toward an anxiety about learning: "We cling to a long-accepted theory, just as we cling to an old suit of clothes. Any new theory, like a new pair of breeches, is sure to have hardfitting [sic] places . . . such being our habitual state of mind, it may well be believed that the perusal of the new book 'On the Origin of Species by Means of Natural Selection' left an uncomfortable impression, in spite of its plausible and winning ways."[35] Gray compares Darwin's work to that of Galileo: the public initially denounced Galileo's theory that the Earth revolves around the sun. Eventually, with time and considered reflection, the majority of the public accepted the idea.[36] So it is with Darwin's theory, though a percentage of the population still does not adhere to the theory of evolution.

Born in Shrewsbury, Shropshire, England, in 1809, Charles Darwin loved to be outdoors enjoying nature almost before he could walk. He enrolled in Edinburgh University with the goal of following his father's and grandfather's famous footsteps into medicine. There, Darwin learned that the brutality of surgery and the sight of blood turned his stomach—not especially good qualities for someone in the medical profession in those days.

During a brief period, Darwin also thought about becoming an Anglican pastor and studied religion—but mostly botany—under the tutelage of Reverend John Stevens Henslow. Intrigued by his protégé's keen interest in the outside world, Henslow suggested that Darwin take a position as naturalist on an expedition commanded by Captain Robert Fitzroy aboard a rebuilt brig quaintly named the HMS *Beagle*.[37] Darwin had always dreamed of traveling the world, and, though Captain Fitzroy had offered to cover his accommodations in return for his services as a naturalist, Darwin insisted on paying a fair share of the meal expenses.[38] Little could anyone have imagined at the time that the pairing of Charles Darwin and the *Beagle* would live on in history as the boy and the ship that would shock the world and forever alter the science of human history.[39]

Casting off under damp and dreary skies but not particularly rough seas, the *Beagle*—with a crew of seventy-three men, including young, untested Charles Darwin—sailed out of Plymouth Harbor on the morning of December 27, 1831.[40] Becoming seasick almost immediately—a malady that would curse nearly all his days at sea—Darwin started to have second thoughts about being on the voyage.[41]

In 1835—after almost four years of exploring the world's oceans—the *Beagle* reached the Galapagos Islands, one of Earth's most remote and least explored archipelagos.[42] Exploring terra firma, of course, was right in Darwin's wheelhouse, and he bounded ashore at each stop with an unbridled exuberance for making new discoveries—a little like telling a child he could keep whatever he found in a gigantic toy store. One of Darwin's fondest memories in the Galapagos was hopping on top of a giant tortoise and trying to keep his balance as the gentle reptile lumbered through hillsides covered with volcanic rubble.[43]

Returning home in 1836 after nearly five years at sea, Darwin turned his meticulously crafted notes into a book entitled *The Voyage of the Beagle*.[44] Still in print today, this colorfully written book, infused with occasional flashes of wit and humor, perfectly captures the essence of the *Beagle*'s voyage and Darwin's onboard adventures.

Darwin crystallized his theories about evolution while observing the genetically isolated populations of animals living on the Galapagos Islands. He was especially intrigued by the finches that he found colonizing each separate island.[45] Prior to his time on the *Beagle*, several mentors shaped Darwin's budding theories on evolution. Jean Baptiste Lamarck planted, in Charles's head, the notion that humans evolved from a lower species via adaptations.[46] The two men differed in their view of how these adaptations came to be: Lamarck hypothesized that they happened during an individual's life while Darwin postulated that animals were born with the adaptations that led to reproduction and species survival. Another mentor for Charles was Thomas Malthus, whose studies on population economics led to Darwin's idea of "survival of the fittest," in which the species most well adapted outcompeted other species for limited resources.[47] And, notably, Charles's grandfather, Erasmus Darwin, shared his views on speciation with Charles; Erasmus eventually published his own book on the subject.[48]

After arriving back home and studying his collected bird specimens with the help of a few professional ornithologists, Darwin noticed that each Galapagos Island's finch population had a beak that was a different size and shape than that of the finches on the other islands. Moreover, the beaks of each isolated finch population appeared specially adapted to the different food species found on its island.[49]

How could this happen, Darwin wondered? In his mind, the only explanation was that the finches on each island had evolved beaks that were best suited to eating the food that was most available on that island and were thus being naturally selected to reproduce.[50] Darwin's paramount publication, where he combined many of the aforementioned theories with his studies on the *Beagle*, was to come later—much later, in fact. While his finches formed a substantial part of the backbone for his theory of natural selection, Darwin was not anywhere near ready to go to publication with his ideas in 1839.

Many explanations have been proposed to identify Darwin's reasons for waiting so long to come forward with his evolutionary theory. Some suggested that he was waiting for other scientists to produce findings that would help verify his beliefs. Others suggested that Darwin was worried he would be ostracized by the Anglican Church and his friends and family. John van Whye, Lecturer on the History of Science, Darwin, Wallace, and Evolution at the National University of Singapore, asserts that not only is there no clear evidence for these theories, they are "overwhelmingly contradicted by the historical evidence."[51] Professor van Wyhe continues, "Darwin was understandably very busy and began his species book when he had completed work in hand, just as he had intended all along."[52]

At first, Darwin's beliefs that animals and humans shared a common ancestor shocked the Anglican Church and Victorian society to the core. By the time of his death in 1882, however, Darwin's evolutionary imagery had spread through all of literature, science, and politics.[53] Although professedly an agnostic, Darwin and his evolutionary theory were finally vindicated when he was buried in London's Westminster Abbey—the ultimate British accolade.[54]

> ***It is not the most intellectual or the strongest species that survives, but the species that survives is the one that is able to adapt to or adjust best to the changing environment in which it finds itself.***
>
> **—Charles Darwin**[55]

So what can an earnest climate scientist learn from studying the life of past environmental hero Charles Darwin? First, Darwin had the gumption and stamina to stand by what he believed in his heart and mind to be true. Then he steadfastly maintained these beliefs and worked diligently to prove their veracity, even when he knew it would subject him to a storm of professional debate, and even ridicule. He boldly and unflinchingly published his controversial theories and then lived to see them widely accepted. Steadfastly standing firm in the face of withering criticism and proving what is not only true but is also right is one of the most strenuous tests of heroism on the planet.

THE GREENHOUSE EFFECT IDENTIFIED: AGAIN!

In 1861, an Irish physicist by the name of John Tyndall demonstrated that water vapor mixed with atmospheric gasses, becoming the second person—after Jean Baptiste Joseph Fourier, as we discussed earlier—to describe what became known as the "greenhouse effect." Tyndall found that "aqueous vapor is a blanket more necessary to the vegetable life of England than clothing is to man."[56] The British government honored Tyndall more than a century later by naming a climate research organization after him—the Tyndall Center. Unfortunately, Tyndall—like Fourier before him—made no connection between human activities and Earth's warming.

SCOURGE OF AMERICAN HISTORY: THE CIVIL WAR

From 1861 to 1864, the ugly pall of the US Civil War descended and covered all but the most western and newest states in bloody battlegrounds that split the Union into two separate nations and threatened to undo all the human rights gained during the American Revolution.[57] The senseless carnage killed 618,000 American soldiers. That is the most American deaths of any war in our nation's history, including more than the totals of World War I, World War II, the Korean War, and the Vietnam War combined.[58] The war's end also left in rubble and ruins much of the seven southern states that had initially seceded from the Union to form the "Confederate States of America."

The only good thing to come out of all this death and destruction was the abolition of slavery, although we have trouble justifying such horrors for getting rid of something that should never have existed in a free country with equality for all in the first place.

As you might expect, most people were not devoting a lot of thought to environmental and conservation matters during the Civil War. However, President Abraham Lincoln can be celebrated as a bona fide American hero for more than one reason.

President Lincoln was, of course, most famous for his tremendous social victories in fighting slavery and issuing the Emancipation Proclamation that freed all slaves on January 1, 1863. But few people know that this lanky, bearded man also contributed to the embryonic environmental movement by establishing the US Department of Agriculture in 1862 to assist the nation's farmers—which at that time made up 90 percent of the population.[59] He also founded the National Academy of Sciences in 1863 and then signed a bill that made the Yosemite Valley a state park in 1864.

Fortunately, while the Civil War was raging, a few individuals still managed to carry the ball for the protection of our dwindling natural resources.

GEORGE PERKINS MARSH: TELLING IT LIKE IT WAS

Cherub-faced, with granny glasses and a slight paunch, George Perkins Marsh would today be called the ultimate environmental nerd—a real tree-hugger. However, Marsh was the first true environmentalist with the guts to stand up and say, "Hey folks, we are really making a mess of things here on Earth!" In 1864, he published *Man and Nature*,[60] followed by a revised edition in 1874 entitled *The Earth as Modified by Human Action: Man and Nature*.[61] Taken collectively, these two books are widely regarded as the first modern discussion of our planet's environmental problems.[62]

Born in 1801 in Woodstock, Vermont, Marsh grew up in an egalitarian household full of the trappings of wealth and prosperity, and he attended the finest schools—Philips Exeter Academy, Dartmouth College, and Vermont Law School. Possessing boundless energy, endless enthusiasm, and immense intelligence, Marsh was definitely a Renaissance man.[63]

During his eighty years, Marsh held many positions. At various times, he was a newspaper editor, lawyer, mill owner, sheep farmer, lecturer, politician, and diplomat. A master of linguistics, he knew twenty languages. He wrote a definitive book on the origin of the English language and often was referred to as the foremost Scandinavian scholar in North America. Always using his creative mind, he invented tools and designed buildings—most notably the Washington Monument.[64] In his "spare time," Marsh served his country in several important capacities, including as a member of the US House of Representatives from Vermont (1843–1849), Minister to the Ottoman Empire (1850–1853), and Ambassador to Italy (1861–1882).

As we discussed earlier, the United States was dominated by rapid westward expansion in the second half of the nineteenth century. The California Gold Rush and the massive economic upheaval spawned by the start of the Industrial Revolution fueled this burgeoning expansion. No one felt much of a sense of environmental accountability to the American landscape until 1864, when Marsh published *Man and Nature*.

Man and Nature advocated a new way for evaluating human progress. Marsh realized that natural-resource use—for energy production, forest products, hydropower, fisheries stocks, and the like—was essential to sustain economic progress. But he also warned that unrelenting and unmitigated overuse of our natural treasures would lead to significant problems down the road.

Given his unique—at the time—understanding of Earth and its processes, Marsh was the first person to document systematically how human activity could have a cumulative and destructive effect on ecosystems and on the ability of those ecosystems to support human culture. Before Marsh came along, humans assumed that nature existed outside of human culture and was unchanged by human acts and works. Most appallingly, the basic belief was that nature was infinitely capable of providing the resources that human economy extracted from it.[65] Marsh worked to change these basic beliefs, delineating ways that human actions on Earth could be having negative effects on the world's natural resources and climate.

To demonstrate his points, Marsh conducted extensive surveys of the benefits of natural forests, including their capacities to moderate local and regional climates. In this frighteningly portentous passage based on his research, Marsh writes:

> Even now . . . we are breaking up the floor and wainscoting and doors and window frames of our dwelling, for fuel to warm our bodies and seethe our pottage. [As a result, our planet is] fast becoming an unfit home for its noblest inhabitant. . . . Another era of equal human crime

> and human improvidence . . . would reduce it to such a condition of impoverished productiveness, of shattered surface, of climatic excess, as to threaten the depravation, barbarism, and perhaps even extinction of the [human] species.[66]

We can only be left to believe that Marsh—writing more than 150 years ago—could see the handwriting on the wall for what we are now facing from the threat of climate change. Watching the "Doomsday Clock" (see chapter 21 for more on this) now ticking ever closer and closer to midnight, we can bear witness to the premonitory truth of Marsh's words.

As earnest climate-change analysts, we all need to pay special attention to George Perkins Marsh's beseeching writing about working toward a harmonious blend of human activities and ecosystem health. Even while the United States still contained an enormous bounty of natural wealth, Marsh emphasized that we should be paying close attention to the effects our actions were having on the planet and working diligently to improve the sustainability of our lifestyles. Marsh's words can be used to eloquently drive home the point that concerns about climate change have been around for a very long time.

THE HOMESTEAD ACT: A PRECURSOR OF ENVIRONMENTAL DISASTER

Owing in large part to the Homestead Act of 1862—which provided each family with 160 acres of land to plow and farm as they saw fit—westward expansion was still spreading steadily, like a soaking ink blot, across the American landscape. Access to virgin prairie was so easy that intrepid settlers tended to plow, plant, and harvest until they depleted the fertility of the land through a combination of soil erosion (due to lack of control devices such as contour plowing and planting cover crops to keep the soil in place) and poor farming practices (no crop rotation). They knew that when the land became unproductive, they could just pack their families up, move further west, claim more land, and start all over again.

In a situation with an eerie similarity to today's climate-change crisis, the major faux pas of the pioneering farm families was believing that their resources were boundless and because of this doing nothing to protect the resources they were using. Seventy years later, in 1932, these same laissez-faire attitudes created the worst environmental disaster the United States has ever experienced—the Great Dust Bowl (which we will discuss in chapter 13).

DR. ERNST HAECKEL INTRODUCES ECOLOGY TO THE WORLD

Ironically, four years after we obliterated nearly 2 percent of the US population during our Civil War, a word now famous for emphasizing the importance of biodiversity made its first appearance in the lexicon of biologists. In 1866, Ernst Haeckel, a German zoologist and master of many other scientific and artistic endeavors, introduced the concept of ecology onto the world scene.

While Haeckel is officially credited with coining the term, we owe the origin of the word *ecology* to the ancient Greeks. In fact, we can understand the word most easily by looking at its Greek roots—*oikos* meaning *house* and *logia* meaning *study*. So to best understand the word *ecology* envision observing everything that is going on in your house, including everything from the interactions of your family with each other to how each family member is using everything in the house. This visualization of your household will give you a good handle on the meaning of ecology.

Now apply this idea to what is happening in the natural world. Ecology is about the ongoing relationships among all living things and their associated nonliving elements on Earth.

For example, if a bird captures and eats a worm, that is an ecological interaction. If the bird in turn is eaten by a fox, that is another ecological interaction. When the fox stops by a pond to wash his meal down with a drink of water that is still another ecological interaction. Taken collectively, all the ecological interactions occurring in all the ecosystems on Earth form the science of ecology—the term Dr. Haeckel officially introduced to the world so long ago.

Why is this so important to know? The study of ecology has become the basis for analyzing and evaluating everything that is going on in the natural world, including the influence of humans, popularly known as anthropogenic effects. By looking at how our activities are affecting individual ecosystems—with an emphasis on increases or decreases in biodiversity (the number of different species of organisms living in an ecosystem or ecosystems)—we gain an understanding of how and why our behaviors need to be modified.

Understanding the need for a paradigm shift is of critical importance today because if we take a hard, honest look at Earth's ecology, we will quickly see that what we are doing does not paint a pretty picture. Ice sheets melting, oceans warming, coral reefs dying, species going extinct—these are all symptoms of broken ecosystems; our existence is predicated on their survival, so we need to prioritize healing them.

Chapter 10

THE RISE OF INDUSTRIAL AMERICA

Progress or Perish (1870–1900)

THE SECOND INDUSTRIAL REVOLUTION: CLIMATE CHANGE GETS ITS START

While the Wild West may have been reaching its pinnacle, the latter half of the nineteenth century in the United States was also a time of innovation, invention, and rapid growth. The Second Industrial Revolution—also known as the Technological Revolution—corresponded to this period and lasted up until World War I. The First Industrial Revolution had begun in the eighteenth century with the textile industry, but the Second Industrial Revolution transformed the country to an even greater extent. It began in the 1860s, when Henry Bessemer invented a process to produce steel more quickly and cheaply, and culminated in the introduction of production lines and the mass production of goods.[1]

Historian Richard White perfectly described what was going on during this period:

> If a Western Rip Van Winkle had fallen asleep in 1869 and awakened in 1896, he would not have recognized the lands that the railroads had touched. Bison had yielded to cattle; mountains had been blasted and bored. Great swaths of land that had once whispered grass now screamed corn and wheat. Nation-states had conquered Indian peoples, slaughtering some of them and confining and controlling most of them. Population had increased across much of this vast region, and there were growing cities along its edges. A land that had once run

largely north-south now ran east-west. Each change could have been traced back to the railroads.[2]

This phase of American history was also marked by significant economic development featuring the introduction of power-driven machinery. Factories even made machines for other factories. All these new workshops needed coal to power their machinery, and they made many new products from natural resources like iron. As a result, new mines opened across the country. Moreover, all these new industries also needed workers.[3]

Daily life was changing quickly throughout the United States, owing to landmark advances in technology and transportation. The telegraph and telephone made it possible to communicate directly with relatives and friends who lived in distant states. Inventors like Lewis Latimer and Thomas Edison made the delivery of electric power and light a reality for homes and businesses from coast to coast. Meanwhile, the world's first automobiles were rolling off Detroit's assembly lines and steam-driven engines and trains lumbered across tracks nationwide.[4]

Millions of Americans took advantage of the flurry of new jobs now available in the factories and mines. The transition from a rural agrarian to urban laborer lifestyle was fast and furious. Demand for new mass-produced products—such as light bulbs, textiles, and telephones—caused the manufacturing sector to soar.[5]

The owners of these new factories and mines became incredibly wealthy, heralding the advent of affluent capitalists dominating the American landscape—something that continues to this day. These men—the Carnegies, Vanderbilts, and Rockefellers—soon dominated all aspects of American life, including economics, politics, and social lifestyles.[6]

The Industrial Age allowed more people to make consistent salaries—albeit mostly based on poor pay—while attaining higher standards of living. But it also meant a significant increase in the demand for natural resources, in particular for coal to run the factories and for raw materials to produce the manufactured goods. In general, higher living standards led to the destruction of more forests to make way for expanding cities and to provide lumber for construction.

In the minds of many climate scientists of today, this Second Industrial Revolution marked the start of humans having significant anthropogenic effects on the rate of greenhouse gas (GHG) emissions into the atmosphere and increasing global air temperatures. The main culprit in this onset of global warming was the massive increase in coal burning that was now needed to fire all the furnaces in the factories springing up throughout the United States and Europe like mushrooms after a summer rain.

The Industrial Age also ushered in the Age of Consumerism, which continues to permeate the fabric of American life to this day. Less wealthy Americans clung to an erroneous notion of entitlement, and pursued a lifestyle of consumption that they could not well afford. Despite this unrealistic sense of potential wealth, the American population—for the most part—kept pursuing the American dream dutifully during the Second Industrial Revolution. They were caught up in desperately striving for wealth and prosperity, all the while giving little thought to the health of the environment or protection of natural resources. As far as most Americans were concerned, the riches of the natural world were still inexhaustible and fully available to help humans continue to conquer Earth.

Economic expediency ruled the day, and the damages to the environment—such as the obliteration of billions of passenger pigeons and millions of American bison—were enormous. Fortunately, a diverse array of environmental triumphs and environmental heroes also took the stage during the last half of the nineteenth century and started a minor backlash that at least got some people thinking about what was happening to their natural world.

DEDICATION OF THE WORLD'S FIRST NATIONAL PARK

On March 1, 1872, the United States saw the most momentous event—to date—in the history of national resource conservation. On that date, President Ulysses S. Grant—relying on geographic and topographic information from a survey expedition led by geologist Ferdinand V. Hayden, as well as the large-format photographs of William Henry

Jackson and the iconic landscape paintings of Thomas Moran—made the Yellowstone Territory the world's first national park.[7]

FOUNDING OF THE US GEOLOGICAL SURVEY (USGS)

From the viewpoint of outstanding federal service to the American public, President Rutherford B. Hayes made a major contribution to the environmental movement when he authorized the US Geological Survey (USGS) on March 3, 1879. The primary task of the USGS was to study the landscape of the United States, including its natural resources—biology, geography, geology, and hydrography—and the natural hazards that threaten these resources. Based on our professional experience, the USGS has always been one of the most valuable and serviceable federal agencies in terms of preparing and disseminating invaluable information like topographic quadrangles, river/stream hydrographs, groundwater mapping, aerial photography, and geologic substrata.

THE ADVENT OF RENEWABLE ENERGY

In 1882, the world first ventured into the use of renewable energy with the dedication of the first hydroelectric power plant on the Fox River in Appleton, Wisconsin. This hydropower venture was a rather rudimentary facility, generating just enough electricity to power the plant itself and a few nearby buildings. But by the turn of the twentieth century, with power production exponentially increased through the use of dams, inexpensive hydropower was providing a significant portion of the US energy production. This was particularly true in rural areas, where power generated by rivers and dams spurred industrial growth that could not otherwise have occurred.[8]

WILLIAM TEMPLE HORNADAY RESCUES THE AMERICAN BISON

In an iconic image of the preeminent success of his life's work, William Temple Hornaday—another one of our past environmental heroes—is holding a leash and looking down lovingly at a newborn American bison (more commonly called the buffalo). We owe Hornaday a deep debt of gratitude for personally saving this symbol of the western American landscape from almost certain extinction.

Hornaday was born in 1851 in Avon, Indiana, and educated at Oskaloosa College (now Iowa State University). While working as a taxidermist in the 1870s, he had the opportunity to join a series of scientific expeditions. Traveling extensively throughout the United States and the world—to Florida, Cuba, the Bahamas, South America, India, Sri Lanka, the Malay Peninsula, and Borneo—Hornaday gained quite a reputation as a marksman in hunting big game animals. He also applied his taxidermy skills to create what he called *life groups*—featuring animals in their natural settings—for museums across the country. In 1882, Hornaday's high-quality animal displays vaulted him into the position of chief taxidermist of the United States National Museum, at the distinguished Smithsonian Institution in Washington, DC.[9]

In this post at the Smithsonian, Hornaday took it upon himself to investigate what he had heard about the dwindling herd of American bison on the western prairies. He sent hundreds of letters to ranchers, settlers, explorers, and homesteaders all over the American west.

What he heard back painted an appalling and depressing picture. As Hornaday wrote to George Brown Goode, his superior at the Smithsonian, "In the United States the extermination of all the large herds of buffalo is already an accomplished fact."[10] His diligence in collecting and reporting this discouraging information led to a trip that forever changed his life and set a milestone in the history of North American wildlife management.

In 1886, Hornaday traveled to Montana's Musselshell River to observe a few remnant bison herds for himself and collect museum specimens before the species went extinct. The fact that he knew what to expect did not diminish the deep distress he felt at seeing that the vast herds of buffalo had vanished and only a few animals still survived in widely scattered groups. To counter his anguish over what he had seen in Montana, Hornaday returned home and, at the still-young age of thirty-six, immediately transformed his work orientation to focus on saving the bison from extinction. To initiate this effort, he acquired live bison that he brought to Washington, DC, and placed on display behind the Smithsonian's administration building (nicknamed "The Castle" for its unusual architectural design).[11]

Hornaday's strategic decision to display live bison proved to be sheer genius on two levels. First, the live exhibit was much more popular than the museum's encased bison group display and soon familiarized thousands of Americans—who had never traveled to the West—with the magnificence of these wildlife icons and the imminent threat of their disappearance forever. Second, this created the groundswell of public support Hornaday was seeking and opened the door for funding to ensure the bison's long-term preservation. It also led to the creation of the Smithsonian's National Zoological Park, with Hornaday serving as the first director.

Hornaday followed up his successful work at the Smithsonian in 1889 with the publication of *The Extermination of the American Bison*, a book that proved very popular and generated increased public support to save the species.[12] Then, in 1896, he received the ultimate honor when he was appointed director of New York City's Bronx Zoo, where he remained for the next thirty years. Now—thanks in large part to Hornaday's efforts—the Bronx Zoo is the foremost zoo in the United States, with a long history of emphasizing the importance of saving American native wildlife.

Throughout his tenure at the Bronx Zoo, Hornaday used his impressive skills as an articulate orator and influential writer to produce hundreds of newspaper and magazine articles and more than twenty books.[13] His works led to the passage of important conservation and wildlife protection legislation. In particular, his unceasing efforts battling against old-fashioned bureaucrats and obstinate politicians led to the passage of the 1911 Fur Seal Treaty and, most notably, the 1918 Federal Migratory Bird Treaty Act, which still protects all migratory birds in the United States.[14]

By 1918, the buffalo was no longer in danger of extinction, thanks in large part to Hornaday's diligent efforts. Today, the National Wildlife Federation carries on his legacy by helping to ensure that free-roaming buffalo herds will forever be found across the American landscape. In the process of dedicating his life to preserving the American bison, Hornaday also earned the title "Founder of the American Conservation Movement."[15]

Climate-change activists can learn a great deal from studying William Temple Hornaday's biography. First, he dedicated his life to a cause and then figured out how to create the groundswell of public support needed to accomplish his goal. The positive techniques he used to accomplish his objective are also admirable. Instead of emphasizing a doomsday outcry for the American bison, Hornaday first turned the public on to the beauty of these burly beasts and then kept emphasizing that it was not too late to save them from extinction. This is exactly the same approach we need to emphasize with climate change: while the livability of our magnificent planet is in serious jeopardy, it's not too late to save it—if we all act together right now.

YOSEMITE NATIONAL PARK BECOMES A REALITY

As we discussed previously, in the midst of the Civil War, Abraham Lincoln turned his attention for a moment to something brighter—the establishment of the first parkland ever set aside by the federal government for preservation and public use.[16] He signed the legislation for the Yosemite Grant, protecting the Yosemite Valley and Mariposa Grove of big trees, on June 30, 1864.[17] This set the precedent for designating land for recreation and public use, leading to the establishment of Yellowstone as the first national park in 1872.[18] In 1890—after decades of persistently badgering the US government and Congress—renowned naturalist, and another of our past environmental heroes, John Muir finally got his wish when Yosemite National Park received full federal protection as one of the United States' most iconic landscapes.

The government established Yosemite as protected land, but John Muir and many of his California alpine-loving friends realized that an organization was needed to ensure preservation of the natural beauty in the surrounding Sierra Nevadas as well.[19] These men formed the Sierra Club on May 28, 1892, to promote recreation in the Sierras by making the area more accessible and well known.[20]

We will talk more about Yosemite Park, John Muir, and his Sierra Club in the next chapter.

CREATION OF THE ADIRONDACK PARK PRESERVE

Next, in 1892, the state of New York set aside 2.8 million acres of the Adirondack Forest Preserve to be protected as "forever wild."[21] Impetus for creating this park for the people actually came from one of our previous heroes, George Perkins Marsh, whose book *Man and Nature* we discussed earlier in this chapter.

Today, the Adirondack State Park encompasses the northeastern lobe of Upstate New York. This gives it a total area of more than six million acres, making it the largest publicly protected land area in the contiguous United States. In fact, the Adirondack State Park is greater in size than Yellowstone, Everglades, Glacier, and Grand Canyon National Parks combined![22]

ARRHENIUS PROVIDES FIRST EVIDENCE OF HUMAN-CAUSED CLIMATE CHANGE

In 1896, Swedish physicist/chemist, Svante Arrhenius, became the first person to show that fossil-fuel combustion was most likely having an effect on warming the Earth's climate.

We talked more about Arrhenius's work and how much it influenced scientific thought in chapter 2.

HARRIET LAWRENCE HEMENWAY: HIGH SOCIETY GOES TO BAT FOR BIRDS

The next past environmental hero we will discuss requires a bit of a backstory to set the stage. In 1886, Frank Chapman—founder of *Audubon* magazine—decided to take a stroll from his uptown Manhattan office to the heart of the fashion district on 14th Street. Along the way, Chapman, a talented birder, counted a total of 174 birds comprising 40 species, including woodpeckers, orioles, bluebirds, blue jays, terns, and owls.[23] A pretty impressive array of birds for the middle of New York City—right? Hardly. You see the problem was that all the birds Chapman counted that day adorned the hats that sat on the top of women's heads.

In the late nineteenth century, America's hat craze was in full swing. To satisfy this overwhelming fashion demand, millinery companies needed countless colorful and flowing bird feathers. This spelled doom for millions of North American wild birds, which were slaughtered in droves.[24] The greedy practice became so lucrative that plume hunters would often wipe out all the birds in a rookery (nesting area)—taking just the feathers and leaving eggs to rot and newly hatched chicks to starve to death. The going feather rate was $32 per ounce—more valuable than gold at the time.[25]

In 1896, nearly five million birds, representing fifty different species, were killed for fashion. Entire populations of shorebirds and wading birds—including herons, egrets, spoonbills, gulls, and terns—along the Atlantic Coast were wiped out. But this despicable situation was about to change in a very dramatic way. Enter Harriett Lawrence Hemenway!

Most successful environmental organizations owe their starts to individuals with a deep and abiding respect for and dedication to the natural world. This is certainly the case with the Massachusetts Audubon Society. For much of her life, through early adulthood, Harriet Lawrence Hemenway lived a life of luxury and privilege at the pinnacle of Boston society, in a family dominated by accomplished men.

Although she escaped to watch birds along the Charles River whenever she could, Hemenway spent most of her time as a prominent socialite, moving gracefully through all the right places while always decked out to the nines in the latest trendy fashions. That was until she sat down on a cold winter day in 1896 and read a newspaper story that made her blood boil and caused all hell to break loose within the confines of polite Boston society.[26]

When Hemenway read about thousands upon thousands of magnificent wading birds being slaughtered just for feathers to decorate women's hats, she knew she had found her life's calling. From that day forward, she became a true "Champion of Conservation" and girded herself for the battle she knew would soon follow with the male-dominated world.

Hemenway's portrait by the famous artist John Singer Sargent shows an arrestingly handsome woman, with a deep-set gaze that practically shouts out "don't trifle with me . . . no matter who you are!" Soon after she read the bloodcurdling account of entire rookeries being wiped out in Florida, many of the cocky men who thought they ruled the roost in Boston were being called on the carpet to atone for the sins of the millinery trade.

The first thing Hemenway did was to contact her cousin Minna B. Hall.[27] Together, they organized a series of ladies' teas, with the intent of discussing much more than the latest social goings-on. Hemenway and Hall first told the women about what was happening to wild birds just to assuage society's haute couture needs. Then they beseeched their guests to start refusing to buy hats with bird feathers and to rally everyone else they knew to do the same.[28]

Their strategy worked like a charm. Using their social networks as a springboard, Hemenway and Hall reached out to hundreds of scientists and businesspeople and soon had gathered enough support to establish the Massachusetts Audubon Society (MAS) in 1896.[29] The MAS was the oldest Audubon Society in the country, leading to the creation of many state-level Audubon societies, which eventually united into the National Audubon Society in 1905.[30] (As an ironic exception, the MAS still prides itself on being totally independent of all other Audubon societies, including the National Audubon Society.)

In less than a year, the MAS had applied sufficient pressure to convince the Massachusetts legislature to outlaw the wild bird feather trade in the Commonwealth of Massachusetts.[31] Within two more years, bird lovers in New York, Connecticut, Pennsylvania, Tennessee, Maine, Iowa, Texas, Colorado, and the District of Columbia had followed Hemenway and Hall's lead. Women in all these states started societies dedicated to ending the feather trade and then convinced male civic leaders and local scientists to join the cause.[32] On average, women accounted for about 80 percent of the membership of each Audubon Society, including 50 percent of the leadership roles. In 1900—just four years after Hemenway and Hall started their work—Congress passed the Lacey Act, which provided the necessary legal teeth for prohibiting the interstate shipment of wild species killed in violation of state laws. By 1905, operating off this federal legal benchmark, thirty-three states had moved to pass their own versions of the Lacey Act and the millinery trade of wild bird feathers—while still breathing slightly—was on life support.

The death knell finally started ringing in 1911. First, New York State passed the Audubon Plumage Bill—a legal triumph that banned the sale of plumes of all native birds and shut down the domestic feather trade in the state. Then the 1913, the Weeks-McLean Law prohibited the spring hunting and marketing of migratory birds while the Underwood Tariff Act banned all importation of feathers except for purposes of scientific research or education.[33] These two laws placed all migratory birds nationwide under federal jurisdiction—finally ending the wild bird plume trade in the United States for good.

In the final analysis, America's Audubon Societies played the critical role in changing people's attitudes toward killing birds for their feathers. And it all started because Harriet Lawrence Hemenway read an article that upset her, took to a venue that she knew well—high society tea parties—and started the ball rolling.

There's an old adage that the best advice anyone can give to an aspiring writer is "just write about what you know." Whether the goal is protecting wild birds or deciding how to most effectively deal with climate change, this principle can be modified only slightly to "just work with who you know." Each of us has a unique sphere of influence—family, friends, and community members we connect with on a regular basis. Using our zeal for climate solutions within our sphere of influence might start with local changes but could also spill into national changes—you might be amazed by what you can accomplish!

US SUPREME COURT SEALS THE DEAL: UPHOLDING THE 1918 MIGRATORY BIRD TREATY ACT

In 1920, the US Supreme Court upheld the most powerful piece of legislation ever passed to protect wild birds. Still in effect today, the Migratory Bird Treaty Act of 1918 (which updated and replaced the 1913 Weeks-McLean Law)[34] gives all migratory birds full federal protection. This statute makes it illegal to take any action that could either directly harm (i.e., pursue, hunt, take, capture, kill, or sell) a migratory bird or indirectly impact its nesting habitat. This protection is extended to both live and dead birds, including any bird parts, feathers, eggs, and nests.[35]

More than one thousand species are currently on the list of birds protected by the Migratory Bird Treaty Act.[36] Writing for the majority in this landmark decision, Justice Oliver Wendell Holmes declared, "Without such measures, one could foresee a day when no birds would survive for any power—state or federal—to regulate."[37]

Chapter 11

WELCOME TO THE PROGRESSIVE ERA (1901–1919)

INDUSTRIAL AGE BACKLASH

When the twentieth century arrived, the "hell-bent-for-leather, get-rich-or-bust" mentality mellowed somewhat, along with the end of the lawless Wild West attitudes that officially closed the "cowboys-and-Indians" era on the US frontier. America finally breathed a deep breath, sat back, and took a fresh look at its disappearing natural resources:

> Owing to a general backlash against the excesses and waste of the Industrial Age, Americans attempted to cleanse the nation of the real and perceived evils that had resulted from unrestricted economic growth. This basic attitudinal change was driven by the fact that the "never-ending" West had [finally] ended. Realizing that the ideals of unlimited natural resources and untamed wilderness were about to disappear forever came as sobering revelations to the American populace.[1]

Along with this newfound concern about the loss of natural resources came a movement aimed at cleaning up the massive social problems dragging the nation down. The Progressive Era that dominated this period emphasized responses to the economic and social problems—racism, poverty, greed, violence, and class warfare—that rapid industrialization had worsened in America.

Most Progressives were college-educated people who lived in cities. They rejected Social Darwinism—the theory of the strong (in terms of wealth, power, and property ownership) being superior to the weak, and therefore more valuable to society (according to Social Darwinism, the poor were unfit and therefore should not be aided).[2] Progressives also believed that most of society's ills could be solved by providing everyone with three essentials for a happy life: a good education, a safe and healthy environment, and a productive workplace. Although it started as a social movement, Progressivism soon took on an active political component.[3]

Unfortunately, much of the talk about doing a better job of conserving natural resources and protecting the environment was just rhetoric—primarily spewing from the mouths of Progressive politicians who were trying to gain votes. On the plus side, however, the roots of the conservation movement were established in the midst of the Progressive Era.

It is important to note here that the conservation movement at the time was not rooted in protecting natural resources for their intrinsic value, which was instead known as preservation—a far different concept. At the time, conservation called for the wise use of natural resources, managing them for the benefit of all current Americans, while also maintaining their essence for future generations to use and enjoy.

In other words, conservation was based on sustainability—not simply preserving natural resources just as they existed with no human economic benefits. Examples of conservation included harvesting a forest for its timber value and then planting a new forest in its place, catching fish in streams and then restocking the streams on a seasonal basis, and hunting ducks with bag limits in place to restrict the total killed and then designating new habitat (wildlife refuges) to produce more young waterfowl.

THEODORE "TEDDY" ROOSEVELT: THE CONSERVATION PRESIDENT

The government became the agency of choice to manage conservation during the Progressive Era because the bulk of the citizenry believed that private corporations and organizations were too corrupt, self-serving, and greedy to be trusted. When Theodore "Teddy" Roosevelt—another of our past environmental heroes—assumed the presidency after the assassination of William McKinley in 1901, Progressives and conservationists finally had the friend they needed in the White House.

Usually depicted as immaculately dressed in a three-piece suit, with his intensely honest stare framed by a bushy mustache and chained glasses, Roosevelt epitomized what was known at the time as a true "man's man." While his unfailingly politeness and courtesy earned him consistent favor with women, his reputation as a Rough Rider adventurer and big-game marksman always garnered heroic accolades from men.

In an ironic twist for such a lifelong outdoorsman, Roosevelt was born in 1858 in the middle of America's most populous city—in a Manhattan brownstone—and home-schooled as a sickly child. But he didn't let these inconveniences keep him away from what he wanted to do. At the age of seven, he formed a local nature club with some of his cousins, and they quickly started badgering every creature they could find creeping and crawling in their urban stomping grounds. Within a few years, the boys had collected, analyzed, and mounted enough specimens to start what they called the "Roosevelt Museum of Natural History."

In another ironic twist of fate, a dual tragedy in Roosevelt's life significantly enhanced his love of the rugged outdoors. On the same day, February 14, 1884, both his mother and his wife tragically died.[4] To deal with the unbearable grief he was feeling over this twin loss, Roosevelt packed up, left New York City, and moved to the Dakota Territory for two years. While there, he left his infant daughter in the care of his elder sister while he worked as a cowboy and cattle rancher in the peaceful solitude of the American West's wide-open spaces.[5]

Soon after he took office, Roosevelt began earning his reputation as the "Conservation President." As an avid adventurer and lover of nature, he dedicated himself to protecting both wildlife and natural resources. He realized that dramatic action would be required to prevent the precious natural resources and transcendent landscapes of our country from disappearing as quickly as the American bison—leaving future generations without a legacy of natural splendors.[6]

As president, Roosevelt was faced with a quandary that was a carryover from the banner years of the Industrial Age. He knew that the deep-pocket entrepreneurs—who controlled the bulk of the nation's wealth—were so busy trying to make more money that they did not have time to worry about protecting the country's fast-disappearing natural resources. Roosevelt also knew that these same power players did not have any desire to give up the "man's dominion over nature" philosophy that had governed their lifestyles for decades. But he sincerely believed that the long-term happiness of most Americans was directly associated with how intelligently the country managed its natural resources.

On the surface, these conflicting ideologies presented a major dilemma. But good old "Uncle Teddy" had a plan. He believed that—with enough foresight—natural resources could be used, economically and recreationally, while simultaneously being conserved for the long-term. In other words, Roosevelt was onto something that would become the world's first plan for sustainable resource management, and he set about proving that his plan could work.

Administration of the nation's vast tracts of national forests perfectly epitomized Roosevelt's beliefs on sustainable management. In his opinion, the national forests had to serve multiple purposes. While they had to provide a broad range of recreational opportunities—including everything from hunting and fishing to hiking, mountaineering, and birdwatching—the national forests also needed to pay their way.[7] This meant portions of the forests had to be selectively logged and sold for building construction and pulp. The trick was to accomplish this sustainably, in a manner that provided some income while not detracting from the full visitor enjoyment of these resources. For example, federal lands where income-producing timber harvests occurred had to be immediately replanted with trees for use by future generations. As we will read later, this forest management concept formed the basis for the operation of the US Forest Service under Gifford Pinchot.[8]

During his presidency, Theodore Roosevelt protected more than 230,000,000 acres of public land. This is an area equivalent to the total size of all of the states that form the Eastern Seaboard of the United States—from Maine to Florida.[9] Included in these newly protected lands were the first 51 federal bird reservations (now national wildlife refuges), the first 18 national monuments, 5 new national parks, the first 4 national game preserves, the first 24 reclamation—or federal irrigation—projects, and 150 national forests.[10] Many of these federal designations were bitterly opposed by commercial interests. For example, the nation's first national wildlife refuge, on Florida's Pelican Island in 1903, raised the hackles of the millinery trade, since it was specifically established to thwart the acquisition of wild bird feathers.[11]

Roosevelt also showed his pragmatic side when he created the Bureau of Reclamation in 1902 to manage water resources in the seventeen western states. Reclamation's goals were to provide a mix of economic benefits—hydropower, irrigation, and flood control—while also maintaining a number of recreational activities. Western residents—many of whom were just settling into new villages, towns, and cities—received the huge benefit of the boating, fishing, and water skiing provided by federal water projects.[12]

As an adjunct to his political career, Roosevelt published more than twenty-five books on such diverse subjects as biology, geography, history and philosophy. In addition, he wrote a four volume biography/autobiography entitled *The Winning of the West*.[13]

Theodore Roosevelt died in his sleep at the age of sixty on January 6, 1919, at his Long Island estate, Sagamore Hill, after suffering a coronary embolism. On January 16, 2001, President Bill Clinton posthumously bestowed the Congressional Medal of Honor on Roosevelt—the highest award for military service in the United States—for his part in the Battle of San Juan Heights during the Spanish-American War more than one hundred years earlier.[14]

> ***Conservation means development as much as it does protection. I recognize the right and duty of this generation to develop and use the natural resources of our land but I do not recognize the right to waste them, or to rob, by wasteful use, the generations that come after us.***
>
> **—Theodore Roosevelt**[15]

President Teddy Roosevelt had an idea about how to get the nation's wealthy elite and middle and lower classes working together for a common cause. Then he applied his unwavering dedication—some might even call it his "Bull Moose" stubbornness—to make his plan work. Thanks to Roosevelt's gritty combination of foresight and fortitude, our nation put into place the world's first plan for sustainable resource management and long-term conservation. Roosevelt's accomplishments more than a century ago are even more apropos today, as we work toward bringing a divided nation together and deciding how to live sustainably with the goal of combating climate change.

GIFFORD PINCHOT AND THE US FOREST SERVICE

Fortunately for President Roosevelt, Gifford Pinchot—another of our past environmental heroes—came along at just the right time to help him with the daunting task of sustainably managing our nation's millions of acres of national forests. Nobody—including Roosevelt—epitomized the ideals of the progressive conservation movement more than Pinchot.

A tall, dapper man—always sporting a world-class handlebar mustache and a gentlemanly countenance—Gifford Pinchot was born in 1865 into a very wealthy family in Simsbury, Connecticut. His family's money gave him a top-flight education, and he graduated from the famed Philip Exeter Academy in New Hampshire and then moved on to Yale University in New Haven, Connecticut.

When he entered Yale in 1885, his father asked him this simple question: "How would you like to be a forester?" At the time, no American had ever made forestry a profession. Pinchot replied that he "had no more conception of what it meant to be a forester than the man in the moon. ... But at least a forester worked in the woods and with the woods—and I loved the woods and everything about them.... My Father's suggestion settled the question in favor of forestry."[16]

Inspired by this paternal conversation, Pinchot enrolled in the L'Ecole Nationale Forestiere in Nancy, France, since education in forestry did not exist in the United States at the time. He returned a year later "fired with enthusiasm for managing forests as a crop."[17] But he quickly realized that land development in the United States was out of control. He wrote, "When I got home at the end of 1890 ... the nation was obsessed by a fury of development. The American Colossus was fiercely intent on appropriating and exploiting the riches of the richest of all continents."[18]

Pinchot's observations were right on the mark. During the so-called Gay Nineties, the American public still believed that the abundance of US forestland was inexhaustible. The forestry practices in vogue at the time were cut, slash, level, and leave. State and federal governments typically gave no consideration to replanting to restore the forests for future use. In fact, forest managers considered wasting timber to be a virtue, not a crime, while they thought that second growth (sustainable) timber management was just a delusion of fools. Essentially, the resources were so plentiful and the country was so rich, that managers didn't want to waste their precious time or energy-saving timber for the future. In Pinchot's words, "What talk there was of forest protection was no more to the average American than the buzzing of a mosquito and just about as irritating."[19]

Spurred on by these strong feelings about abysmal land management practices, Pinchot jumped into his chosen profession—as America's first professional forester—with unabashed enthusiasm. In 1892, he accepted a position as resident forester on the Biltmore Estate in North Carolina. Due to Pinchot's leadership in sustainable forestry management, the Biltmore Estate is now known as the Cradle of American Forestry. Over the next thirteen years, Pinchot worked in a variety of other positions, helping to transform forestry management and natural resource conservation into a nationwide movement.[20]

Then, in 1905, Pinchot became the first Chief Forester of President Roosevelt's newly created US Forest Service (USFS) within the Department of Agriculture. Once in office, Pinchot used his energy, maverick philosophy, and dynamic personality to transform the management of forestland across the United States permanently. He diligently worked toward wisely using the nation's forests for the benefit of humans, not just preserving them for nature's sake. Pinchot believed that "the object of our forest policy is not to preserve the forests [just] because they are beautiful . . . or because they are refuges for the wild creatures of the wilderness. The forests are to be used by man. Every other consideration comes secondary."[21]

Despite his belief that forest use had to be the first priority, Pinchot's primary goal was to prove that forestry could produce timber for harvest while also maintaining the quality of forests for future generations. With this philosophy, he was the first person to coin the term "conservation ethic" and one of the first practitioners of what is now known as managing for resource sustainability.

Emphasizing Pinchot's two primary driving philosophies—"the greatest good for the greatest number over the long run" and "conservation coupled with wise use of natural resources,"[22] the redefined USFS soared to great new heights. Under his administration, the number of forest reserves—later called National Forests—grew from 60 units, covering 56 million acres, in 1905 to 150 separate management areas, covering 172 million acres, in 1910.[23]

Pinchot believed that multiuse management was the best way to proceed. With this notion in mind, he extended federal regulation to all resources—including forestry, grazing, water power dam sites, mineral rights, and recreational activities—within national forests boundaries. This management approach still abides throughout today's USFS.[24]

Maintaining his close friendship with Roosevelt, Pinchot also served on several of the president's commissions, including the Commission on the Organization of Government Scientific Work, the Commission on Public Lands, the Commission on Departmental Methods, the Inland Waterways Commission, and the Country Life Commission.[25] He was also the primary founder of the Society of American Foresters, which first met at his home in Washington, DC. Pinchot died of leukemia in New York City on October 4, 1946, at the age of eighty-one.[26]

> ***Without natural resources life itself is impossible. From birth to death, natural resources, transformed for human use, feed, clothe, shelter, and transport us. Upon them we depend for every material necessity, comfort, convenience, and protection in our lives. Without abundant resources prosperity is out of reach."***
>
> **—Gifford Pinchot[27]**

Gifford Pinchot was as dedicated to his beliefs as anyone who ever lived—perhaps even to a fault. In a love story for the ages, Pinchot maintained a secret affair with Laura Houghteling—a socialite and the jewel of his life—whom he met in 1893. At first blush, this bond may not sound that unusual, but deeper investigation reveals the intrigue. Ms. Houghteling died less than one year after she first met Pinchot, but Pinchot wrote letters to her and kept diaries describing their imagined—or perhaps very realistic, at least in his mind—relationship for twenty years after her death. He remained faithful and celibate during this whole time, not marrying until he was forty-nine.[28]

Pinchot's long-standing dedication to the things he loved, alongside his commitment to transforming the nation's forest-management system from wanton slash-and-burn policies to wise, sustainable, long-term use is a model for a climate-change hero. All along the way, Pinchot battled both corporate and political resistance to changing the way things have always been done. But his dedication never wavered, despite great obstacles. His legacy of making wise short-term use of natural resources to foster their long-term protection endures to this day and provides a prototype for achieving sustainable management of Earth's energy sources.

JOHN MUIR AND HIS BATTLE FOR YOSEMITE PARK

The battle between conservationists and preservationists continued after Roosevelt and Pinchot left office. The preservationists believed the conservationists were exposing the nation's natural resources to widespread overuse and eventual destruction, while the conservationists categorized the preservationists as idealistic amateurs—precursors to today's tree-huggers—who had their heads in the sand when it came to economic reality. A rancorous dispute over the proposal to dam the Tuolumne River and flood Hetch Hetchy Valley in Yosemite National Park brought the head-to-head conservationist/preservationist struggle to a fever pitch.

Let's start by looking at the need for putting a water supply reservoir in Yosemite National Park in the first place. You've probably heard about the historic earthquake that struck the city of San Francisco in 1906. Blamed for more than three thousand deaths and for devastating

80 percent of the urban area, the "Great Quake" still ranks as one of the worst natural disasters to ever hit the United States. But you might not know that the widespread fires sparked by the quake revealed the woefully poor quality of the city's water infrastructure. There simply was not enough water available to successfully douse the flames, which then roared unabated throughout the entire city.

For two years before the earthquake struck, the city leaders had, in fact, been struggling with the question of how to meet the city's water needs while protecting the area's natural resources. From the purely human perspective, Hetch Hetchy Canyon—located 167 miles west of San Francisco—offered the perfect topographic configuration for constructing a dam and reservoir that would provide a long-term solution to San Francisco's water requirements. However, the Hetch Hetchy Valley rivaled the Yosemite Valley in terms of spectacular scenery, abundant wildlife, and marvelously varied recreational opportunities. Building the dam and reservoir would negatively impact the valley.

As so often happens following a calamity, the severity of the earthquake and fires flipped the ongoing debate in favor of damming the Tuolumne River at Hetch Hetchy. In 1908, the US Department of the Interior—which had previously denied a permit—granted the city of San Francisco's application for development rights of the Tuolumne River in Hetch Hetchy Canyon. With the permit firmly in hand, the process of planning for construction of the Hetch Hetchy Dam would soon follow—or so the city fathers thought. What they hadn't planned on was a feisty fellow named John Muir and his legion of devoted followers from the newly formed conservation organization known as the Sierra Club.

John Muir was born in 1838 to a very strict and religious family in Dunbar, East Lothian, Scotland. Muir spent his boyhood alternating between two incongruous pursuits—playground fighting and nature walks with his grandfather.[29] John's playful and adventurous spirit led him to crossing the harbor by hopping from boat to boat and to rock-climbing feats on the walls of Dunbar Castle[30]—foreshadowing his future escapades on the granite walls of Yosemite National Park. In 1849, seeking a stricter religious foundation than he had in Scotland, Muir's father moved the family to Fountain Lake Farm near Portage, Wisconsin.[31] Partly to escape the religious fervor in his home, young John rose early for clandestine visits to neighbor's houses to read poetry in the summer and built inventions in the darkness of winter.[32] His love of nature led to regular Sunday afternoon ramblings around his home in in Wisconsin. These pursuits often resulted in lashings from his father who believed any activities that did not involve Bible study were a waste of time. An itinerant Presbyterian minister, Muir's father insisted that Muir memorize the Bible, and had no patience for anything besides religious or practical books.[33]

From the time he hit Wisconsin soil, Muir's life as an adventurer took off in earnest. While he did not acquire his distinctive long, thick beard, tousled hair, and crooked walking stick until adulthood, his tireless ramblings took him all over the Wisconsin countryside throughout his college days at the State University in Madison.

Though Muir delighted in his studies at the university, and they formed the basis for many of his passions, including geology and botany, he left before receiving a graduate degree.[34] Instead, he went to work at a steam-powered factory in Indianapolis,[35] pursuing his dreams of being an inventor, which were dashed when a file slipped and punctured his right eye.[36] This may have been one of the best things to happen to our country's natural resources, despite its toll on Muir. After he regained sight, his biggest misgiving was that he had almost lost the ability to investigate God's creation. With little haste, he embarked on a thousand-mile "stroll" from Indianapolis to the Gulf of Mexico, sailing to Havana, Cuba, and later Panama, then cruising up the West Coast to San Francisco, California.[37] Finally, in 1868, at age thirty, he made his way to the Yosemite Valley, the place that was to become the land of his lasting legacy and unrelenting devotion.[38]

In 1869, after working as a sheepherder in the California Sierra Nevada high country for a season—"The Range of Light," as he referred to it[39]—Muir took a job building a sawmill in the Yosemite Valley. He spent almost all of his free time exploring the peaks, nooks, and crannies of what was to become Yosemite National Park. During this time, he theorized that the Yosemite Valley had been carved by glaciers and then unconditionally surrendered to nature.[40] Muir believed that humankind was just one part of a world where everything was interwoven like a massive spider web. He continued to maintain his belief in God, though, expressing the opinion that God is revealed only through personal communication with the natural world.[41]

In 1873, after four years in the Sierra High Country, Muir moved to the city of Oakland. He began publishing articles in leading literary publications like *Atlantic Monthly*, *Overland Monthly*, *Scribner's*, and *Harper's Magazine*, describing the ecstasy he felt in the natural world. These narratives soon made Muir nationally famous and helped him build strong coalitions throughout the government, corporate, and political worlds. He combined these strong contacts with his widespread national fame and critical acclaim as a speaker, activist, and proposal writer to become our nation's most accomplished and important land preservationist. Muir's love of the western high country gave his writings a unique spiritual quality. His readers—who included presidents, congressmen, and ordinary people—were inspired by his words and often moved by his enthusiasm to take action to protect the country's natural places.

Muir became the public voice for protecting the high country around the Yosemite Valley. In 1890, Yosemite was named a national park, which set the stage for the nation's national park system. In fact, the creation of Sequoia, Mount Rainier, Petrified Forest, and Grand Canyon National Parks can all be attributed to Muir's efforts.[42] In 1892, as a further testament to his growing reputation as the nation's leading conservationist, Muir and a cadre of his devoted followers founded the Sierra Club to "do something for wildness and make the mountains glad."[43] Muir served as the first president of the Sierra Club until his death in 1914, and the group is now one of the leading environmental organizations in the world.

Muir's greatest cause célèbre came during a 1903 three-night excursion to the Yosemite Valley with President Teddy Roosevelt. Many authors have called this sojourn the camping trip that changed America, and for excellent reasons. First, Muir successfully persuaded Roosevelt to transfer the spectacular Yosemite Valley and the equally magnificent Mariposa Grove of giant sequoias from state park status to federal protection, adding them to the portion of Yosemite National Park that had been created in 1890.[44]

This trip, away from the tightly wound vagaries of Washington, DC, and the White House, had a profound and lasting impact on national conservation policies throughout the rest of Roosevelt's presidency. Of his Yosemite escape with Muir, Roosevelt fondly remembered, "It was like lying in a great solemn cathedral, far vaster and more beautiful than any built by the hand of man."[45] Roosevelt also said of the park, "There can be nothing in the world more beautiful than the Yosemite, the groves of the giant sequoias and redwoods . . . our people should see to it that they are preserved for their children and their children's children forever, with their majestic beauty all unmarred."[46]

The epic battle over Hetch Hetchy, which we talked about earlier, was, by far, the most titanic and traumatic struggle of Muir's life. He wrote, "These temple destroyers, devotees of ravaging commercialism, seem to have a perfect contempt for Nature, and instead of lifting their eyes to the God of the Mountains, lift them to the Almighty Dollar. Dam the Hetch Hetchy! As well dam for water tanks the people's cathedrals and churches, for no holier temple has been consecrated by the heart of man."[47]

The preservationists that battled with Muir drew withering fire from the citizens of San Francisco who believed these "tree-huggers" were destroying the city's long-term growth. Muir and the Sierra Club published numerous articles in magazines and newspapers throughout the country, gaining widespread support. The issue was finally decided in 1913, however, when Congress passed a bill to allow the building of the dam and the flooding of the Hetch Hetchy Valley. John Muir was heartbroken. A pristine jewel of the land that he had spent the best years of his life exploring and writing about was about to be desecrated forever. Many of Muir's devout followers believed that approval of the Hetch Hetchy Dam cost him his life. After the overwhelming *yes* vote, Muir became severely stressed, increasingly depressed, and died a year later of pneumonia.

Despite this, Muir's bountiful life and his legacy as America's first true preservationist offers much to today's climate-change heroes. Often called the "Father of Our National Park System,"[48] Muir lived life to the fullest and, through his writing and speaking, made others aware of the joy he found in his famous western cathedrals: "Climb the mountains and get their good tidings. Nature's peace will flow into you as sunshine flows into the trees."[49]

But Muir also realized the importance of fighting for the protection of these iconic landscapes as a way to honor the entire natural world: "When we try to pick out anything by itself, we find it hitched to everything else in the universe."[50] In the end, he gave his life—battling corporations, city officials, and federal bureaucrats—trying to save a sacred piece of his hallowed cathedral in the High Sierras.

In spite of the loss of the Hetch Hetchy battle, Muir's legion of preservationists gained quite a bit of traction during the second half of the twentieth century. For one thing, they learned that using the print media—magazines and books—could be a very effective tool for rallying public support to their side. Operating in a similar fashion, the climate-change community should take advantage of every type of media available—particularly emphasizing social media these days—to spread awareness of the critical importance of maintaining our seashores, coastal wetlands, coral reefs, and oceanic icefields for all future generations of Americans to behold.

CREATION OF THE NATIONAL PARK SERVICE

No doubt spurred on by articles describing America's fabulous scenery—particularly in the West—thirteen new national parks were created by 1916, including California's Sequoia and General Grant (1890), Washington's Mount Rainier (1899), Oregon's Crater Lake (1902), South Dakota's Wind Cave (1903), Colorado's Mesa Verde (1906), Montana's Glacier (1910), and Colorado's Rocky Mountain (1915). These new parklands included properties that protected a combination of historic and archeological features, as well as natural resources.[51]

Responding to a multifold increase in total park visitation—from 69,000 in 1908 to 335,000 in 1915—Congress created the National Park Service (NPS) in 1916 to manage these priceless preserves. As a new federal bureau in the Department of the Interior, the NPS now assumed the ultimate challenge of earning the utmost trust of conservationists and preservationists alike.[52]

The nation's thirty-five national parks and monuments—along with those yet to be created—were placed in the hands of carefully chosen staffs of dedicated scientists, wise land managers, and proven federal servants. As stated on the NPS website, "the Service thus established shall promote and regulate the use of the Federal areas known as national parks, monuments and reservations . . . by such means and measures as . . . to conserve the scenery and the natural and historic objects and the wild life therein and to provide for the enjoyment of the same in such manner and by such means as will leave them unimpaired for the enjoyment of future generations."[53] Based on our years of professional experience working with the NPS and delighting in visiting our parks and enjoying their natural splendors, we will say that the NPS has definitely lived up to this challenging commitment.

ANNA BOTSFORD COMSTOCK AND HER HANDBOOK OF NATURE STUDY

Anna Botsford Comstock provides us with another strong example of a past environmental hero who used her exceptional writing skills to spawn sincere interest in caring for the natural world. She was an ardent conservationist long before most people in the United States even knew the word.[54] In her 1912 book, *The Handbook of Nature Study*,[55] she was also way ahead of her time in stressing the importance of the interactive relationships—both biological and nonbiological—that work together to form what we now call ecosystems.

Born in 1854, Comstock grew up on a farm in Otto, New York, where she spent a lot of time outdoors with her Quaker mother, who taught her about the elements of the natural world—such as insects, birds, wildflowers, and trees. In 1874, she enrolled at Cornell University, where she met and fell in love with her husband-to-be, John Henry Comstock, an entomology professor. She then withdrew from school and spent several years drawing exquisite insect illustrations for her husband's books.[56]

In the mid-1890s, Comstock finally had the chance to break out and shine like the star she was destined to be. The New York Society for the Promotion of Agriculture asked her to help introduce a nature-study program—

the first of its kind in the United States—into local schools in Westchester County, New York. At first, many parents and teachers resisted the idea of teaching about the outdoors, arguing that it was frivolous, unproductive, and a waste of time. Despite these objections, Comstock's environmental education initiative soon grew into a nationwide teacher-education program administered by Cornell University, and her career was off and running.[57]

Comstock stayed busy promoting her nature-study program by producing study guides and instructional booklets for teachers to use across the country. Her works emphasized taking young people outdoors where they could see and experience relationships between people and the natural world for themselves. In so doing, she left her mark on countless generations.[58]

During the early 1900s, Comstock and her husband opened the Comstock Publishing Company, which had as its motto: "Nature through Books." It was here that she wrote and illustrated a series of books, including *Ways of the Six-Footed* (1903), *How to Keep Bees* (1905), *The Pet Book* (1914), and *Trees at Leisure* (1916). But her tour de force remained her nearly nine-hundred-page tome, *The Handbook of Nature Study*, which is now a famous sourcebook for teachers. Since its original publication, this guidebook has gone through twenty-four editions and has been translated into eight languages.[59]

Throughout her landmark work, Comstock continually emphasizes the rewards of direct observation of the natural world, writing, "I want to cultivate the child's imagination, love of the beautiful, and sense of companionship with life out-of-doors."[60] Expressing a departure from today's de rigueur reliance on indoor electronic tethers, it's no wonder that the *Handbook of Nature Study*—with its emphasis on leaving the classroom and going outside to actually see, breathe in, and experience the great outdoors—still remains so popular with grade schools teachers.

Comstock retired from full-time teaching in 1922 but continued to lecture. A 1923 poll by the League of Women Voters named her one of America's twelve greatest living women.[61] Comstock died of cancer in Ithaca, New York, on August 24, 1930.

Outside of the classrooms, Comstock's work as a conservationist remained largely unknown and unappreciated until the US environmental movement started to gather steam in the 1960s and 1970s. Then in 1988 she was named to the National Wildlife Federation's esteemed Conservation Hall of Fame, where she is now forever lauded as the "Mother of Nature Education."[62]

While she was not as outspoken as conservation leaders such as contemporaries like Theodore Roosevelt or John Muir, Anna Botsford Comstock heightened the national environmental consciousness by reaching out to America's youth. Since climate change holds the greatest peril for future generations of Americans, we must all involve our children and grandchildren in finding solutions.

THE PASSING OF MARTHA: THE WORLD'S LAST PASSENGER PIGEON

In a landmark tragedy for the US environmental movement, the last passenger pigeon—named Martha in honor of our original first lady Martha Washington—died at the Cincinnati Zoo on September 1, 1914. Perhaps no other event in US history so clearly illustrated the fact that humans were dramatically impacting the natural environment. When European colonists first arrived, the passenger pigeon was the most abundant wild bird in the United States, and possibly in the world. Flying in tightknit flocks numbering in the billions, passenger pigeons once darkened the skies across the United States for days on end. As one of the United States' most iconic representative of the horrors of species extinction, Martha's stuffed remains were displayed at the Smithsonian Institution from June 2014 until September 2015 in an exhibit entitled "Once There Were Billions."[63]

WORLD WAR I: THE GREAT WAR

In 1914, World War I—also known as "The Great War"—broke out after the assassination of Archduke Franz Ferdinand, heir to the throne of the Austro-Hungarian Empire, and his wife in the Bosnian capital of Sarajevo. For four long years, the miserable and bloody trench warfare and close-quarters assaults resulted in more than 37 million military and civilian casualties (killed, wounded, and missing-in-action). When the war started, the United States, with just a small army and a pitiful navy, stayed out of the European conflict.

Ironically, just months after winning the election with the slogan "He Kept Us Out of the War," President Woodrow Wilson declared war in 1917 when Germany refused to stop their U-boats from sinking neutral nonmilitary ships—most notably the *Lusitania*, in 1915, which resulted in 1,200 deaths, including 128 Americans. Overall, World War I sent more than 100,000 American troops to their deaths. The Great War finally ended on November 11, 1918, after the Germans realized they could not withstand the seemingly endless onslaught of fresh American troops and signed an armistice. To this day, many historians believe the world could have avoided the outbreak of the war if prudence and common goodwill had found a voice after Ferdinand's assassination.

In retrospect, World War I proved to be a watershed event in US history. With orders for war-damage repairs flooding in from all over Europe, American manufacturers became wealthy and American industrial might began to lead the world. The international financial system set up its headquarters in New York City. Overall, the war catapulted America into a role as the world leader in both economic and military strength. This war—which caused so much heartache and devastation throughout the world—established the military-industrial complex that rules our country to this day. Our rise to world power proved especially critical some two decades later, when an even greater world war—involving the twin scourges of Nazi Germany and Imperial Japan—loomed on the horizon.

World War I's establishment of the United States' dominant military-industrial complex has a significant place in today's climate-change debate. As Naomi Klein emphasizes in her book *This Changes Every Thing*—which we will talk about more in chapter 21—solving climate change may very well require a significant revision in the way the United States conducts business, both domestically and throughout the world.[64]

NORTH AMERICA'S HIGHEST PEAK GETS ITS DUE

Back on the home front, after eleven years of lobbying and wrangling between Alaska and Congress, preservationist Charles Alexander Sheldon, along with George Bird Grinnell and the Boone and Crockett Club, achieved an important goal. On February 26, 1917, President Wilson signed legislation designating Alaska's Mount McKinley National Park. Since then, the name of the park and the mountain—the tallest peak in North America—has been changed to Denali, meaning "the High One," the native Athabaskan term—for the mountain.

The Progressive Era had brought much awareness to the conservation and preservation movements. Much had changed in only a few years. The next decade, however, would struggle to continue the forward momentum begun by people like Gifford Pinchot and John Muir.

Chapter 12

WATCH OUT

Here Come the Roaring Twenties (1920–1929)

DECADE OF ENVIRONMENTAL DECAY

Just when everything was starting to go swimmingly well for the preservationist ideal—the belief that nature should be protected just for itself and not for use by humans—along came the decade of environmental decay known as the "Roaring Twenties." The end of World War I and a resistance to the implementation of Prohibition brought out a "let the good times roll" mentality about living the high life.

Celebrating the good times meant speakeasies with door slits and secret passwords, bathtub gin, machine gun–toting gangsters, and general social mayhem. Lifestyle excesses focused on America's newfound love affair with the automobile—making the entire nation immediately accessible—plus a sudden abundance of electrically powered devices like the radio and kitchen appliances such as refrigerators, ovens, and toasters.

The presidential administration of Warren G. Harding went right along with the populist liberalism that characterized the 1920s, and, for the most part, previous concerns about natural-resource preservation and conservation issues flew right back out the window. In fact, the most prominent events that occurred under Harding's command were scandals—most notably the Teapot Dome scandal in Wyoming, in which oil reserves were illegally leased for donations of cash, bonds, and even a herd of cattle. When Calvin Coolidge succeeded to the presidency after Harding's sudden death in 1923, the scandals went away but the indifference toward environmentalism in the United States did not.

IZAAK WALTON LEAGUE: AMERICA'S FIRST MAINSTREAM CONSERVATION ORGANIZATION

Although the US environmental movement was hanging on by a thread as thin as spider's silk, some notable environmental events still managed to happen during this period of extreme human excess. In 1922, charismatic advertising executive Will Dilg established the Izaak Walton League of America (the IWLA) named after the author of *The Compleat Angler*.[1] As the first mainstream conservation organization with mass membership—over 100,000 supporters by 1924—the IWLA promoted natural resource protection and outdoor recreation across the nation.[2]

The decade of the 1920s also brought the emergence of two extremely significant past environmental heroes. One is very near and dear to our hearts—Budd's especially as a nature photographer. The other is responsible for protecting the pristine quality of natural resources on extensive tracts of public lands.

ANSEL ADAMS: LANDSCAPE PHOTOGRAPHER AND CONSERVATIONIST SUPREME

Choosing a professional career is typically a difficult decision for anyone to make. For years, I (Budd) aspired to be a full-time nature photographer, making a living by sharing my love for the amazing breadth of beauty that gave life to each of my footsteps in the great outdoors. After a few years of trying, I realized that there was simply too much competition and that I should just be happy with the occasional opportunities I had to see and capture nature in its rawest and most inspiring forms.

That is why we will forever remain in awe of the man who had the requisite skills and desire to be both a concert pianist and a master photographer. Fortunately for legions of nature lovers and conservationists throughout the United States, he left the performance halls and proceeded to become the greatest landscape photographer the world has ever known.

Ansel Adams was born in the Western Addition of San Francisco in 1902—just a few years before the city's Great Quake. While the four-year-old Adams was uninjured by the quake's main shock, an aftershock sent him tumbling into a garden wall. The face-first smash broke his nose, and he never bothered to have it surgically repaired.[3]

First visiting Yosemite National Park in 1916—only two years after John Muir's death and three months before the founding of the National Park Service—Adams stood mesmerized by the landmark's iconic splendor. While music was still his primary passion and planned profession at this time, this first experience with Yosemite at age fourteen planted the sparks that burned brightly in his brain for the future. From that day forward, he joyfully explored the natural world—especially his beloved Yosemite—while attempting to capture black-and-white replicas of the grandeur he saw at every turn.

In 1927, Adams got the break that changed his life forever when he was named the Sierra Club's official trip photographer. His role in the Sierra Club grew rapidly and the group's organized hikes and talks became instrumental to his career development, as the *Sierra Club Bulletin* published Adams's first photographs and writings.[4] By suggesting proposals for improving parks and wilderness areas, Adams was soon deeply involved in the club's environmental activities. He quickly became widely known as both a talented artist and a passionate advocate for his beloved Sierra Nevada Range and its majestic Yosemite National Park.[5]

Between 1929 and 1942, Adams expanded his repertoire, focusing on detailed close-up photographs, as well as on large-format images of everything from mountains to factories. He spent a great deal of time in New Mexico socializing with other well-known artists such as Alfred Stieglitz, Georgia O'Keeffe, and Paul Strand. During this period, Adams also joined photographers Dorothea Lange and Walker Evans in their commitment to effecting social and political change through art. After the internment of Japanese people in the United States during World War II, Adams photographed life in the camps for a photo-essay on wartime injustices.[6]

Adams first used his images for environmental purposes when the Sierra Club sought the creation of a national park in the Kings River region of the Sierra Nevada. He published a limited-edition book entitled *Sierra Nevada: The John Muir Trail*, while lobbying Congress for a Kings Canyon National Park.[7] His text and exquisite black-and-white photographs helped influence both Interior Secretary Harold Ickes and President Franklin Roosevelt to create Kings Canyon Park in 1940.[8]

Adam's iconic landscape images—especially those of the American West—have since inspired millions of people. His photographs and writings—including more than forty books—helped expand the National Park System and the nascent Sierra Club. His collective works have been hailed as providing the foremost record of what many of our national parks were like before the advent of tourism. While he pursued his art, Adams also tirelessly advocated for balancing progress with the maintenance of the peace and solitude that can only be found in undeveloped natural areas.

As a past environmental hero, Adams was a visionary figure in nature photography and wilderness preservation. He wrote, "I hope that my work will encourage self-expression in others and stimulate the search for beauty and creative excitement in the great world around us."[9]

In 1968, the Interior Department awarded Adams the Conservation Service Award, their highest civilian honor, "In recognition of your many years of distinguished work as a photographer, artist, interpreter and conservationist, a role in which your efforts have been of profound importance in the conservation of our great natural resources."[10] Then in 1980 he was awarded the Presidential Medal of Freedom, for "his efforts to preserve this country's wild and scenic areas, both on film and on earth. Drawn to the beauty of nature's monuments, he is regarded by environmentalists as a national institution."[11] Ansel Adams died from cardiovascular disease at the Community Hospital in Monterey, California, on April 22, 1984.

Why was Ansel Adams revered by Americans as few other artists or conservationists had ever been? Author William A. Turnage offers this explanation, "More than any other influential American of his epoch, Adams believed in both the possibility and the probability of humankind living in harmony and balance with its environment."[12]

Ansel Adams was a master of more than just landscape photography. He used his magnificent works of art to draw people into his thoughts. Once he had the public's attention, he communicated—in his writings and speaking engagements—the message that if we are not careful, we could easily lose all of our nation's unparalleled natural splendor.

Adams's skill of transforming enjoyment of art into support for a cause is critical to converting climate-change deniers and fence-sitters. If he were alive today, Ansel Adams would certainly be in the forefront of our heroic charge to design and implement climate-change solutions that would allow the harmony and balance of humans and the environment to continue.

ROBERT "BOB" MARSHALL: WILDERNESS WAS IN HIS BLOOD

There is a revered—many would say mythical—landscape in northwestern Montana that features 1.5 million acres of pristine lakes, crystal rivers, snow-capped mountains, and evergreen valleys. This magnificent wilderness jewel is known simply as "the Bob."[13]

Wilderness had always been in Dr. Robert "Bob" Marshall's blood. While he did not live to see the completion of his namesake wilderness area—a fitness fanatic, he ironically died far too early, at age 38—his legacy as a conservationist and humanitarian will forever stand strong in Montana, as well as throughout wilderness areas scattered across the United States.

Born on January 2, 1901, to Louis Marshall—a wealthy civil rights lawyer and philanthropist—in New York City, Marshall spent his summers at the family home, named "Knollwood," on Lower Saranac Lake in the heart of the Adirondack Mountains.[14] As a teenager and young adult, he roamed all over his beloved mountain group, eventually becoming a member of "46ers Club" by climbing all forty-six of the range's peaks that are above four thousand feet.[15]

After graduating from the Ethical Culture School in New York, Marshall attended Columbia University, the New York State College of Forestry at Syracuse, and then Harvard University. From there, his national wanderings took him to the Northern Rocky Mountain Forest Experiment Station in Missoula, Montana, where he worked from 1925 to 1928. He went back to school in 1928 at Johns Hopkins University to get a doctorate degree in plant physiology—one of three doctorate degrees he earned in his lifetime.

In February 1930, Marshall's most poignant essay, "The Problem of the Wilderness," was published by the *Scientific Monthly* and became one of the most influential works in conservation history.[16] In this piece, Marshall wrote that there are many reasons for preserving wilderness beyond just its intrinsic value as landscape untrammeled by human activities: "Fundamentally, the question is one of balancing the total happiness which will be obtainable if the few undesecrated areas are perpetuated against that which will prevail if they are destroyed."[17] Today, many wilderness historians credit "The Problem of the Wilderness" as being a seminal call to action, while setting the stage for the passage of the 1964 Wilderness Protection Act more than thirty years later.[18]

While much has been made—and rightfully so—about Marshall's work for wilderness preservation, his life's interests extended far beyond such pursuits as hiking, mountain climbing, fishing, and horseback riding. Although he came from a very wealthy family, he had a deep and abiding—bordering on the spiritual—concern for those who were less fortunate than he was. In particular, Marshall cared about the indigenous groups of people he met during his extensive travels in pursuit of wilderness preservation.

For example, in 1930 and 1931, while living among the native Koyukuk in the remote Alaskan village of Wiseman, Marshall got the inspiration for writing his first book, *Arctic Village*.[19] The book became a bestseller in 1933 and earned him $3,600—a princely sum of money back then. After the book came out, instead of pocketing all the money for himself, Marshall gave away half of the profits to the Koyukuk he had known during his stay in Wiseman.

Then, throughout his career as a federal public lands administrator, Marshall made equal rights for all people—no matter their race, sex, creed, or religion—a personal priority.[20] He worked diligently to involve all Native Americans in the management of their tribal forests and mountain ranges. After realizing that their customs and freedoms of religious worship had been stripped away, Marshall fought to provide them with the level of equality he felt all Americans should enjoy. He wanted to provide every American with equal access and visitation rights to all recreational areas on federal lands. To accomplish this, he proposed subsidized travel so families with lower incomes could experience the beauty of the American landscapes firsthand.[21]

Marshall believed that truly wild lands had intrinsic values for the human spirit, providing places where people could go to find themselves and rekindle their souls. While serving as Chief Forester for the Bureau of Indian Affairs in 1935, Marshall, along with Aldo Leopold (whom we'll meet in chapter 14) and others, established the Wilderness Society, an organization dedicated to protecting the wilderness and inspiring Americans to care for wild places.[22] Then, as Assistant to the Chief of the US Forest Service—a position he held until he died—Marshall devoted much of his energy to the development of the National Forest's wilderness system.

Marshall's name and strong promotion of America's wilderness preservation system will forever remain a driving force for land conservation today. When he passed away in 1939, he bequeathed one-fourth of his $1.5 million estate to the Wilderness Society—thus assuring the organization's future in the name of wilderness preservation. Today, the Bob Marshall Award is the highest honor bestowed by the Wilderness Society.

> *[Wilderness] is the song of the hermit thrush at twilight and the lapping of waves against the shoreline and the melody of the wind in the trees. It is the unique odor of balsams and of freshly turned humus and of mist rising from mountain meadows. It is the feel of spruce needles under foot and sunshine on your face and wind blowing through your hair. It is all of these at the same time, blended into a unity that can only be appreciated with leisure and which is ruined by artificiality.*
>
> **—Robert "Bob" Marshall[23]**

> *How many wilderness areas do we need? How many Brahms symphonies do we need?*
>
> **—Robert "Bob" Marshall[24]**

Throughout Bob Marshall's remarkable—albeit far too short—life, he demonstrated an amazing blend of caring deeply about both natural preservation and equal human rights for all people. This is exactly the type of dual thinking that will allow us to bring climate-change solutions home to all individuals on Earth. Citizens of developing countries must be fully involved in every decision and action that affects them. The environmental justice movement must be for the entire world and not bypass developing nations in favor of catering to the wealth and power of developed countries.

Chapter 13

WOE IS US

The Great Depression and the Dust Bowl (1930–1939)

BLACK TUESDAY AND THE COLLAPSE OF WALL STREET

Some semblance of genuine concern for the environment finally returned to Washington, DC, with the Herbert Hoover administration in 1929. With his campaign slogan promising a "chicken in every pot and a car in every garage,"[1] Hoover won in a landslide election over Alfred Smith, who was the first Roman Catholic to run for president.

Unfortunately, this federal promise of economic prosperity and the refreshing of natural resource ideals was very short-lived, owing to the collapse of the stock market on October 29, 1929. Infamously known as "Black Tuesday," the Wall Street panic wiped out millions of investors. The "deepest and longest-lasting economic downturn in the history of the Western industrialized world,"[2] the Great Depression lasted for a decade (1929–1939).[3]

FDR COMES TO THE RESCUE

Our next past environmental hero is Franklin Delano Roosevelt (FDR), who was an indubitable American hero on three other major fronts—health care, economic issues, and military prowess—in addition to land conservation. Born in 1882 on his family's estate in Hyde Park, New York, FDR was given the best schooling, attending the prestigious Groton

School in Massachusetts, Harvard University, and Columbia University. As an adult standing more than six feet tall, Roosevelt was handsome, lean, and athletic, with deep blue eyes, dark wavy hair, and a strong jaw. He was also charismatic and genuinely interested in people.[4]

In 1921, while visiting his summer home on Campobello Island, on the border of New Brunswick, Canada, and the state of Maine, Roosevelt contracted poliomyelitis, otherwise known as infantile paralysis, the dreaded—at the time—disease called "polio." Perhaps Roosevelt's most heroic feat was how he dealt with this debilitating condition with dignity for the rest of his life. Roosevelt worked very hard to make it appear he wasn't really disabled, but he was never again able to stand without aid. Unfortunately, in those days, physical disability was considered a dehumanizing disgrace. In 1938, he started what would come to be known as the March of Dimes, a fundraising organization that led to the development of the Salk vaccine that eventually wiped out polio in the United States.

From an economic perspective, Roosevelt's election in 1932 proved to be the saving grace for the United States and the US environmental movement. FDR's program for relief, recovery, and reform, known as the New Deal, included a significant expansion of the role the federal government played in the economy. New Deal policies introduced an array of social programs—Social Security, the Wagner Act, and the Fair Labor Standards Act—that still form the backbone of the federal government's provisions for its citizens. FDR's economic policies were so monumental that they overshadowed the positive environmental changes he made. According to environmentalist Benjamin Kline, "If it were not for the ravages of the Great Depression, Roosevelt may have ranked among the most successful of environmental leaders in American history."[5]

Roosevelt's alphabet soup of other federal programs he instituted included the Agricultural Adjustment Act (AAA), the Civilian Conservation Corps (CCC), and the Tennessee Valley Authority (TVA). The AAA paid subsidies to farmers for not planting crops—intentionally leaving fields fallow—and not slaughtering livestock. This subsidy program prevented surplus agricultural products—like wheat, corn, livestock, and dairy products—from flooding the market, which would have driven the price of these commodities down and put many farmers out of business.

Although always controversial, the federal price subsidy program still exists today to maintain the price of agricultural products.

The CCC put young men to work on federal lands all across the nation. Many of the hundreds of infrastructure improvements created at this time—including bridges, dikes, reservoirs, roads, trails, and shelters—are still in use today in our national parks and monuments, national forests, and national wildlife refuges.[6] FDR's concepts for combining conservation and development were perhaps best epitomized by his TVA projects, which were established to bring water supplies, flood control, and inexpensive and renewable hydropower—as well as recreational amenities—to underserved areas of the southern United States.[7]

Several new national parks—Olympic, Shenandoah, Kings Canyon, plus the groundwork for Grand Teton—were also created during FDR's administration. Additionally, he oversaw two momentous milestones in wildlife management—the ban on killing predators in national parks and the establishment of the US Fish and Wildlife Service in 1940.[8]

Today, there are more than 560 national wildlife refuges across the country, with at least one in every US state and territory. Wildlife refuges attract nearly 50 million visitors every year, who come in pursuit of both active and passive recreational activities—including wildlife-watching, hunting, fishing, photography, hiking, canoeing, kayaking, and environmental education.[9]

During Roosevelt's unprecedented third term in office, he assumed the role of Commander-in-Chief throughout the World War II conflict with Nazi Germany and Imperial Japan. We'll discuss this war in more detail in the next chapter, but it's important to note that, throughout the war, Roosevelt served as the American counterpart to Great Britain's Winston Churchill. Roosevelt stayed actively involved in all military activities and even overrode the decisions of field commanders when he deemed it necessary.[10]

Roosevelt also worked to create a "Grand Alliance" against the Axis powers—Germany, Italy, and Japan—through a peacekeeping organization that is today known as the United Nations. Many historians believe that the additional stress of World War II—on top of managing his health and the nation's economic problems—proved too much for

Roosevelt. He died a few hours after experiencing a massive stroke at his Little White House in Warm Springs, Georgia, ironically on the eve of the United States' complete military victory in Europe and just months before the victory over Japan in the Pacific Theatre.[11]

> *We seek to use our natural resources not as a thing apart but as something that is interwoven with industry, labor, finance, taxation, agriculture, homes, recreation, and good citizenship. The results of this interweaving will have a greater influence on the future American standard of living than the rest of our economics put together.*
>
> **—Franklin Delano Roosevelt**[12]

Despite FDR's diligent remediation efforts, most of America's concerns about environmental protection went out the window—along with just about everyone's dutifully scrimped-and-saved-for life savings—during the Depression decade of the Thirties.

On a positive note, however, many of the ideas FDR put into practice to bring the United States out of the Great Depression are relevant to solving the climate-change crisis. For example, as with the AAA's price subsidies for agriculture, subsidizing fossil-fuel companies not to extract new reserves could maintain their financial statuses while forcing them to develop renewable energy sources to meet the customer demands on their systems. Also, a new federal agency along the lines of FDR's TVA could be established. This new agency would be specifically tasked with solving the climate-change crisis by promoting and managing the concurrent reduction in fossil fuels with the expansion of renewable energy supplies. Finally, a CCC-like organization could put hundreds of scientists and laborers to work immediately on designing and constructing the sources and infrastructure required to deliver renewable energy supplies to every corner of the United States.

THE DUST BOWL OBLITERATES THE NORTH AMERICAN PLAINS

During the 1930s, nowhere was the concept of "nature strikes back" more evident than on the soil-ravaged and agriculturally pillaged plains of the

United States and Canada. The Great Dust Bowl occurred because poor farming practices caused huge portions of what had been luscious and golden prairie grasslands to dry up and blow away with the wind.

Eager settlers had moved west to the prairies, lured by advertisements promising a Garden of Eden. They brought with them farming techniques that had worked in the Northeast but that were incompatible with the prairie ecology.[13] In what was called "the Great Plow Up," settlers dug deeply into the virgin topsoil and pulled up the thickly rooted native grasses that had held the soil in place and trapped moisture through periods of severe drought.[14] Recent rapid advancements in farming equipment in the early twentieth century—notably gasoline-powered tractors and massive combine harvesters—allowed more and more of these arid, native grasslands to be converted to wheat fields.[15]

When drought came and the crops dried up, the unanchored soil turned to dust and blew away, forming huge dark clouds that blackened the sky for miles around.[16] These choking billows of dust—named "black blizzards"—traveled cross-country, reaching as far as New York, Philadelphia, and Washington, DC. In total, the drought and erosion of the Dust Bowl affected 100 million acres (more than 150,000 square miles—roughly the size of Montana, our fourth-largest state) across states like Oklahoma, Texas, Kansas, Colorado, and New Mexico.[17]

In response to the social and environmental horrors of the Dust Bowl, Congress passed two significant pieces of legislation. First, the Taylor Grazing Act of 1934 defined designated areas where livestock grazing could occur on federal lands—thus minimizing the rampant overgrazing that was occurring. Next, in 1935, the Soil Conservation Act led to the establishment of the Soil Conservation Service (SCS)—now the Natural Resource Conservation Service (NRCS)—a federal agency tasked with advocating good farming practices that would minimize the potential for future wind and water erosion of farming topsoil.

Meanwhile, bankrupt and forced to abandon their homes and farms to foreclosure due to loss of crops, at least one-quarter of the plain's homesteaders migrated to California and other states.[18] They arrived at their destinations with great hope, only to find situations that were little better than those they had left.[19] In the minds of many, the Dust Bowl still ranks as the worst and most prolonged natural disaster the United States has ever experienced. In his 1939 novel, *The Grapes of Wrath*, John Steinbeck painted a poignant image of what it was like to experience the Dust Bowl:

> Carloads, caravans, homeless and hungry; twenty thousand and fifty thousand and a hundred thousand and two hundred thousand. They streamed over the mountains, hungry and restless—restless as ants, scurrying to find work to do—to lift, to push, to pull, to pick, to cut—anything, any burden to bear, for food.[20]

JOHN STEINBECK: PROLIFIC AUTHOR AND STAUNCH CONSERVATIONIST

John Steinbeck—another one of our past environmental heroes—is, in our opinion, one of America's greatest writers. We claim this not just because of Steinbeck's empathy with the human spirit but also because of the ardent environmentalism he displayed in his works. His perpetually exasperated look, with his thin mustache overtopping his ever-present cigarillo—hid his strong support for everyday working people and their living conditions.

From *The Grapes of Wrath*[21] to *Cannery* Row,[22] Steinbeck wrote scathingly about America's degradation of our environment. He lashed out at the overharvesting of fishery stocks, at harmful farming practices, and—most emphatically—at the global disaster of human overpopulation. In his novel *Sweet Thursday*,[23] Steinbeck warns that "Man, in saving himself, has destroyed himself."[24]

Throughout his writings, Steinbeck staunchly preached conservation and, in his last work, *America and Americans*,[25] he put forth the hope that we could learn to not "destroy wantonly."[26] If he were still alive today, we strongly suspect his next nonfiction book would be something along the lines of *Dark and Angry Skies: Fighting the World's War Against Climate Contaminants*.

Born in Salinas, California, in 1902, John Steinbeck lived in a modest family home in the midst of a prosperous farming community. This setting formed the background for many of his novels and the basis for his characters, who strongly identified with the land. Beginning in early adolescence, Steinbeck had a penchant for the pen.[27] In high school, he would hide away in his attic room and write short stories that he would send to magazines under pseudonyms without a return address. In later years, he sheepishly admitted that "he was scared to death to get a rejection slip, but more afraid of getting an acceptance."[28]

In 1919, Steinbeck enrolled at Stanford University but never graduated, although he took courses for five years, on and off. It was during this time, however, that he became interested in science and biology—beginning his lifelong concern with environmental issues. While working for a fish hatchery in Tahoe City, California, he met and married Carol Henning in 1930. The couple moved to a cottage owned by his father on the Monterey Peninsula, and Steinbeck wrote, "Financially we are in a mess, but spiritually we ride the clouds. Nothing else matters."[29]

In 1932, with the acceptance and publication of *Pastures of Heaven*,[30] a loosely connected collection of short stories about the Salinas Valley, Steinbeck's writing career took off in earnest. *The Grapes of Wrath*, which many consider his best work, won both a Pulitzer Prize and a National Book Award in 1939.

Steinbeck's ties to environmentalism and ecology—before each became a watchword in the United States—have been acknowledged and described by many scientists. In the book *Steinbeck and the Environment*, authors Clifford Gladstein and Mimi Reisel Gladstein write, "Literary works often precede and foretell the articulation of philosophical concepts. And lovers of the natural world have been among the most devoted readers of John Steinbeck. Maybe it is because they see in his works strong identification with and respect for tillers of the soil and harvesters of the sea as well as an abiding reverence for the earth in its pristine state."[31]

Writing in the same book, Lorelei Cederstrom takes things a step further, "In his depiction of the fertile earth and the lives of those who have depended on her for abundance, John Steinbeck in *The Grapes of Wrath* presents a visionary foreshadowing of the universal ecological disaster that looms so prominently on the horizon today."[32]

Steinbeck continued to write in his later years, publishing many highly acclaimed and widely read books—*Burning Bright* (1950),[33] *East of Eden* (1952),[34] *The Winter of Our Discontent* (1961),[35] and *Travels with Charley: In Search of America* (1962).[36] In 1962, he received the Nobel Prize for Literature, which was awarded for his "realistic and imaginative writings, combining as they do sympathetic humor and keen social perception."[37] Steinbeck died of heart disease far too young—at the age of sixty-six—on December 20, 1968, at his home in New York City.

The best two pieces of advice any writer can ever receive are "be observant of the world around you" and "write about what you know" (as we mentioned earlier in the book). Based on his prolific and heartfelt works, Steinbeck was a master at both of these directives. Writers today who are working to counter climate change would do well to study Steinbeck's books. We need to learn how to emulate his skill at expertly portraying the essence of the social, business, and environmental inequalities he saw happening around him. Applying the same burning passion that Steinbeck displayed in his writing will certainly help us rally the world to our climate-change cause.

ROGER TORY PETERSON: FATHER OF THE FIELD GUIDE

Who would you say the local townspeople in the small city of Jamestown, New York, voted for in 1975 as their most famous native daughter or son? If you guessed the world-famous television star and comedienne Lucille Ball, you'd be wrong. Although Lucille Ball was indeed born in Jamestown, the townsfolk chose Roger Tory Peterson as their best-known resident.

This man, with his handsomely-craggy face and—by many accounts—somewhat obstinate personality—was then known to some 20 million birdwatchers around the nation as simply "Peterson"—as in "quick, look it up in your Peterson."[38] In 1934, naturalist and wildlife artist Roger Tory Peterson published his first of many field guides, which would together sell millions of copies throughout the United States.

Born on August 28, 1908, in Jamestown, Peterson was known to his teachers and other townsfolk as a constant troublemaker—always pulling pranks and raising hijinks. Due to his budding genius, he skipped two grades; the school kids called him "Professor Nuts Peterson" since he was atypically (for a kid his age) absorbed in insects, bugs, and biota.[39] Peterson's true bird-loving epiphany occurred at age eleven when he joined the local Junior Audubon Bird Club. Once he latched onto his first birding field guide and pair of binoculars, his wayward days ended and a life-long passion began.

Peterson was also interested in art, and, as a young adult, he enrolled in the Art Students League in New York City and also studied at the National Academy of Art Design. While studying in New York City, Peterson was influenced by a number of serious young birders—including Joe Hickey, Allan Cruickshank, and John Aldrich.[40]

In his early twenties, while he was struggling to make a living as an artist, Peterson's two life passions began to come together. He started to notice that each bird species had distinctive field marks that clearly set it apart from other—even closely related—species. He realized from these initial observations that he had discovered a way for birds to be quickly identified—even at a distance.

Barely able to contain his excitement, Peterson set to work drawing and writing his first book, *A Field Guide to the* Birds,[41] which was published in 1934. Since then—through more than forty-seven total reprintings—over seven million copies of Peterson's two primary birding field guides—*Guide to the Eastern Birds*[42] and *Guide to the Western Birds*[43]—have been sold.[44] Peterson continued to boost public interest in wild birds by writing articles for popular publications, bridging the gap between professional ornithologists and amateur backyard birdwatchers.

Peterson's 1948 book, *Birds over America*,[45] also demonstrated his depth and breadth as a dedicated conservationist. While primarily describing birds and birdwatching across North America, this book is notable in the fields of both ecological principles and environmental ethics. In particular, *Birds over America* features the interconnected web of all living things, how hunters and farmers affect conservation, the ominous threats posed by invasive species, and the importance of protecting endangered species and their critical habitats.

For the 50 million folks who regularly watch birds in the United States, Peterson is now known as the "Father of the Field Guide." Birders everywhere now carry these indispensable pocket-sized books, which provide vital field-mark clues for accurate bird identification. The bird paintings included in his first field guides also earned Peterson the title of "The Audubon of the Twentieth Century."[46]

Peterson's entire series of field guides—which included identification books on everything from amphibians to butterflies, fish, reptiles, wildflowers, and even seashells—fostered an appreciation for the natural world and helped set the stage for ramping up the US environmental movement during the sixties and seventies. It became impossible to find someone who was interested in the natural world who didn't have at least one Peterson Field Guide on their home or office bookshelves. In 1984, Peterson founded the Roger Tory Peterson Institute of Natural History in Jamestown in order to further educate people about nature and conservation.

Although never a professed activist, Peterson always maintained his dedication to environmental protection and his dogged opposition to environmental hazards throughout his career—for example, the use of DDT.[47] In fact, when Rachel Carson's 1962 *Silent Spring* sounded its warning about pesticides' threat to bird habitats, the Peterson Field Guide Series had primed the country to care about birds; Roger's field guides were a tool that helped people connect with and cherish living things.[48] The Roger Tory Peterson Institute has this to say about Peterson's contributions, "Birdwatching soon became a hobby—even a sport—for vast numbers of people. From identifying birds emerged a love of birds, and from there it was only a short step to a more inclusive passion for nature."[49]

Peterson continued to write and publish an array of field guides and other books after moving to Old Lyme, Connecticut, where he died in 1996. As any birder or naturalist will agree, Peterson's many accolades—including the Gold Medal of the New York Zoological Society, the American Ornithologists' Union Brewster Medal, the Linnaean Society of New York's Eisenmann Medal, and the Order of the Golden Ark of the Netherlands—are all well deserved.[50] The most meaningful award he received was the Presidential Medal of Freedom—our nation's highest civilian honor—which was presented by President Jimmy Carter in 1980. Enjoying the great outdoors would just not be the same without the Peterson Field Guides in our pockets and packs.

In 1994, Peterson invited a group of world-famous outdoor photographers, editors, vendors, and authors to his Institute in Jamestown for the founding of the North American Nature Photography Association (NANPA). NANPA is now the world's only organization devoted solely to practitioners of outdoor photography.

I (Budd) am very proud to say that I am a charter member of NANPA—having joined in 1995 at the first Nature Photography Summit in Fort Myers, Florida—and I have also served on NANPA's board of directors from 2009 to 2015.

Roger Tory Peterson—a troubled youth with a passion for nature, but no clear direction—started a revolution in the field of environmental education that has continued to this day. He brought the study and understanding of the natural world out of the halls of academia and into the farms, fields, meadows, and mountains of North America.[51] This "boots on the ground" approach is exactly the type of educational effort that we need to create the groundswell of public support in the climate-change movement.

> ***The philosophy that I have worked under most of my life is that the serious study of natural history is an activity which has far-reaching effects in every aspect of a person's life. It ultimately makes people protective of the environment in a very committed way. It is my opinion that the study of natural history should be the primary avenue for creating environmentalists.***
>
> **—Roger Tory Peterson**[52]

DING DARLING: THE MAN WHO SAVED DUCKS

Our next past environmental hero had the best name in the history of the US environmental movement and—like Roger Tory Peterson—also used his creative ability in a masterful way. But this time, the craftsmanship was combined with offbeat humor, to get the public's attention and build support. Jay Norwood "Ding" Darling combined an artist's eye with a humorist's ear and, over fifty years, produced prize-winning political cartoons that amused the American public while galvanizing their support for landmark conservation initiatives.

Born in Norwood, Michigan, in 1876, Darling always considered himself an Iowan—even though his middle name derived from his place of birth. After he moved with his parents to Sioux City, Iowa, in 1886, he reveled in getting out and exploring the edge of the American frontier. Rambling about through unspoiled prairie teeming with seemingly limitless wildlife gave Darling a life-long passion for protecting nature's bounty.

An affable and energetic man, who often wore a bow tie and who bore a strong resemblance to news anchor Walter Cronkite—"the most trusted man in America"—Darling began drawing his political cartoons in 1900 while working for the *Sioux City Journal*. In 1906, he moved on to the staff of the *Des Moines Register*, where he began signing his cartoons with his nickname "Ding." He had invented this catchy pseudonym when he was in college, combining the first initial of his last name with its last three letters.[53] By 1917, his work was syndicated across the country through the *New York Herald Tribune*.[54]

Appearing in 130 daily newspapers, Darling's witty and insightful drawings entertained an audience of millions. Using his satirical pen to focus public attention on environmental concerns and resource conservation issues earned him Pulitzer Prizes in both 1923 and 1942.[55]

While he always claimed that he was a conservationist as "only a hobby,"[56] Darling's monumental accomplishments in the conservation arena belie that statement. In 1934, as Chief of the Bureau of Biological Survey—predecessor of the US Fish and Wildlife Service—he dramatically cut waterfowl bag limits and seasons in order to bolster dwindling waterfowl populations. This action earned him his reputation as "the man who saved ducks."[57]

Always an articulate speaker and tireless activist, Darling convinced President Franklin Delano Roosevelt to call the first North American Wildlife Conference in 1936—a landmark meeting that emphasized the need to have a permanent organization advocating for the protection of wildlife and wild places throughout the United States. Out of this conservation confab, the General Wildlife Federation (GWF)—forerunner of today's National Wildlife Federation—was born, and Darling became its first president. The GWF provided a long-term public platform for reining in the uncontrolled exploitation of wildlife—one of Darling's life concerns.[58]

Darling's other focus was the wanton destruction of vital waterfowl habitat, and he initiated the Federal Duck Stamp Program to do something about that concern. Proceeds from the sale of duck-hunting stamps—the first of which Darling drew himself—went into a fund set aside specifically for purchasing wetlands to preserve waterfowl nesting and migratory habitat.[59]

Darling was also in large part responsible for establishing the network of National Wildlife Refuges that now spread across the country. The Federal Aid in Wildlife Restoration Act of 1937 (also known as the Pittman-Robertson Act), which provides money to states for the purchase of game habitat and helps fund wildlife research through a tax on sporting firearms and ammunition, also owes its existence to Darling's work.[60]

For many years, Darling owned a winter home on Captiva Island in South Florida. Thanks to the efforts of many of his island neighbors and the J. N. "Ding" Darling Foundation, the adjacent Sanibel National Wildlife Refuge—which had been protecting wildlife habitat since 1948—was renamed the J. N. Ding Darling National Wildlife Refuge and officially dedicated to him in 1978. Today, "the Ding" is a mecca for both birdwatchers and photographers living throughout South Florida and around the world.[61]

I (Budd) have enjoyed the distinct avian pleasure of visiting "the Ding" many times every year, starting in the winter of 1995. Every morning just before sunrise, serious photographers armed with telephoto lenses the size of bazookas and life birders toting their Swarovski binoculars line up to be the first to see what new birds have arrived on the refuge. What many of them may not realize is that the refuge's namesake is the primary reason the birds are there in the first place.

> ***Land, water and vegetation.... Without these three primary elements in natural balance, we can have neither fish nor game, wildflowers nor trees, labor nor capital, nor sustaining habitat for humans.***
>
> **—Jay Norwood "Ding" Darling**[62]

Jay Norwood "Ding" Darling used his fame as a cartoonist as a springboard to help create two American wildlife management institutions—the US Fish and Wildlife Service and the National Wildlife Federation. But, more importantly, he also found creative ways to finance his two favorite conservation causes—wildlife research and waterfowl management. Fees collected from the very people who participated in—and therefore benefited from—sport hunting were used to purchase habitat and protect nesting areas. America could apply the same fundraising approach to combat climate change. Fossil-fuel consumers could be charged a user fee and then the collected money—no doubt in the billions of dollars—could be applied to research facilities for renewable energy power generation and distribution.

THE CALLENDAR EFFECT PROVES GLOBAL WARMING

The year 1938 also brought the first definitive measurements showing that global warming was occurring. Relying on background data showing CO_2 concentrations had risen over the previous century, Guy Callendar, a British engineer, proved temperature increases occurred over the same period.[63] As we discussed in chapter 2, his correlation of increasing CO_2 levels with rising temperatures was labeled the "Callendar Effect" and was—for the most part—dismissed by meteorologists and the general public.

The 1930s had had their ups and downs in terms of environmental issues and the next several decades would prove to have their own variety of environmental challenges.

Chapter 14

ENVIRONMENTAL CONCERNS MOVE TO THE BACK BURNER (1940–1959)

WORLD WAR II AND TWO DAYS THAT CHANGED OUR WORLD FOREVER

During the first half of the 1940s, most of the world was preoccupied with the ongoing horrors of World War II. The horrific actions of Adolf Hitler, the worst dictator the world has ever known, and the Nazi Party that followed him, meant that no one in the United States could pay any serious attention to environmental matters. The Nazis' unthinkable plot to exterminate an entire civilization of people remained foremost in everyone's minds. After the momentously stirring—albeit sadly prolonged and costly—victory of the Allied Forces on the beaches of Normandy, the mentality of the US population began to return to normal . . . or almost normal, at least.

Historian Donald Worster wrote perhaps the most poignant passage about what would happen next in the US environmental movement:

> The Age of Ecology began in the desert outside of Alamogordo, New Mexico, on July 16, 1945, with a dazzling fireball of lights and a swelling mushroom cloud of radioactive gases. One kind of fallout from the atomic bomb was the beginning of widespread, popular ecological concern around the globe.[1]

Suddenly the United States was the most powerful, feared, and—in some countries—loathed nation on Earth. As physicist J. Robert Oppenheimer, civilian leader of the Manhattan Project and "Father of the

Atomic Bomb," sadly noted on that frightful July day, "Now I am become Death, the destroyer of worlds."[2]

On two days—August 6 and 9, 1945—the disparity of world opinion about whether the United States should be considered the "Savior of the World" or the "Doomsday Purveyor of Death" increased dramatically. On these two days, the United States dropped atomic bombs, "Little Boy" and "Fat Man"—the first and hopefully the last to ever be deployed against human beings—on the Japanese cities of Hiroshima and Nagasaki, respectively.[3] The bombs obliterated huge sections of the cities and killed a total of at least 450,000 people (including men, women, and children)—and over the long run—possibly twice that many.[4]

While these atomic bombs certainly achieved their short-term result—World War II was finally over when Imperial Japan surrendered just six days after Nagasaki was bombed—their lingering effect extended for decades. The era of the Cold War—which focused world attention on the nuclear arsenals of the United States and Russia—had officially begun.

The scientific advances that came from the war years changed the way everyone in the United States lived. Who among the baby boomers can ever forget curling up under our school desks to practice air raid drills. Or seeing our parents perform their civilian duties by going to little tin huts on hilltops to watch the skies for Russian bombers.

World War II also had—at first, at least—a profound effect on environmental philosophy. The standard thought process for the economic community was that any environmental problem could be overcome by the application of science. If we could figure out how to blow up an entire city in one fell swoop, how could we possibly be stymied by something as insignificant as logging an old-growth forest or damming a salmon-choked whitewater river?

This period of scientific supremacy in the US environmental movement ushered in an attitude of "why worry about the future?" Americans operated under the approach of just doing what needed to be done right then and letting future generations figure out their own problems. Unfortunately, this same thought process still permeates contemporary society and has been one of the primary reasons for the refusal to address climate change in any significant manner.

Despite the presumption that science would overcome all environmental issues, Mother Nature had other opinions. In 1947, a noxious layer of pollution enveloped the steel-mill dominated town of Donora, Pennsylvania, killing twenty people and making another six thousand residents violently ill.[5] Then, in 1949, automobile-generated pollutants—or vehicular smog, as it was later called—caused a massive crop loss in the Los Angeles area.[6] But the worst—in the form of DDT and other deadly chemical pesticides invented specifically for the agricultural industry and sprayed everywhere with reckless abandon—was still to come. We will discuss DDT in detail in chapter 15.

As a child on sultry summer evenings after World War II, I (Budd) vividly remember gleefully running with my sister and cousins through clouds of foul-smelling chemicals—in all likelihood DDT—spraying from the back of Panama City, Florida, tanker trucks. Since the acrid fumes killed hordes of hungry mosquitoes—and most likely every other type of insect around—the adults just watched placidly from their lawn chairs and sipped juicy cocktails while we repeatedly subjected ourselves to the poisonous gasses. Back then, we had no inkling whatsoever that these noxious fumes would soon put many of our most iconic wild bird species on the brink of extinction.

THE BUREAU OF LAND MANAGEMENT: CREATED TO CONTROL GRAZING RIGHTS

While the minds of most Americans were still preoccupied with recovering from four years of world war, some significant environmental events did take place in the late 1940s. The year 1946 saw the creation of the US Bureau of Land Management (BLM)—another federal agency created to help western ranchers better manage their livestock herds and maximize their incomes in the process. In fact, throughout its existence, the BLM has been routinely criticized for helping ranchers too much—for example, by leasing them grazing rights on federal land at way below the going rate.

MARJORY STONEMAN DOUGLAS: TAKE HEED, GREEDY LAND DEVELOPERS

The late 1940s were significant in the life of another of our past environmental heroes—this time a feisty little woman with the heart of a lion and the tenacity of a wolverine. Marjory Stoneman Douglas looked more like a wealthy socialite—in her characteristic Panama hat and horn-rimmed glasses—than an outdoor lover, but, as the old saying goes, looks can be deceiving.

Douglas worked diligently to turn a lifelong passion for doing the right thing into her own personal environmental justice movement, which culminated in helping preserve one of the most unique ecological areas on the face of the earth. In 1947—after more than thirty years of fighting with politicians, bureaucrats, and local land developers—she published her landmark book, *The Everglades: River of Grass.*

In her masterpiece—which has been favorably compared to Rachel Carson's *Silent Spring* and Aldo Leopold's *A Sand County Almanac*—Douglas lovingly described the unfathomable beauty and untold treasures encompassed by Florida's Everglades. And she did so at a time when most Americans—especially those who had recently moved to South Florida—considered the Everglades just worthless square miles of mosquito- and snake-infested wastelands. Not only did *River of Grass* spark a movement to protect the Everglades from uncontrolled filling, land development, and wanton destruction, but it also opened the eyes of the rest of the country to see and appreciate the many critical functions—including flood control, water quality protection, aquifer recharge, and wildlife habitat—provided by wetlands.

Ironically, Douglas was born in 1890 in Minneapolis, Minnesota, and grew up in Taunton, Massachusetts, two places and environments that were about as far removed as possible from the South Florida landscape she grew to love and cherish.[7] During her undergraduate years at Massachusetts's Wellesley College, Douglas earned straight As and was voted "Class Orator"—a prophetic title, as wealthy landowners and their corrupt politicians throughout South Florida were about to learn.[8]

In 1915, Douglas moved to South Florida and began working as a columnist for her father's newspaper, the precursor to the *Miami*

Herald. Combining skillful writing with a firebrand personality, she quickly gained local notoriety by getting embroiled in battles over racial inequality, feminism, and resource conservation—long before these issues became the focus of the national spotlight.[9]

The state of Florida's outlandishly uncontrolled land development practices provided the perfect grist for Douglas during her early years as a columnist/poet for the *Miami Herald*. She regularly wrote editorials urging protection of Florida's unique regional character and arguing against the rampant development that was threatening to destroy these irreplaceable natural resources.[10]

As an aside here, if you're interested in learning about the worst possible way to conduct land development, research the settlement history of South Florida, starting around 1900. The monumental comedy of errors—including "improving" (i.e., straightening) hundreds of miles of cool, meandering streams and diking the sheet flow of massive Lake Okeechobee, costing thousands of lives—is simply beyond belief.

The litany of egregious environmental impacts continued through Walt Disney's filling of hundreds of acres of pristine wetlands in the 1960s. Many people will tell you that Disney World—because of the rampaging development it fomented—is the worst thing that ever happened to Central Florida. In *National Geographic*, T. D. Allman says that Disney World and SeaWorld have transformed Orlando into "Exhibit A for the ascendant power of our cities' exurbs: blobby coalescences of look-alike, overnight, amoeba-like concentrations of population far from city centers" and "a place whose specialty is detaching experience from context, extracting form from substance, and then selling tickets to it."[11]

Carl Hiaasen, bestselling novelist and columnist for the *Miami Herald*, in his book *Team Rodent: How Disney Devours the World*, suggests that Disney World changed the face of an entire state. He writes, "Three decades after it began bulldozing the cow pastures and draining the marshes of rural Orlando, Disney stands as by far the most powerful entity in Florida; it goes where it wants, does what it wants, gets what it wants ... the worst thing Disney did was to change how people in Florida thought about money ... suddenly there were no limits. Merely by showing up, Disney had dignified blind greed in a state pioneered by undignified greedheads."[12] Even today, environmental atrocities are still running wild throughout South Florida. Again quoting Carl Hiaasen: "The Florida in my novels is not as seedy as the real Florida. It's hard to stay ahead of the curve. Every time I write a scene that I think is the sickest thing I have ever dreamed up, it is surpassed by something that happens in real life."[13]

Now back to the head-turning exploits of Ms. Douglas: Marjory Douglas was small, but her diminutive size belied her zeal for standing up to the power interests in South Florida. In his introduction to her autobiography *Voice of the River* (1987), writer John Rothchild describes Douglas's appearance in 1973 at a public meeting in Everglades City: "Mrs. Douglas was half the size of her fellow speakers and she wore huge dark glasses, which along with the huge floppy hat made her look like Scarlet O'Hara as played by Igor Stravinsky. When she spoke, everybody stopped slapping [mosquitoes] and more or less came to order. . . . Her voice had the sobering effect of a one-room schoolmarm's. The tone itself seemed to tame the rowdiest of the local stone crabbers, plus the developers, and the lawyers on both sides. I wonder if it didn't also intimidate the mosquitoes. . . . The request for a [US Army] Corps of Engineers permit was eventually turned down. This was no surprise to those of us who'd heard her speak."[14]

When it came to the Everglades, Douglas took on all comers, including greedy land developers who wanted to drain and fill the "worthless swamp" to political hacks and power brokers who would bend over backward to "make things work out" for a little extra money under the table.[15] For her ceaseless efforts to block land development in the Everglades and maintain its vital sheet-flow water source emanating from Lake Okeechobee, Douglas made many enemies in South Florida. But, in the process, she also earned a great deal of respect, as verified by her well-deserved nickname, "the Grande Dame of the Everglades."

Douglas's relentless campaigning for South Florida finally paid off in 1947—just after publication of *River of Grass*—with the establishment of Everglades National Park and then again twenty-two years later, in 1969, with the founding of the conservation organization Friends of the Everglades. Her tireless efforts as a conservationist earned her many awards. In 1986, the National Parks and Conservation Association established the Marjory Stoneman Douglas Award "to honor individuals who often must go to great lengths to advocate and fight for the protection of the National Park System."[16] Finally, in 1993, she was awarded the Presidential Medal of Freedom—America's highest civilian honor.[17]

Living to the remarkable age of 108, Douglas passed away in 1998 in the Coconut Grove neighborhood of Miami. Even near the end of her life, she was still advocating in support of the ongoing federal and state efforts to restore the hydrologic sheet flow out of Lake Okeechobee that was historically the lifeblood of the Everglades. Upon her death, an obituary in the *Independent*, a British national morning newspaper, stated, "In the history of the American environmental movement, there have been few more remarkable figures than Marjory Stoneman Douglas."[18] Per her request, Douglas was cremated and her ashes were scattered across Everglades National Park—her beloved "river of grass."[19]

> ***There are no other Everglades in the world. They are, they have always been, one of the unique regions of the earth—remote, never wholly known. Nothing anywhere else is like them.***
>
> **—Marjory Stoneman Douglas**[20]

No one in the history of the US environmental movement has ever been more dedicated to a singular conservation issue than Marjory Stoneman Douglas. She took on all the major power brokers that South Florida could throw at her during a time when support for natural resource conservation was minimal. Remarkably, she won her war when the Everglades National Park was established just one month after her landmark book, *River of Grass*, was published. Douglas's bold fight to stand up for what she believed—even in the face of withering resistance and daunting opposition—is the exactly the type of feistiness and resilience that is required to successfully take on Big Oil and win the climate-change battle.

THE FIERCE GREEN FIRE OF ALDO LEOPOLD

In 1949, the majority of the US population still held fast to the religious fervor that maintained the indomitable belief that the early colonists first brought with them. As we discussed earlier, the dominant thinking was that humans had a God-given right to exert dominion over all of nature's creatures. The general idea was that the beasts of the forests, fields, rivers, and streams were put there to serve human needs. Not taking advantage of this natural bounty was still considered a sacrilege and an affront to human integrity almost 175 years after our nation was founded.

Ironically, the man who was to raise the greatest challenge, to date, to this deeply held belief was an avid outdoorsman himself. As a young adult, Aldo Leopold hunted and fished in his native Wisconsin countryside with boundless zeal.[21] Then, on a hunting trip to Arizona, he had an experience that changed his life forever, after he shot a female wolf:

> We reached the old wolf in time to watch a fierce green fire dying in her eyes. I realized then, and have known ever since, that there was something new to me in those eyes—something known only to her and to the mountain. I was young then, and full of trigger-itch; I thought that because fewer wolves meant more deer, that no wolves would mean hunters' paradise. But after seeing the green fire die, I sensed that neither the wolf nor the mountain agreed with such a view.[22]

Born in January 1889 in Burlington, Iowa, Leopold was an outdoorsman almost before he could walk. His dad, a German immigrant and woodcrafter, regularly took the young Aldo on nature forays around the Iowa countryside and, during the summers, in Michigan's Les Cheneaux Islands in Lake Huron. Attending Yale University, Leopold became one of the first graduates of the Yale School of Forestry, which had been created through an endowment from Gifford Pinchot, the first Chief Forester of the US Forest Service.[23]

The experience with the old wolf changed Leopold from a man who took great pleasure in killing animals to one who often relished just watching and documenting what they did. In his younger years—before the "fierce green fire," he was seldom seen without a hunting jacket on and a high-powered rifle slung over his shoulder. After he shot the wolf—although he didn't give up hunting—he was more likely to have a long-stem pipe poking thoughtfully out of the corner of his mouth and a pair of binoculars hanging around his neck.[24] Forty years later, in 1949, this dramatic change in Leopold's personal thinking led to the publication of *A Sand County Almanac*, a book that also changed the thoughts of the entire US environmental movement and ushered in a totally new field, the science of wildlife management.

Acknowledged by many as the "Father of Wildlife Conservation" and one of the most influential conservation thinkers of the twentieth century, Leopold was also one of the early leaders of the American wilderness movement.[25] As we have previously discussed, earlier forms of conservation were based almost solely on economics and benefits to humanity.

In his stirring essay, "The Land Ethic," which concluded *A Sand County Almanac*, Leopold described his belief that everything on Earth was interrelated and that humans and nature existed in a harmonious relationship.[26] These beliefs—which were the precursor to the modern concept of ecology—stressed that humans were just part of the overall global ecosystem. Plus, as the most intelligent component of this global ecosystem, humans had the responsibility for being the caretakers of all living things on Earth. Leopold also stressed that all living things were owed the right to a healthy existence. These concepts were not only incredibly innovative but were also far ahead of their time.

Leopold wrote *A Sand County Almanac* from a refurbished chicken coop—which he called simply "the Shack"—located on a farm he was restoring in the sand counties of Wisconsin, near Baraboo. Today, with more than two million copies in print and having been translated into twelve languages, this book is one of the most beloved and respected books about the environment ever published.[27]

Leopold died in 1948, just a few months before the publication of his most famous work, from a heart attack while fighting a forest fire on the property of one of his neighbors.[28] Despite this, his legacy and writings still live on—both spanning and blending the disciplines of forestry, wildlife management, conservation biology, sustainable agriculture, restoration ecology, private land management, environmental history, literature, education, esthetics, and ethics.

Although Aldo Leopold did not live long enough to hear much—if anything—about global warming, his "land ethic" views form the basis of the rationale for combating climate change. If he were alive today, he would most certainly have taken a firm stance on the issue, arguing that as an integral—and supposedly harmonious—part of the natural world, humanity must not only take responsibility for the warming climate, but we must also take the lead in combating it. He would emphasize how studying our place within, rather than outside of, nature will allow us to most effectively see how we are influencing these changes. Once we have this understanding, we can make the necessary corrections to counteract the looming crisis. As Mike Dombeck, Emeritus Professor of Global Environmental Management at Wisconsin–Stevens Point, recently wrote, "As a society, we are just now beginning to realize the depth of Leopold's work and thinking."[29]

> *Like winds and sunsets, wild things were taken for granted until progress began to do away with them. Now, we face the question whether a still higher "standard of living" is worth its cost in things natural, wild, and free.*
>
> —**Aldo Leopold,** *A Sand County Almanac*[30]

> *The last word in ignorance is the man who says of an animal or plant: "What good is it?" To keep every cog and wheel is the first precaution of intelligent tinkering.*
>
> —**Aldo Leopold,** *A Sand County Almanac*[31]

THE RISE OF SUBURBIA AND FURTHER DECLINE OF ENVIRONMENTAL CONCERN

For the most part, the 1950s skittered along with most Americans still pursuing their way to the prosperity epitomized by sparkling new homes festooned with immaculate green lawns, two-car garages, and all the latest appliances from General Electric and Bendix. The song about the row of little houses made out of "ticky-tack" certainly seemed apropos for defining the new American dream. Meanwhile, many people who were trying to introduce an environmental ethic back into the American consciousness were fearing the possibility of being called up in front of the McCarthy Commission for advocating "Un-American Activities."

As they often contradicted the prevailing emphasis on finding and living "the good life," environmental considerations essentially disappeared from the American psyche. The majority of Americans just off-handedly assumed that natural resources were there to serve their recreational needs and provide family enjoyment. Indifference and callous misuse of the environment was rampant. Litterbugs were everywhere. Piles of rubbish accumulated along the nation's roadsides and in recreational areas. People blindly tossed their garbage out of their car windows as they drove along.

Most national parks, monuments, and forests suffered severe overuse as more and more vacationers fled to these scenic outposts to escape the drudgery of their urban jobs. People wanted to be entertained, without any thought of being sensitive to the native flora and fauna in these areas.

Unfortunately, most federal recreational areas were not sufficiently staffed to deal with the hordes of seasonal visitors. Because of this, the quality of the natural resources on federal lands across the nation was being severely impacted—often, irreversibly so. Something desperately needed to be done. But no one in a position of leadership volunteered to step forward—or even seemed to know exactly what actions to take.

Even the federal government seemed to be making all the wrong decisions and taking all the wrong actions. President Dwight D. Eisenhower, retired general and World War II hero, appointed a Secretary of the Interior who became known as "Giveaway McKay." In fact, during his tenure as Interior Secretary, Douglas McKay repeatedly tried to turn federal energy projects over to the private sector, abolish national wildlife refuges, and transfer federal game reserves to state agencies.[32]

FORMATION OF THE NATURE CONSERVANCY: A POSITIVE STEP FORWARD

While strides forward in the environmental and conservation movements in the 1950s were rare, there were a few hopeful moves. Formed in 1951, the Nature Conservancy (TNC) remains one of the most active and successful nonprofit land-management organizations in the United States. To date, TNC has successfully protected more than 15 million acres of land nationwide. TNC's strategy of improving and then selling the most developable portions of land to pay for saving the most feature-rich and valuable areas is one of the best natural resource acquisition and preservation strategies ever devised.

In fact, TNC's strategy could be expanded to become a very effective component of the climate-change solution. Funding for TNC's operations could be appropriated either through finder's fees on new fossil-fuel extractions or user's fees on those still using fossil fuels—or better yet by doing both. Then TNC could use these monies to acquire undeveloped land—preferably with dense forest cover—to be set aside as CO_2 sequester zones. Alternatively, TNC could use these funds to buy land for siting renewable energy generation facilities—wind farms, solar fields, geothermal plants, and so on—plus rights-of-way for the new infrastructure required to transport this renewable energy.

THE BUREAU OF RECLAMATION: WATERING THE WEST

The creation of a new federal agency, the US Bureau of Reclamation, paved the way for a dam project in the Great American West that would set the stage for the raging environmental battles that would characterize much of the 1960s and 1970s. The US Bureau of Reclamation ("BuRec"—surreptitiously referred to in some western circles as the "Bureau of Wreck-the-Nation") was created as a separate bureau in 1907 for the expressed purpose of "Watering the West." Under congressional directives, the BuRec's administrators attacked their task with great vigor—knowing that the only thing that could hold back westward expansion was a lack of water in the parched western landscapes.

The agency's hydrologists and engineers soon began eyeing the magnificently carved river valleys of many of the West's great rivers—especially the mighty Colorado and its major tributaries. In fact, the BuRec's first great success—at the time considered one of the human-made Wonders of the World—was Hoover Dam. Completed in 1936, this mega-monolith stands 726 feet tall and—when full—backs up 248 square miles of water to create the Lake Mead National Recreation Area.[33]

ECHO PARK DAM: A LANDMARK VICTORY FOR CONSERVATION

In the 1950s, Colorado's Dinosaur National Monument confined the confluence of two of the Colorado River's major and most magnificent tributaries—the Yampa and the Green Rivers. At the time, the fantastic palette of scenic, archaeological, and paleontological beauty in this remote corner of Colorado was known only by the sparsely settled local communities. In the minds of the BuRec's chief bureaucrats, however, these characteristics offered them exactly what they were looking for—another Hoover Dam.

The BuRec's engineers began work by focusing their attention on Echo Park—a remote canyon that snaked its way around an eight-hundred-foot high sandstone monolith called Steamboat Rock.[34] Steamboat Rock stood in the riverbed just below the confluence of the Yampa and Green Rivers and, topographically, provided the ideal location for constructing a massive dam. The channel's narrow trough here would require a relatively minimum concrete span, while backing up the water in both channels of the two major rivers.[35] The resulting long expanse of open water would provide plentiful hydroelectric power and boundless recreational opportunities—serving as the perfect complement to other downstream dams and reservoirs that were also on the engineers' drafting tables.[36]

All systems were go for this Echo Park Project, or so the Washington politicians and BuRec bureaucrats thought, until they met up with the dissenting minds of the Sierra Club led by the "Archdruid" himself—Executive Director David Brower—and leaders of several other conservation groups. Far from becoming another showcase for the western water controllers, the proposed Echo Park Dam became one of the most momentous victories ever recorded in the history of the US environmental movement.

The opposition to Echo Park actually first started when a group of downstream river guides in Utah got wind of what was being bandied about upstream. The guides moved quickly to rally support for their cause and invited Brower and the other conservation leaders to join them for a free trip to see exactly what would be lost if the canyon was flooded.

After spending a few days with the guides, Brower and his fellow conservationists oversaw the formation of an alliance that challenged the BuRec's claims about the project's major benefits and limited impacts. One of the opposition group's catchphrases was, "We're not opposed to development. We're not opposed to dams. We're just opposed to dams in national parks."[37] We discuss this in greater detail in chapter 16.

Facing such difficult to refute logic of the absurdity of a dam in a national park, alongside withering opposition led by a master organizer like Brower, Congress—after a prolonged and controversial battle—eventually removed Dinosaur National Monument and Echo Park from the act that created the Colorado River Storage Project. But unfortunately for environmentalists the story of Echo Park doesn't end there. In their crusade to preserve Echo Park and Steamboat Rock, Brower and his band of activists accepted a tradeoff—a loss downstream that would anguish them and others for years to come. The deal involved an agreement not to oppose a dam on the Colorado River in a little-known place called Glen Canyon.

Still smarting from the damming of Hetch Hetchy Valley in Yosemite National Park, the Sierra Club's focus in the 1950s was squarely on national parks and monuments, but Glen Canyon had no such protected status. The rationale was that since Glen Canyon wasn't in a national park, it probably wasn't all that valuable. But this time, the Sierra Club was wrong—very wrong—and Brower deeply regretted his compromise decision. The problem was that no one had really taken the time to check Glen Canyon out, and by the time they did it was too late.[38]

The Glen Canyon Dam drowned more than one hundred miles of spectacular canyon, a rippling desert wilderness that few had ever seen. Salt Lake City river guide, Richard Quist, called Glen Canyon "a sprawling labyrinth of wonder" and said the canyon's ancient, still-colorful pictographs were "stunningly beautiful." He also described stumbling onto the canyon's long-abandoned pit houses as being among his greatest childhood adventures.[39]

Brower later wrote a foreword to a 1963 book called *The Place No One Knew*, which honored Glen Canyon and featured the stunning nature photographs of Eliot Porter.[40] But by the time Glen Canyon's beauty and wildness became more widely known, the dam fighters couldn't backtrack. The dam was completed in the fall of 1963, and Lake Powell was filled to capacity by 1980—drowning Glen Canyon presumably forever.

As journalist Brandon Loomis describes it, "The odds of defeating Glen Canyon Dam would have been long even if the coalition had turned its attention to land outside national parks. The dams weren't equals—Echo Park would have held about a quarter of what Lake Powell can—and Glen Canyon Dam was always intended to anchor the upper Colorado River's water storage system."[41]

Despite the less-than-perfect victory, Echo Park became "a symbol of wilderness," and the battle to save it was documented in a 1994 book by the same name.[42] The project's showdown between Congress, the BuRec, and the Sierra Club—led by David Brower—and other opposing conservation organizations was also widely acknowledged to be a major steppingstone that led to passage of the Wilderness Act of 1964.[43]

DAVID BROWER: THE ARCHDRUID HIMSELF

What happens when you combine the athleticism of a world-class mountaineer, the charisma and good looks of a movie star, and the passionate leadership of a dedicated head of state? You get David Brower, another of our past environmental heroes, and a man who singlehandedly founded more environmental conservation/activism groups than any other person in US history. Included among Brower's organizations were the Friends of the Earth, the League of Conservation Voters, and the Earth Island Institute. In addition to his prominence in the successful fight against the Echo Park Project and other BuRec dams in the 1950s, Brower became one of the most prominent figures in the creation and expansion of the US environmental movement in the late 1960s and early 1970s.[44]

David Brower was born in Berkeley, California, on July 1, 1912, and first visited the High Sierras and Yosemite National Park when he was only six years old. He had a passion for mountaineering and made first ascents of more than seventy peaks in the western United States. During World War II, he served in the legendary 10th Mountain Division, where he repeatedly used his rock climbing skills to lead daring assaults on enemy positions. Not surprisingly, Brower's outdoor skills eventually led him to the Sierra Club, where he served as the first executive director (ED) from 1952 to 1969.[45]

It was in this position that Brower first achieved national fame—leading the opposition to several Bureau of Reclamation dam proposals in Dinosaur National Monument (now Park)—as we have previously discussed—and the Grand Canyon. Brower's advocacy as Sierra Club ED also helped establish nine national parks and seashores—including three in his native California—Redwoods National Park, Kings Canyon National Park, and Point Reyes National Seashore.[46] Under his direction, the Sierra Club began publishing their now-famous large-format coffee table books, which combined mind-blowing outdoor photography with poignant and powerful conservation messages.

Brower was also one of the first environmental activists to feature full-page advertisements in prominent newspapers as a way of shifting public opinion and building grassroots support for his causes. Owing to his tireless drive and passion, the Sierra Club's membership boomed—from 2,000 members (mostly in California) to 77,000 across the United States—during his seventeen years as executive director.[47] Despite this success, his leadership of the organization was viewed as extreme and controversial. Notably, Brower spent time and money aggressively lobbying against the construction of dams in the Grand Canyon; this lobbying backfired and cost the Sierra Club its tax-deductible status in 1967 and David Brower his leadership position two years later.[48]

Writing in *Publishers Weekly*, environmentalist and author Paul Hawken commented that "no single person created more ways and means for people to become active and effective with respect to the environment than David Brower."[49] Russell Train, head of the Environmental Protection Agency under Presidents Richard Nixon and Gerald Ford, added, "Thank God for Dave Brower; he makes it so easy for the rest of us to look reasonable."[50]

Brower earned international respect because of his passion for Earth and its inhabitants. He was nominated for the Nobel Peace Prize three times (in 1978 and 1979, and then jointly with professor Paul Ehrlich in 1998). In 1998, Brower also received the Blue Planet Prize for his lifetime achievements.[51] His successful advocacy for many environmental causes led acclaimed author John McPhee to publish a series of articles on Brower, and then a bestselling book called *Encounters with the Archdruid*.[52]

> ***I believe that the average guy in the street will give up a great deal, if he really understands the cost of not giving it up. In fact, we may find that, while we're drastically cutting our energy consumption, we're actually raising our standard of living.***
>
> **—David Brower, The Archdruid**[53]

> ***Polite conservationists leave no mark save the scars upon the Earth that could have been prevented had they stood their ground.***
>
> **—David Brower, The Archdruid**[54]

If he were still alive today, David Brower would certainly be in the forefront of the climate-change battles. He never gave up on finding innovative ways of getting his environmental messages across to nonbelievers and naysayers. He was one of the most forceful—and also controversial—conservation activist in the history of the US environmental movement. His brazen willingness to take and then defend positions that he believed in—even when they meant getting fired as executive director of the Sierra Club—represent exactly the type of person we need to assume the leadership role in solving the climate-change crisis. Brower's skills for organizing activist groups and then devising methods for rallying public support need to be carefully studied and then applied to our cause.

GILBERT PLASS: FATHER OF MODERN GREENHOUSE GAS THEORY

Several more events significant to the history of climate change occurred in the 1950s as well. First, in 1955, climate researcher Gilbert Plass performed detailed computer analyses showing that doubling CO_2 concentrations in the atmosphere would increase global temperatures by 3 C to 4 C (5.4 F to 7.2 F).[55] We talked more about Plass and his work in chapter 2.

HANS SUESS AND ROGER REVELLE SUPPORT PLASS

Next, a pair of scientists, chemist Hans Suess and oceanographer Roger Revelle, proved that seawater would not—as previously believed—absorb all the CO_2 that enters our atmosphere. In fact—foreshadowing the peril we face today—Revelle wrote that, "Human beings are now carrying out a large scale geophysical experiment of a kind that could not have happened in the past nor be reproduced in the future."[56] We talked more about both Suess and Revelle in chapter 2.

THE KEELING CURVES

Finally, in 1958, Dr. Charles David Keeling began a project that continues to this day—the Keeling Curves. Using systematic measurements taken at Mauna Loa in Hawaii and in Antarctica, Keeling provided the first unequivocal proof that CO_2 concentrations in our atmosphere are rising. Using the most sophisticated technology available at the time, Keeling produced concentration curves for atmospheric CO_2.[57] Again, you will find more information about the Keeling Curves in chapter 2.

Though our heroes often battled against the government as it threatened to devour our environment in the 1940s and 1950s, their efforts would soon be rewarded. The 1960s brought a whole decade where "hippies, peace, and harmony" were the trends and environmental issues were one of the main causes du jour.

Chapter 15

FULL SPEED AHEAD

Preparing for the Environmental Years (1960–1968)

THE SIXTIES: DECADE OF SOCIAL UPHEAVALS

From the Beat Generation to hippie communes; civil rights marches; anti–Vietnam War demonstrations; and sex, drugs, and rock and roll, the 1960s were known as the decade of social upheaval, civil unrest, and cultural shifts. In particular, many Americans finally became aware of environmental concerns. The severe impacts human activities were having on land, water, and air could no longer be ignored or tolerated.

Spurred on by the replacement of consumerism and materialism with free-spirit simplicity and lifestyle mellowing, Americans began demanding that better care be taken of Earth's natural resources, advocating for clean air, clean water, enhanced biodiversity, plus more open land, green spaces, and recreational amenities. "Progress at all costs" was renounced in favor of finding a balance between comfortable living and preserving, instead of constantly altering, the natural environment. In fact, the wild social and political radicalism of the sixties was the best thing that ever happened to the US environmental movement.

RACHEL CARSON: HERO ABOVE ALL

Nowhere was this new environmental thinking more noticeable than with one person who—figuratively speaking—stood head and shoulders above her peers. Avoiding the limelight, Rachel Carson preferred instead to nestle down behind her microscope in her government lab, while allowing her writings to tell the public about her keen insights into the natural world.

If asked to select the most pivotal environmental accomplishment of the twentieth century, it would be difficult not to vote for the publication of Carson's book *Silent Spring*.[1] Not only did this landmark publication lead to the eventual eradication of DDT—a pesticide that had high environmental and toxicological effects, exemplified by the deaths of massive numbers of peregrine falcons and their offspring—the book also saved many of our iconic birds of prey from extinction. Thanks to *Silent Spring*, bald and golden eagles, peregrine falcons, brown pelicans, and ospreys still fill our skies with their dramatic flights. Furthermore, Carson's writing had an substantial effect on the passage of the National Environmental Policy Act, and establishment of the US Environmental Protection Agency.[2]

Born in 1929 in the rural river town of Springdale, Pennsylvania, Rachel Carson inherited her lifelong love of nature from her mother. Her education included stints at Massachusetts's famed Woods Hole Oceanographic Institute (then Marine Biological Laboratory) and Baltimore's prestigious Johns Hopkins University, where she earned her master's degree in zoology in 1932.[3] Carson was enrolled in a doctoral program at Johns Hopkins but dropped out due to financial hardship in 1934; she was the sole breadwinner for her mother, father, and sister, Marian, who was weakened by diabetes.[4]

Most people don't realize that Carson was actually quite famous as an author before *Silent Spring* came along. While employed as a scientist and editor with the US Fish and Wildlife Service, she wrote three books describing the environmental treasures of our oceans,[5] including her prize-winning and bestselling, *The Sea Around Us*.[6]

In *Silent Spring*, Carson's revelations about the myriad horrors of synthetic pesticides such as DDT scared money out of people's pocketbooks and forced letters to Congress out of their pens. Prophetically summarizing what may be said about our contemporary society after 2050, Carson wrote that, "We have allowed these chemicals to be used with little or no advance investigation of their effect on soil, water, wildlife, and man himself. Future generations are unlikely to condone our lack of prudent concern for the integrity of the natural world that supports all life."[7]

As you might expect, the 1962 publication of *Silent Spring* was met with a bitterly negative reaction from the corporate world. The resistance started with the small-time farmers and moved all the way up to the mega-giants—the Monsantos—of the agricultural world. This antagonistic pushback also went sideways to the corporate chemical conglomerates—the DuPonts and the Union Carbides—who were used to getting whatever they wanted wherever and whenever they wanted it.

In particular, the chemical companies threatened to sue Carson over her "inflammatory statements" in *Silent Spring*. They argued that her "outlandish opinions" were crippling American agriculture while also threatening human health. As a chemical-industry spokesman bluntly stated at the time, "If man were to follow the teachings of Miss Carson, we would return to the Dark Ages, and the insects and diseases and vermin would once again inherit the Earth."[8]

Carson's most vehement critics tried to push her to the far left of the political spectrum, arguing that she was consorting with unsavory parties who were trying to undermine American agriculture and free enterprise. While the word "Communist"—the most potent of insults in the decade following the scourge of McCarthyism—wasn't used directly, it was certainly implied. Monsanto even published and distributed five thousand copies of a brochure called *The Desolate Year*, which parodied *Silent Spring* by describing a bleak world wracked by famine, disease, and uncontrolled hordes of insects, which existed because chemical pesticides had been banned.[9]

But nothing—not even the debilitating effects of the cancer that was ravaging her body—deterred Carson from defending the content of her now-famous book. She endured all of the witheringly negative personal attacks—being labeled as everything from radical and disloyal to hysterical and unscientific[10]—with dignity and professionalism, assuring the American public that what she was saying deserved to be heard and heeded.

In 1971, eight years after Carson's death in 1964, the US Environmental Protection Agency (EPA) finally banned DDT nationwide, though US corporations continue to be among the largest producers of the chemical for use in other nations.[11] Today, *Silent Spring* is credited with generating a groundswell of grassroots activism that eventually opened the door for the massive environmental movement of the late 1960s and early 1970s.

Critical reaction to *Silent Spring* in 1962 also set the stage for the battles between industry and environmentalism that have persisted ever since. No matter the subject (air pollution, water pollution, endangered species protection, etc.), industry always follows the same basic strategies: question the science, attack the scientists' credibility, and warn of extreme financial loss. In fact, from this battle over deadly pesticides to the fights over the carcinogenic effects of tobacco in the 1970s to today's feuding with the fossil-fuel companies over climate change, the same "experts" (as featured in the book, *Merchants of Doubt*, see part 3, chapter 21) have been brought forth to perpetuate their bloated rhetoric and denigrate the scientific findings.[12]

While the battle plans aren't always successful, there's no denying that the United States has become cleaner and healthier since the publication of *Silent Spring*. But, as we all now know, the fight is far from over—something the current polarized debate over climate-change solutions certainly demonstrates.

Our new climate-change heroes would do well to use Rachel Carson as their role model for advancing their causes. She believed in her goal of ridding Earth of deadly pesticides and persisted in her mission in the face of powerful opposition and outright personal humiliation from chemical/agricultural conglomerates that were used to getting their own way.

Indeed, the fervor and passion of people like Rachel Carson are necessary to push through the significant political, cultural, and lifestyle changes that will be required for counteracting the ongoing climate-change crisis. Using a persuasive piece of literature to turn the tide of public opinion is also an effective strategy for our current heroes to emulate.

MURRAY BOOKCHIN: THINKING AHEAD OF HIS TIME

The year 1962 saw the publication of another landmark environmental warning book as well—*Our Synthetic Environment*, from the pen of social ecologist Murray Bookchin.[13] In his writings, Bookchin calls for major socioeconomic shifts that are also now being put forth as key activities for controlling climate change, most notably by Naomi Klein in her book *This Changes Everything*.[14] In a precursor to Klein's thoughts, Bookchin advocates for "a reordering and redevelopment of technologies ... based on wind and solar power, methane generators, and possibly liquid hydrogen that will harmonize with the natural world."[15]

He also writes that society should be decentralized into autonomously functioning community units, with economies that serve the basic needs of individuals instead of the broader-based corporate financial interests of a region, nation, or the world. In the minds of both Bookchin and Klein, such a shift in socioeconomic structure would provide the basis for implementing much-needed new technologies for energy production and wealth redistribution.[16]

JFK: AN ENVIRONMENTAL SAVANT GONE TOO SOON

Spurred on in large part by the publication of *Silent Spring*, a flurry of other environmental milestones occurred throughout the 1960s. First, in 1960, a young, dynamic leader and progressive thinker—John Fitzgerald Kennedy—was elected to the White House.

For an environmental activist, JFK had everything that was needed to transform the nation into a caring and considerate country where natural resources were involved. Under his administration, Kennedy set the stage for the Land and Water Conservation Fund, which would, in 1968, begin using federal revenues generated by offshore oil drilling to acquire more land for federal and state recreational areas.[17]

Unfortunately, the world never really got the chance to see just how much of an environmental savior and conservation savant JFK would actually be. He was dealing with the more imminently pressing issues of the nuclear arms race and the associated Cold War with Russia when Lee Harvey Oswald's bullets tore through him on what began as a bright and promising November day in Dallas's Dealey Plaza.

No one who was alive that day will ever forget where we were and what we were doing when Walter Cronkite reported that JFK was dead. Hearing that this brilliant, movie-star handsome man, who had promised to do so much good for our country, was suddenly gone forever just didn't seem true at first. It was almost too much to bear, even for a carefree sixteen-year-old boy—which I (Budd) was at the time.

STEWART UDALL: CONSERVATION MASTER

Fortunately for the ever-burgeoning masses of environmentalists and conservationists in the United States, JFK had nominated Stewart L. Udall—another past environmental hero—as Secretary of the Interior. With this ambitious Arizonan leading the way, it was suddenly off to the races for a host of conservation causes and the US environmental movement. Early in his eight-year tenure as Interior Secretary, Udall initiated the first White House Conference on Conservation since the administration of President Theodore Roosevelt.[18]

Born January 31, 1920, in the small community of St. Johns, Arizona, Stewart Udall was raised in a family with strong ties to the Mormon Church. In fact, he served as a Mormon missionary in Pennsylvania and New York before receiving his bachelor's and law degrees from the University of Arizona, where he was also one of the school's first sports stars as a point guard on the basketball team.[19]

As Secretary of the Interior in both the Kennedy and Johnson administrations, Udall crafted federal land acquisitions like a master landscape painter working with broad brush strokes. His brilliant artistry resulted in the acquisition of 3.85 million acres of new holdings, including four new national parks—Canyonlands in Utah, Redwoods in California, North Cascades in Washington State, and Guadalupe Mountains in Texas. Udall's legacy of protection also included six new national monuments, nine new national recreation areas, twenty new historic sites, and fifty-six new national wildlife refuges.[20]

Of all the federal lands that Udall dealt with, establishing the nation's first national seashores proved to be the most difficult. Because of the typically high cost of and demand for coastal property, Udall faced vehement opposition based on the assumption that taking coastal property out of private hands would prove disastrous for the local economies.[21]

Still he held fast to his goals, eventually establishing a host of national seashores—including Cape Hatteras on the Outer Banks of North Carolina; Assateague Island, with its hundreds of wild horses, in Maryland and Virginia; Point Reyes in Northern California; and Cape Cod in Massachusetts. And, lo and behold, instead of being financial money pits each of these coastal recreational treasures became havens for tourists and economic bonanzas for both state and municipal governments.[22]

Udall also helped achieve the passage of many of our landmark conservation laws, including the Wilderness Act of 1964, the Water Quality Act (1965), the Solid Waste Disposal Act (1965), the Land and Water Conservation Fund Act (1965), the Endangered Species Preservation Act (1966), the National Historic Preservation Act (1966), and the Wild and Scenic Rivers Act (1968).[23]

Throughout his career, Udall left a monumental legacy as a guardian of America's natural beauty. In his bestselling 1963 book—*The Quiet Crisis*—he called for a nationwide "land conscience" to conserve America's wild places, warning of the dangers that pollution and wanton waste posed to our natural resources. Udall's stated goal in *The Quiet Crisis* was "to outline the land-and-people story of our continent."[24] Elaborating further, he wrote, "We cannot afford an America where expedience tramples upon esthetics and development decisions are made with an eye only on the present."[25]

Nancy Pelosi, then Speaker of the House, had this to say about Stewart Udall after his passing in 2010: "Mr. Udall was one of the greatest champions in our nation's history for conservation. As Interior secretary he championed the burgeoning environmental movement, protected the treasures that are our parks, seashores and wildlife refuges, worked for energy independence, and ensured the arts remain a central part of civic life."[26]

President Barack Obama paid tribute to Udall as well: "As secretary of the interior, Stewart Udall left an indelible mark on this nation and inspired countless Americans who will continue his fight for clean air, clean water, and to maintain our many natural treasures."[27]

> ***Cherish sunsets, wild creatures and wild places. Have a love affair with the wonder and beauty of the earth.***
>
> —Stewart Udall[28]

The climate-change movement certainly craves public-sector leaders like Stewart Udall. We need people in the seats of power who can see the big picture and understand how all the pieces—from key land purchases to implementation of pollution controls to development and allocation of new energy resources and equitable distribution of wealth and power—must fit together to provide a roadmap for the future that is both realistic and achievable.

SCENIC HUDSON DECISION: SETTING A LEGAL PRECEDENT FOR ENVIRONMENTALISM

On the US environmental movement's overall timeline, the Scenic Hudson decision set a major precedent for private citizens and conservation groups fighting legal battles over controversial federal projects. The case also marked the emergence of environmental law as a legal specialty, launched the environmental movement, and had significant influence on the National Environmental Policy Act (NEPA) when it passed in 1973.[29]

The Scenic Hudson decision involved a battle over building a hydroelectric plant on Storm King Mountain, along the Hudson River near Cornwall, New York. The mountain was named in the nineteenth century by writer Nathaniel Parker Willis, who noticed that he could always tell when a storm was coming by the way clouds gathered around the peak. And, as the Marist College Environmental History Project describes it, "This name would prove true as the mountain range found itself involved in the most contentious environmental legal case our nation has seen."[30]

This proposed action started in 1962, when the Consolidated Edison company (Con Ed)—one of the United States' largest public utilities companies—announced its plan to build the world's largest pumped storage project into the face of Storm King Mountain. Project plans called for a facility that would have a generating capacity of 2,000,000 kilowatts (kW), with an upper reservoir behind the mountain that would be a mile across, and an eight-hundred-foot long powerhouse at the mountain's base. The initial application to construct the plant was filed with the Federal Power Commission (FPC, now the Federal Energy Regulatory Commission) in Washington, DC, in January 1963.[31]

On November 8, 1963, a small group of local residents met at the home of author Carl Carmer to discuss the Con Ed proposal. During this meeting, they formed the Scenic Hudson Preservation Conference (generally known as "Scenic Hudson"). Their goal was to fight the massive project, arguing that it not only posed a serious threat to the scenery and historic importance of Storm King Mountain but also to the Cornwall water supply and the fishing and recreational value of the Hudson River.[32]

The resulting unprecedented seventeen-year legal dispute—lasting from 1963 to 1981—eventually culminated with the defeat of Con Ed's proposal. In the final analysis, this decision established a landmark precedent in the environmental regulatory process. It was the first time a conservation group had been allowed to file suit for protection of a public interest.

THE CLEAN AIR ACT HITS THE HILL

The passage of the original federal Clean Air Act (CAA) in 1963 was the first federal legislation actually aimed at controlling air pollution, and today it has become inextricably intertwined with the climate-change issue. Before the passage of the CAA, there were no regulations on what kinds of substances steel mills or chemical plants or oil refineries could spew into the atmosphere. Since its original passage, major amendments to the CAA—in 1970, 1977, and 1990—have added some real teeth to this legislation, focusing especially on controlling airborne contaminants known to be hazardous to human health, including those that cause smog, acid rain, and holes in the ozone layer.

According to Benjamin Kline, in his book *First Along the River*, the CAA and its amendments accomplished the following:

- The identification of 189 pollutants that cause smog, and the establishment of standards to regulate their emissions.[33]
- The stipulation that factories and power plants must install special smokestack filters—known as scrubbers—to prevent the discharge of ash and other pollutants. They were also required to install antipollution equipment or use alternative fuels in order to limit the release of sulfur dioxide (SO_2), which is linked to acid rain and is produced when coal or oil burns.[34]
- Finally, the CAA identified automobiles as a major polluter and required manufacturers to build cars with catalytic converters, which reduced exhaust fumes, while also stipulating that oil companies must remove lead from gasoline.[35]

Since the CAA plays such a critical role in finding and implementing solutions for the climate-change crisis, we talk a great deal more about it and its regulatory applications and implications—for the auto industry, the ozone layer, and acid rain—in later chapters.

THE LBJs: LYNDON BAINES AND LADY BIRD

After JFK's assassination, many conservationists feared there would be another sharp decline in environmental leadership at the federal level. But much to their relief, Kennedy's successor—Lyndon Baines Johnson (LBJ)—kept things rolling in the right direction.

The environmental highlight of LBJ's administration came in May 1965 when the White House opened its doors to the first Conference on Natural Beauty. Through this event, Johnson kept the environmental movement in the forefront of public attention. Earlier that year, he addressed Congress on the issue, saying, "We must not only protect the countryside and save it from destruction, we must restore what has been destroyed and salvage the beauty and charm of our cities. Our conservation must be not just the classic conservation of protection and development, but a creative conservation of restoration and innovation. Its concern is not but with nature alone, but with the total relation with man and the world around him."[36]

Not to be outdone was the "other LBJ" who was occupying the White House at the time. Lady Bird Johnson is still considered one of the most important women in history of the US environmental movement. Her 1965 national campaign for the beautification of the newly constructed interstate highway system resulted in passage of the Highway Beautification Act of 1965. Her beliefs regarding the importance of national beautification can be summarized in her statement that "where flowers bloom, so does hope."[37] In 1969, Mrs. Johnson founded the Texas Highway Beautification Awards, became a member of the National Park Service's Advisory Board (NPSAB) and went on to serve on the council of the NPSAB for many years.

On her seventieth birthday, in 1982, Lady Bird Johnson founded the National Wildflower Research Center (NWRC)—a nonprofit environmental organization dedicated to the preservation and reestablishment of native plants in both natural and artificial landscapes. Located in Austin, Texas, this facility performs the outstanding dual services of advancing the science of native plant introductions while educating the general public about the benefits of native species.[38]

In December 1997, the NWRC was renamed the Lady Bird Johnson Wildflower Center in honor of Mrs. Johnson's eighty-fifth birthday. In 1999, Department of the Interior secretary Bruce Babbitt honored Lady Bird Johnson by presenting her with the Native Plant Conservation Initiative Lifetime Achievement Award, saying, "Mrs. Johnson has been a shadow Secretary of the Interior for much of her life."[39]

HOWARD CLINTON ZAHNISER: FATHER OF THE WILDERNESS ACT

Sometimes it takes one person with a tenacious spirit and a personal goal to make the world a better place for us all. In 1964, when President Lyndon Baines Johnson signed the Wilderness Act into law, creating the National Wilderness Preservation System, we had one such person to thank for it.

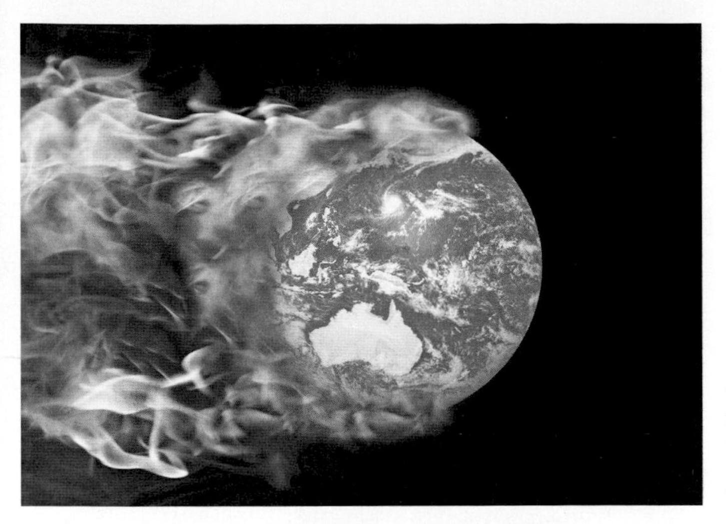

A matter of degrees: How hot is too hot for Earth? (p. 31) © *Reiulf Gronnevik/Shutterstock.*

Industrial Revolution: machine detail—symbolizing the onset of climate change. (p. 112) © *Andreas Krone/ Shutterstock.*

Smoke from a coal plant: industrial pollution filling our skies with greenhouse gases. (p. 112) © *Ungnoi Lookjeab/ Shutterstock.*

Calving glacier: permanent recession of ice masses is one result of a warming global climate. (p. 44) © *Bildagentur Zoonar GmbH/Shutterstock.*

Storm surge and lighthouse: increased storm intensity is happening now as a result of a warming climate. (p. 57) © *CPR62/Shutterstock.*

Skull bone lying on parched riverbed: climate change creates widespread droughts throughout the world. (p. 65) © *24Novembers/Shutterstock.*

Tourists avoiding sea-level rise in San Marco Square, Venice, Italy. (p. 51) © *Yulia Grigoryeva/ Shutterstock.*

Scuba diver looking at bleaching of coral reef in Indonesia. (p. 70) © *fenkieandreas/ Shutterstock.*

Portrait of John James Audubon, the founding father of bird portraits and one of the first conservationists. (p. 88) © *Everett Historical/Shutterstock.*

Portrait of Henry David Thoreau, one of the first to demonstrate how to live in harmony with nature in a modern world. (p. 94) © *Everett Historical/Shutterstock.*

Portrait of Charles Darwin: the man who taught the world about the origin of species. (p. 102) © *Nicku/Shutterstock.*

1998 postage stamp featuring John Muir in his beloved Yosemite National Park; he was also honored on a 1964 stamp. (p. 132) © *catwalker/Shutterstock.*

Sculpture depicting hunger-stricken men in a Great Depression–era breadline. (p. 150) © *Tinnaporn Sathapornnanont/Shutterstock.*

Dust Bowl storm gathering in Rolla, Kansas. (p. 153) © *Everett Historical/ Shutterstock.*

Statue of a multi-faceted hero: President Franklin Delano Roosevelt (FDR) and his dog, Fala. (p. 150) © *Zack Frank/ Shutterstock.*

Hoover Dam on the Colorado River near Las Vegas, Nevada. (p. 175) © *robert cicchetti/ Shutterstock.*

Teddy Roosevelt, protector of wildlife and natural resources, became known as the Conservation President soon after taking office. (p. 125) © *Everett Historical/Shutterstock.*

Iconic view of Yosemite National Park, an American treasure. (p. 118) © *Dan Sedran/Shutterstock.*

Bison herd in Yellowstone National Park, the first area of land to receive protection as a designated national park. (p. 114) © *Richard Wayne Collens/Shutterstock.*

Digital illustration of the extinct passenger pigeon—from billions to none. (p. 140)
© *Nicolas Primola/Shutterstock.*

1981 postage stamp featuring Rachel Carson, savior of countless bird species and author of *Silent Spring.* (p. 182) © *neftali/Shutterstock.*

Portrait of a formerly endangered peregrine falcon: a true conservation success story. (p. 183)
© *Vanessa Belfiore/Shutterstock.*

Denali Peak—North America's highest mountain—rises above the Alaskan taiga. (p. 142) © *Troutnut/Shutterstock.*

Tourists watching Old Faithful Geyser in Yellowstone Park—the National Park Service now protects our most valuable natural resources. (p. 137) © *f11 photo/Shutterstock.*

Sadly, the incredibly dedicated individual who made this possible—writing sixty-six drafts of the legislation over an eight-year period and leading eighteen congressional hearings—was not around to see the landmark act put into force.[40] Howard Clinton Zahniser, the Father of the Wilderness Act and another past environmental hero, died from a heart attack just four months before LBJ signed the act. But anyone who knew this friendly man could sense him there—looking down at the scene and smiling at them—when the president handed the now-famous signing pen over to Zahniser's widow, Alice.

Zahniser—called simply "Zahnie" by all who knew, loved, and respected him—was born in 1906. His father was a Free Methodist minister who moved the family around a lot before settling in the tiny, scenic village of Tionesta on the banks of the Allegheny River.[41] Zahnie graduated from Greenville College, a small Christian institution in Illinois, with a degree in writing. Afterward, he wrote for several papers, including the *Pittsburgh Press*. Throughout his career, Zahnie displayed a humility established in his deeply religious childhood.[42]

After working for twelve years as a writer and editor with the Department of Agriculture's Bureau of Biological Survey, and its successor agency the US Fish and Wildlife Service, Zahniser took a pay cut to serve as executive secretary of the Wilderness Society and editor of the *Living Wilderness* magazine. During his almost twenty years (from 1945 to 1964) in this position, he worked with other conservationists across the nation to lobby for a piece of federal legislation called the Wilderness Act.[43]

Zahniser was a gifted writer and an eloquent speaker, and he passionately shared his love for wild places and arguments for their preservation. He used a variety of media—magazine articles, radio addresses, professional speeches, and congressional testimony—to get his message about the value of wilderness across to the general public and the legislators on Capitol Hill.

Zahniser pushed himself to the breaking point—often going more than twenty-four hours without sleep—to finally get the Wilderness Act passed. In the final tally, this legislation passed in the US House of Representatives by the remarkable vote count of 373–1.[44]

When it first took effect, the Wilderness Act protected 9.1 million total acres in fifty-four areas in thirteen states. By 2015, the number of areas in the Wilderness Preservation System had grown to 765—protecting more than 100 million acres in forty-four states and Puerto Rico.[45] Regrettably, in all of these wilderness areas, there is not a single feature named in honor of Howard Zahniser. We strongly suspect Zahnie wouldn't mind this slight, though—his life's labor was finally in place to benefit the public good.

Never in the limelight like some of his more widely acclaimed contemporaries—including David Brower, Rachel Carson, and Aldo Leopold—Howard Zahniser was always content with staying in the background and subtly pushing his basic message. As journalist Don Hopey describes it: "Zahniser was an early proponent of the ethic that without untrammeled wilderness, mankind would be materially and spiritually impoverished."[46]

Zahniser firmly believed that Congress needed to pass wilderness protection legislation to avoid the preservation of these wild lands from being used as a pawn among greedy politicians who were playing one-upmanship with our nation's public-land legacy.[47] His son Ed is quoted as saying, "It was not just an environmental ethic my father had. He viewed conservation as part of a broad humanism. Thoreau said that in wildness is the preservation of the world, and my father believed that."[48]

> ***I believe that at least in the present phase of our civilization we have a profound, a fundamental need for areas of wilderness—a need that is not only recreational and spiritual but also educational and scientific, and withal essential to a true understanding of ourselves, our culture, our own natures, and our place in nature.***
>
> **—Howard "Zahnie" Zahniser**[49]

E. O. WILSON: FATHER OF BIODIVERSITY

If Mother Nature designed the perfect naturalist, he would look and act exactly like Edward O. ("E. O.") Wilson. Tall, with a long, leisurely stride, hair swept casually to the side, and "aw shucks" good looks, Wilson still appears to be a boy reveling in the natural wonders of his native Alabama than the prolific writer and Harvard University Professor Emeritus that he is.

Wilson is considered to be both the Father of Biodiversity and the Father of Sociobiology. At the same time, he is a past environmental hero to us in a personal way as well. In 2001, Wilson presented our family with the Sudbury Valley Trustees Conservationist of the Year Award at a ceremony in Concord, Massachusetts. Meeting him at that ceremony, where he spoke of the importance of environmental education, shaped my (Mariah's) career decisions. Attending his presentations on campus provided frequent inspiration while I (Mariah) was studying for my master's degree in environmental management at Harvard University.

Born in Birmingham, Alabama, in 1929, Wilson spent almost every waking moment—at least, when he could escape going to school—patrolling the swamps, marshes, and coastlines of South Alabama and the Western Florida panhandle. He attended the University of Alabama, where he received his bachelor's and master's degrees, and then matriculated to Harvard University, where he gained both his doctorate and his home for the rest of his career.

Well-known in scientific circles as the world's foremost authority on ants, Wilson's description of how he came to specialize in insects is typical of his forthright personality and downhome humor. The way he tells it, he was blinded in one eye when he whipped a fish he had caught up out of the water and the spines on the fins caught him in the eye.[50] Also, he was congenitally unable to hear sounds in the upper register. So, since he was blind in one eye and couldn't use binoculars—plus couldn't hear high-pitched bird songs—his first dream, of being an ornithologist, went out the window. Wilson couldn't hear the croaks and calls of amphibians either, so studying frogs also went by the boards. But since he was bound and determined to be a naturalist—he says he never considered doing anything else—he had to find something to study, and insects were about all that was left. At least he could catch the little guys and hold them between his fingers to get a really good look at them.[51]

In hindsight, Wilson's decision to study insects was fortunate for all of us involved with the science of biology. No one has done more with regard to the study and analysis of the biodiversity of life on Earth. His findings have opened our eyes to the mysteries of a planet we thought we knew very well. Wilson's work has also showed the scientific and environmental communities the critical need to stop the mass extinctions that are now proceeding more rapidly than at any other time in human history. He emphasizes that every time we lose a species to extinction we are sacrificing something that may have provided human society with untold medical or sociological benefits.[52]

Through his research and writing, Wilson has developed a large number of concepts and theories that have placed him on an academic sphere attained by few other scientists. Being named a Junior Fellow of Harvard's Society of Fellows opened up the world for Wilson's research. His scientific travels have taken him from Cuba and Mexico to Australia, New Guinea, Fiji, New Caledonia, and Sri Lanka.[53]

In the late 1970s, Wilson became actively involved in the global conservation on biodiversity.[54] His 1984 book, *Biophilia*,[55] explored humanity's attraction to the natural environment and played a major role in shaping the modern conservation ethic. In 1988 he edited the volume *Biodiversity*,[56] which linked his name with the term and earned him a place as the paramount authority on the concept.[57]

Wilson's work has not been without controversy. In his book *Sociobiology: The New Synthesis*, he proposed that the essential biological principles on which animal societies are based also apply to humans.[58] As explained by the Academy of Achievement: "Wilson speculated that hierarchical social patterns among human beings may be perpetuated by inherited tendencies that originally evolved in response to specific environmental conditions. A number of Wilson's colleagues took strong exception, and others condemned Wilson's work on the grounds that it justified sexism, racism, polygamy and a host of other evils."[59] Wilson vehemently denied this was what he had meant, but it didn't stop demonstrators from picketing his talks and, in one case, dousing him with water as he stood on stage.[60]

Throughout the controversy, Wilson staunchly defended his viewpoints and in 1978 published another highly acclaimed work, *On Human Nature*, for which he was awarded the Pulitzer Prize.[61] In this book, Wilson thoroughly examined the scientific arguments surrounding the role of biology on the evolution of human culture and further explained the theories some people found problematic. By the end of the 1970s, scientists generally accepted Wilson's ideas about human behavioral evolution and protesters had stopped picketing at his appearances.[62]

Throughout his time as a professor at Harvard, Wilson continued to write. With fellow myrmecologist Bert Hölldobler, Wilson coauthored the monumental work *The Ants*,[63] earning him a second Pulitzer Prize.

Since retiring from Harvard in 1996 and becoming a Professor Emeritus, Wilson has continued his activism and writing. He has published over a dozen books during this time, and his conservation efforts have benefitted the Columbia University's Earth Institute, the American Museum of Natural History, Conservation International, the Nature Conservancy, and the World Wildlife Fund.

Two of Wilson's books in particular are worthy of special mention. His novel, *Anthill*,[64] tells the story of a boy growing up in rural Alabama who, as an adult, battles against real estate developers to save a piece of wilderness land. The middle section of the book, titled "The Anthill Chronicles," describes the world from the point of view of the ants, who are fighting against a rival colony, and the resulting gargantuan battles resemble a war from the ancient world—the Visigoths versus the Huns, perhaps.[65] The parallels between the insect world and the human world are clear, with struggles for dominance and survival at the forefront. On a much lighter note, Wilson's memoir *Letters to a Young Scientist* should be a must-read for any school child who is considering pursuing biology as a profession.[66]

> ***Destroying a rainforest for economic gain is like burning a Renaissance painting to cook a meal.***
>
> **—E. O. Wilson**[67]

> ***We are drowning in information, while starving for wisdom. The world henceforth will be run by synthesizers, people able to put together the right information at the right time, think critically about it, and make important choices wisely.***
>
> **—E. O. Wilson**[68]

In a question and answer session conducted on April 30, 2012, by Lisa Hymas, senior editor at *Grist*, E. O. Wilson asked, "Why aren't you young people out protesting the mess that's being made of the planet? . . . Why are you not repeating what was done in the '60s? Why aren't you in the streets? And what in the world has happened to the green movement that used to be on our minds and accompanied by outrage and high hopes? What went wrong?"[69]

In his subtle, down-home way, Wilson is hitting the nail right on the head when it comes to doing something that will actually make a difference in counteracting climate change. If we can't get enough people out in the streets, demanding change, then all of our individual efforts collectively just aren't going to get the job done. The sociopolitical history in the United States has proven, time and again, that grassroots organizations and mass demonstrations supported by millions—not hundreds, not thousands, but millions—are what it takes to get the attention of and action from the decision-makers in Washington, DC.

PAUL EHRLICH AND HIS POPULATION BOMB

The next significant book to hit the streets in the late sixties was Paul R. Ehrlich's 1968 bestseller, *The Population Bomb*.[70] While his dire forecasts for climate change hopefully will not come to pass, Ehrlich's theories provided a wake-up call for much of the civilized world. His warnings about a human population that is too large to be sustainably supported by Earth's resources are now being viewed with renewed concern—especially in the face of climate change. While Ehrlich was not the first scientist to raise the alarm about overpopulation issues, his charismatic and media-savvy approach brought the issue to a new level of prominence during the primary US environmental movement years of the late sixties and early seventies.

Part Three

THE US ENVIRONMENTAL MOVEMENT

A Story for the Ages: 1969 through 2016

Chapter 16

THE ENVIRONMENTAL HEYDAYS (1969–1979)

THE MEDIA GETS INTO THE ACT

During the environmental heyday of the sixties and seventies, media, both print and television, began to be used as a primary tool to promote the US environmental movement. One of the most famous examples is the poignant 1971 public service television spot, "Keep America Beautiful," showing a Native American brave canoeing down a river until he arrives at a trash-laden, pollution-spouting urban waterfront, where—after he drags his canoe out of the water—a tossed bag of garbage explodes at his feet.[1] The close-up zoom showing a single tear drop rolling down his cheek is heart-breaking and instantly captured the nation's attention. (As it turned out, of course, the actor was Italian and not Native American, but Americans got the point and responded.)

Throughout the sixties, David Brower, executive director of the Sierra Club, who we talked about back in chapter 14, was still around harassing the Bureau of Reclamation and leading the opposition to federal dams. In 1966, he helped develop an ad campaign attacking a plan to build dams in the Grand Canyon, which would flood parts of the park. One ad read, "Should we also flood the Sistine Chapel so tourists can get nearer the ceiling?"[2]

Public reaction to this ad was enormous and instantaneous, generating almost total support against building dams anywhere in the Grand Canyon. The Washington politicians who had been behind the idea were so incensed that they dispatched an IRS agent to the local Sierra Club office with a letter stating that the group's nonprofit status had been withdrawn. But the Sierra Club had no intention of giving in.[3]

As Tom Turner, editor for Friends of the Earth and Earthjustice, put it, "People in the public may not have known what they thought about the Sierra Club; but they sure knew what they thought about the IRS. Sympathy for the Sierra Club just boiled over and people joined in droves."[4]

The resulting public reaction was so intense that the pressure changed the course of the government's plans. Instead of building the proposed dams, Congress expanded Grand Canyon National Park and prohibited dams from being built anywhere within its boundaries. In the final analysis, the Sierra Club had used a series of simple print ads to change the course of history and become the new face of the US environmental movement.[5]

Making strategic use of the media is certainly something to consider when determining how to build support for our climate-change battle. Everything depends on producing pieces that speak to people and create a national reaction.

As one of the narrators (environmental activist Doug Scott) in the film *A Fierce Green Fire* states, "Every now and then some issue arises that is elevated into the stratospheric focus of public attention, then becomes the symbolic rallying cry for a whole generation of activists."[6] In the 1960s, that issue was the proposed damming of the Grand Canyon; today we need the issue to be solving the climate-change crisis.

ARRIVAL OF THE BROWN CLOUD: RETOOLING THE US AUTO INDUSTRY

The late sixties also featured a number of natural disasters unlike anything the world had seen before, providing even more proof that Earth was in dire need of increased environmental sensitivity and protection. Most prominently, in June of 1969, *Time* magazine published an article about the Cuyahoga River near downtown Cleveland catching on fire.[7] The pictures included with the article depicted the flaming river, an absurdity that even Hollywood movie producers could hardly imagine. In a later article, *Time* reported that the photos included with the article were from a much larger fire in 1952—the 1969 fire was too small and too quickly extinguished to catch on film.[8] Nonetheless, the article ignited more than just the river—the public rallied against pollution.[9]

In many places around the country—places such as Denver, Colorado—newscasters were advising residents to keep their children and elderly relatives inside during the frequent "brown cloud" smog alerts many cities were experiencing.[10] Primarily emanating from automobile engine combustion, the all too frequent brown clouds became a huge concern because they were seriously affecting the health of millions of Americans.

During the early 1980s, we vividly remember often sitting in our living room on cold winter mornings in the foothills of the Rocky Mountains, and looking out at the city of Denver—fifteen miles to the east. Downtown Denver's central phalanx of skyscrapers often looked to be immersed in a bowl of beef bouillon, due to the automobile exhaust fumes trapped by daily temperature inversions.

To counteract this confounding pollution problem as a nation, we took on the all-powerful American automobile industry and slowly forced them to make major changes in the products they delivered to consumers. The first legislated exhaust (tailpipe) emission standards came as an amendment to the Clean Air Act of 1963, focusing on establishing standards for automobile emissions.[11] Then, starting in 1970, the US Environmental Protection Agency (EPA) began progressively tightening national emission standards.[12]

Fuel for our cars changed significantly in the 1970s to adapt to these federal emission modifications. Specifically in 1973, the EPA released a study that confirmed the direct public-health threats posed by lead from automobile exhaust.[13] Regulations for gradually reducing lead in gasoline followed soon thereafter.

Eugene Houdry, a mechanical engineer, invented the catalytic converter in the 1950s to reduce the emissions from smoke stacks and automobiles.[14] Houdry's great concern about air pollution inspired the invention of these "cats" (as the converters were called for short), but they could not be used in automobiles until the lead was removed from gasoline.[15] Lead reacted with the cats and disabled them by forming a coating on their surfaces, so it was necessary to use unleaded gasoline with catalytic converters. When catalytic converters were used, they significantly reduced vehicle emissions and allowed vehicles to meet the tightening emissions standards.

These events—the installation of catalytic converters, lead being removed from gasoline, and a rising concern over air pollution—came together in 1975. At the end of that year, Congress passed the Energy Policy Conservation Act, which set specific miles per gallon (mpg) requirements for cars manufactured annually.[16] The act required that the average fuel economy for passenger vehicles—not including light trucks, such as pickups, minivans, and SUVs—increasingly improve beginning with 18.0 miles per gallon in 1978 up to 27.5 miles per gallon in 1985 and thereafter.[17] We have simplified this story of course—transitioning from leaded to unleaded fuel took time and there were obstacles along the way. But today all cars are equipped with catalytic converters and, at least in America, we fill up with unleaded fuel—which is much safer for human and environmental health—without thinking twice about it. We will talk more about this in our solutions section, but it is important to note how this fuel transition, which seemed insurmountable at first, eventually became widely accepted as normal. We hope that the same transition can be made to renewable energy sources.

In addition to this industry retooling, foreign-made, fuel-efficient cars—led by the Japanese and German auto industries—gained a stronger foothold in the American market.[18] American interest in fuel efficiency began with the rising gas prices brought on by the 1973 oil embargo and Arab-Israeli war, but continued after these crises had been resolved.[19] As a result, Detroit's "Big Three" auto manufacturers—Ford, General Motors, and Chrysler—began manufacturing smaller and more fuel-efficient cars.[20]

Isn't this situation with the US auto industry directly analogous to what we are now facing with climate change? For one thing, we now have all-powerful entities—the Big Oil companies—dominating American lifestyles and finances, just as the "Big Three" US automobile manufacturers did in the sixties and seventies. And we now have a pollution problem that threatens to severely affect the health and welfare of everyone on Earth. So why can't we make the oil companies retool their production lines—just like we did with the automotive production lines—requiring that they extract less fossil fuels and shift to building infrastructure that increasingly creates and transports more renewable energy supplies? We have done it before, so why can't we do it again?

MEETING THE PUBLIC'S DEMANDS: FIX OUR ENVIRONMENT NOW!

Nationwide, our environmental woes continued, with several raptors—including the iconic bald eagle and the spectacular peregrine falcon—nearing extinction due to the lingering effects of DDT poisoning in their food supplies. Then, in 1969, the Santa Barbara oil spill—at the time, the largest in US waters—spewed more than 80,000 barrels of crude oil onto Southern California beaches.

The country's landscape was being destroyed. People everywhere were fed up and angry. As biologist and author Victor Scheffer wrote in *The Shaping of Environmentalism in America*,

> It was the daily commuters who drove with smarting eyes through city smog, the mothers who learned that DDT was present in their breasts and that arsenic from smelter smoke was accumulating in the bodies of their children, the poultrymen who wondered why eggshells broke more easily than they used to, the fishermen who saw trout streams, once pure, now running brown, the farmers who wondered where all the bluebirds had gone, and why the water level in the wells had dropped and why the water now tasted queer.[21]

THE FIRST EARTH DAY LEAVES A MIGHTY MARK

On April 22, 1970, the first Earth Day was held, a sign of the serious interest in environmental protection that was beginning to take hold. The brainchild of Wisconsin senator Gaylord Nelson, Earth Day spotlighted such problems as thermal pollution of the atmosphere, dying lakes, the profusion of solid waste, ruinous strip mining, catastrophic oil spills, and dwindling natural resources. As a pivotal event in the environmental movement, the first Earth Day emphasized that the obsession with industrial growth and consumerism was straining the environment to the breaking point and introduced the idea of living lightly on the Earth.[22]

After the first Earth Day was over, Nelson mulled over what had just occurred—the greatest demonstration of public support for a cause in US history: "No one could organize 20 million people, 10,000 grade schools and high schools, 2,500 colleges and 1,000 communities in three and a half months even if he had $20 million. [Nelson had just $190,000.] The key to the whole thing was the grass roots response."[23]

All concerned climate-change activists would do well to study the unequivocal success of the first Earth Day. The unexpected magnitude of the response to this landmark event clearly shows what can be done when the political and social moods of the country collide, coming together with a message that says, "Let's get something done." Furthermore, the resulting outpouring of federal environmental legislation proves that Congress was listening to what the people were asking for. Such a groundswell of public opinion—all speaking with the same voice—will certainly go a long way toward passing similar laws and regulations to initiate real climate-change solutions both in the United States and around the world.

TRICKY DICK COMES TO THE RESCUE

The public outcry for fixing the environment became so pervasive that when President Richard Millhouse Nixon—arguably the most unpopular leader in US history—took office he was forced to acknowledge that something had to be done about the environment. Nixon made his feelings quite evident during his first State of the Union Address, saying, "The 1970s absolutely must be the decade when America pays its debt to the past by reclaiming the purity of its air and its waters. . . . It is literally now or never."[24] The flurry of environmental legislation coming from Capitol Hill in the first years of the seventies offered solid proof that Nixon and Congress were committed to putting federal funds where their mouths were.

The demand for more national environmental protection was so great that no administration could ignore it, including one staffed primarily by conservative and—as proven by the sordid and shameful Watergate Affair in 1972—criminally unscrupulous politicians.

As an aside here, I (Budd) vividly remember, as a graduate student at Virginia Tech, having dinner at my dad's house in Blacksburg, Virginia, when the Watergate break-in was first reported on the CBS Evening News. Walter Cronkite mentioned it in passing late in the broadcast, essentially as a nonstory—"some minor criminal activity at the Washington, DC, hotel where the Democratic National Convention was being held." My dad immediately looked up from cutting his broiled chicken and—displaying his right-on instincts as a lifelong newspaperman—said, "Boy that sure sounds like that just might be a real story!"

NEPA, CEQ, AND THE BIRTH OF THE EIS

Fittingly, the first federal action to take place during the seventies was the National Environmental Policy Act (NEPA), and, along with it, the president's Council on Environmental Quality (CEQ). While officially intended to be representative of a national policy promoting enhancement of the environment, NEPA, in actuality, turned out to be something quite different. In fact, it quickly became a controversial vehicle for debating the pros and cons of what was nebulously referred to as proposed federal actions that may or may not have significant negative or positive impacts on the quality of the natural or human environments.[25] Got it?

Well, don't feel bad, neither did most of the people who were hired by the federal government to conduct NEPA analyses and then write corresponding Environmental Impact Statements (EISs) summarizing the findings of these analyses. I (Budd) should know—I was one of those people.

I spent my first three years as a "professional environmental scientist" trying to convince Congress to clarify exactly how an EIS was supposed to be written and how the findings should be determined and presented. But we digress . . . at least NEPA represented a legitimate effort on the part of Congress to embed some sort of environmental ethic and consciousness into the land-development business in the United States. And this was, in fact, a big improvement over what previously had been required from an environmental analysis and evaluation standpoint—which was pretty much nothing at all.

RUSSELL TRAIN: THE FIRST GURU OF CEQ

All of this Washington, DC–level hoopla about environmental awareness and protection brought another man to the forefront of the conservation movement. From the day that he first set foot in Washington, past environmental hero Russell Train was destined to be a bright star in the surprising mix of politics, high finance, and natural resource conservation. Meticulous, dapper, and exquisitely mannered, even as a young boy, Train had the combined legal and business acumen to know how to get things done in DC, and *wow* did he ever put those skills to work for national environmental regulation and worldwide wildlife protection.

Born in Jamestown, Rhode Island, in 1920, Train was raised in Washington, where his father, a rear admiral in the US Navy, served as President Herbert Hoover's naval aide.[26] Educated at both Princeton and Columbia Universities, Train started his career as a DC lawyer working in the US Tax Court but quickly realized he could create a niche for himself as an advocate for African wildlife—his first real love.[27] So in 1965 he resigned in the middle of his term and became president and chief attorney of the Conservation Foundation.[28]

Train's success as a conservation attorney was so great that he caught the eye of president-elect Richard Nixon, who decided that his deft combination of innovative ideas and personal passions were exactly what was needed to capture the public's burgeoning concerns—and votes—about environmental protection. Nixon plugged Train into all the right places, and it was actually through Train's advice that Nixon created both the CEQ—to review regulatory activities under NEPA—and the US

Environmental Protection Agency (EPA).[29] Once Train became the first chairman of CEQ, he was off and running as a force to be reckoned with.

His work as head of the CEQ was so effective that Train soon had the moniker "the Father of NEPA," attached to his name. His policy of "look-before-you-leap" became the catch phrase for analyzing the potential impacts of major federal actions before building them.[30] As CEQ chairman, Train also earned the reputation for being the founding father of the United Nations Educational, Scientific, and Cultural Organization's (UNESCO) World Heritage Program.[31]

After being appointed administrator of the EPA in 1973, Train became well known for creating groundbreaking laws and implementing effective enforcement of a host of rules and regulations. With the power delegated to him, Train shaped the world's first comprehensive programs for scrubbing the skies and waters of pollution while safeguarding US citizens from exposure to toxic chemicals.[32]

After leaving government service in 1978, Train moved on to become the first chairman of the World Wildlife Fund (WWF). During his time in leading the organization, the WWF grew from a small, relatively unknown conservation group to a global force for conservation, consisting of a hundred-million-dollar-a-year global network of researchers and technical specialists, recognized by its panda trademark.[33]

For the remainder of his life, Train continued to receive recognition—from serving as Chairman Emeritus of the WWF to being awarded with the Presidential Medal of Freedom by President H. W. Bush in 1991.[34] In 2003, Train published his memoir, *Politics, Pollution, and Pandas*,[35] which is an excellent overview of the birth and growth of the United States' national interest in environmental issues.

> ***I believed strongly that environmental issues needed a sharp, cutting edge in government, one that had high visibility to the public . . . this [the establishment of the EPA as an independent agency] was the view that ultimately prevailed.***
>
> **—Russell Train, *Politics, Pollution, and Pandas*[36]**

Russell Train's organization-building skills—in both the public and the private sectors—should be studied in depth by climate-change leaders and activists. Successfully fostering a conservation cause first depends on having a cohesive, well-staffed, and financially solvent organization in place.

THE FEDERAL ENDANGERED SPECIES ACT

In 1973, the US Congress passed the federal Endangered Species Act (ESA) and placed it under the jurisdiction of the US Fish and Wildlife Service (USFWS). According to conservation writer Paul Rauber, "For the first time in history, the government undertook a concrete, systematic effort to save wildlife from the destructive behavior of humans."[37] Whenever a native plant or animal was brought up for evaluation, the USFWS placed it on a watch list so that its status could be studied.

Then, if a species was deemed to be in jeopardy, the agency made a determination, declaring the species to be either "threatened" or "endangered" by extinction. Since monitoring and protecting individual members of a listed species proved in most cases to be extremely difficult, if not impossible, the ESA sought to protect the critical habitats of each species.

Despite some regulatory rough patches, the ESA was certainly a big improvement over having no jurisdictional protections for wildlife species faced with extinction—save for injunctions based on public demonstrations and federal court interventions. In fact, the ESA has had many notable successes—including such showcase (high visibility) species as the bald eagle, American alligator, the peregrine falcon, the osprey, and the brown pelican—all of which are now common or even abundant in their natural habitats.

THE FEDERAL CLEAN WATER ACT

Aimed at protecting the quality of our nation's waters—everything from major estuaries to rivers, streams, floodplains, lakes, ponds, and bordering wetlands—the Federal Water Pollution Control Act—more commonly referred to as the Clean Water Act (CWA)—was passed in 1972.[38] The CWA's primary goal is preventing the contamination of every body of water or vegetated wetland in the United States by controlling impacts associated with illegal filling, dumping, and pollution runoff. The pollution runoff component deals with both point-source discharges—such as the outflows from factory and sewage plant pipes—and nonpoint-source discharges—meaning the contaminants contained in storm water runoff.

As it turned out—since few polluters and developers were bold or stupid enough to dump contaminants directly into the open water of rivers—the primary regulatory actions under the CWA involved protecting our nation's wetlands.[39] Surprisingly, the CWA also designated the US Army Corp of Engineers (US ACOE) as the first line of regulatory authority over all of the federal wetlands in the United States.

We say surprisingly because—throughout US history—no other entity, public or private, has negatively impacted more wetlands than the US ACOE. In their defense, however, the ACOE's impacts have primarily been associated with wetland dredging, not filling, due to the fact that one of the agency's primary functions is keeping our nation's rivers and harbors open for shipping. Despite this incongruity, the Wetland Regulatory Divisions within the US ACOE that are based in each of the agency's Regional Offices have—in my (Budd's) professional experience—typically done an admirable job of adjudicating wetland boundaries and processing Section 404/401 Wetland Permit Applications for proposed projects under the CWA.

THE COASTAL ZONE MANAGEMENT ACT

Also passed in 1972, the federal Coastal Zone Management Act (CZMA) encourages coastal states in the United States to develop and implement Coastal Zone Management Plans (CZMPs).[40] Initially drafted as a US national policy, the CZMA typically is used to assist states with preservation, protection, development, and "where possible, to restore or enhance the resources of the nation's coastal zone"[41] for current and succeeding generations.

However, again in my (Budd) experience, the CZMA has primarily served as window dressing, since—under it—states are allowed to essentially do whatever they want. Since coastal property is typically the most economically attractive land for development, states have seldom implemented management plans that are overly restrictive to proposed land development. This can be quickly verified by using aerial views on Google Maps to look at the jumbled morass of development that has taken place in many coastal areas—especially on vulnerable barrier islands.

In far too many cases, unrestricted development has removed the first line of defense for coastal communities—the primary foredunes that have naturally built up behind the high-tide lines. The foredunes previously provided a buffer from coastal storms for the secondary dunes, coastal forests, marshes, and estuaries. Now—with these dunes gone—it's not a question of *if* these ill-begotten oceanfront developments will be blown away by a summer hurricane or fierce winter nor'easter but *when*.

The results are sad and devastating—think, for example, of a situation like the Jersey Shore, which was leveled by Hurricane Sandy in late October 2012. Time and time again, you hear folks who lost everything vowing "to never give up" and to build everything "back just like it was before." While the most logical thing to do would be to relocate somewhere that is not so vulnerable to Mother Nature's fiercely unrelenting power, how do you tell that to honest, hard-working families who have just seen their homes and neighborhoods ripped to shreds by uncontrolled storm surges? It's a vexing problem indeed, especially in coastal areas that have already been desecrated beyond any reasonable hope of ecological restoration and recovery any time in the near future.

BARRY COMMONER: THE PAUL REVERE OF ECOLOGY

Several landmark environmental books were released in the seventies as well, with most dealing with the fact that the earth may be about to start spiraling out of control with no hope of recovery. In particular, books by two of our past environmental heroes—Barry Commoner and Donella

"Dana" Meadows—provided early hints that we were on a doomsday track that could culminate in a radically changed Earth.

Although widely criticized in many circles as a doomsdayer, Barry Commoner was way ahead of his time in describing the dramatic consequences of zealous overconsumption, capitalistic greed, and abuse of natural resources. In fact, his writing and speeches predicted the onset of the climate-change crisis we are facing today. In a cover-page feature in 1970, *Time* magazine called Commoner the "Paul Revere of Ecology" and said that "He has probably done more than any other US scientist to speak out and awaken a sense of urgency about the [world's] declining quality of life."[42]

With his thick hair, warm smile, and intense but purposeful gaze, Barry Commoner would be a good choice for one of any environmentalist's three dream guests at a dinner party. Born in Brooklyn, New York, in 1917 to Russian immigrants, Commoner first studied zoology at Columbia University and then at Harvard, where he received his doctorate in biology and ecology in 1941. Commoner was one of the new science of ecology's most provocative thinkers, and he recognized that America's technology boom following World War II was not all good.

As a leading opponent of nuclear testing, Commoner was credited with creating the momentum that led to the passage of the 1963 Nuclear Test Ban Treaty between the United States and the Soviet Union. He also knew we were running the risk of poisoning both the land and ourselves with the preponderance of toxic substances we were spewing across Earth and into our skies. With the publication of his 1971 bestselling book, *The Closing Circle*,[43] Commoner helped launch the environmental movement of the 1970s—being often mentioned with such other notable activists and environmental heroes as Rachel Carson, David Brower, and Aldo Leopold.

The parallels between Commoner's work and beliefs and those of the modern climate-change experts are intriguing. In the 1950s, Commoner first became well known for his emphatic warnings about the hazards of atmospheric testing of nuclear weapons and the fallout generated by these tests. Capitalizing on his newfound fame, he next alerted the American public about the dangers created by the petrochemical industry and toxic substances such as dioxins (persistent organic pollutants, or POPs).[44] These pollutants accumulate in the fatty tissue of animals, and may cause developmental, reproductive, and hormonal problems, as well as cancer.[45]

Laying the groundwork for the environmental justice movement—a catch phrase in the current climate-change debate—Commoner continually emphasized that environmental hazards disproportionately impacted the poor and racial minorities, since dangerous chemicals and associated hazardous conditions are more typically found in rural or blue-collar neighborhoods.[46] Today, climate-change analysts focus on these same points, while also emphasizing that poor people in developing countries throughout the world are the primary sacrificial lambs of the fossil-fuel industry.

Just as do the leading members of today's 350.org—which we'll discuss in chapter 20—Commoner viewed the environmental crisis of the 1960s as a symptom of a fundamentally flawed economic and social system. In his opinion, three primary culprits—corporate greed, illogical government priorities, and the misuse of technology—were driving the world's infatuation with excessive profits and overindulgent lifestyles, threatening to make Earth an unfit place to live.[47]

Commoner continually emphasized the parallels among the environmental, civil rights, labor, and peace movements in the United States, while also connecting the ongoing environmental crisis to world problems of poverty, injustice, racism, public health, national security, and war.[48] These are the same arguments and concerns that are now being analyzed as the primary hurdles to discovering practicable solutions for dealing with climate change.

In the 1970s, Commoner disagreed with Paul Ehrlich's view—as expressed in Ehrlich's book, *The Population Bomb*[49]—that overpopulation, particularly in developing countries, was responsible for depleting the world's natural resources and deepening the earth's environmental problems. In *The Closing Circle*, Commoner introduced the idea of sustainability, now a widely discussed concept but during the 1970s very controversial, and often linked to socialism[50] He emphasized that there is only one ecosphere for all living things and that we needed to maintain it to ensure our long-term survival.

In line with ecological thought, Commoner believed that "what affects one, affects all."[51] Encouraging the now-widespread practice of recycling, he also noted that in nature there is no waste and because of this we can't just throw things away. He advocated designing and manufacturing products that can be reused, thus maintaining the delicate balance between humans and nature.[52] Commoner was one of the first scientists to bring the concept of sustainable living to a mass audience, challenging the petroleum industry and—long before it became politically fashionable—touting solar power as the long-term solution to the world's energy needs.[53]

In his book *The Closing Circle*, Commoner lists his four "laws of ecology," which summarized his overall views:

- Everything is connected to everything else.
- Everything must go somewhere.
- Nature knows best.
- There is no such thing as a free lunch.[54]

A man whose ideas were well ahead of their time, Barry Commoner would have been a prime candidate for the leader of today's climate-change movement. His courage to take stands and express philosophies that were contrary to popular thought are exactly what is needed to get the message across and start implanting the major and significant changes that are required in the social, industrial, and political infrastructures of today's world.

DONELLA MEADOWS AND HER LIMITS TO GROWTH

When it comes to standing behind your beliefs, Donella "Dana" Meadows is at least the equal of Barry Commoner. Meadows had solutions for dealing with climate change years before it became a prominent national and worldwide concern, plus she practiced exactly what she preached. In fact, the groundbreaking book she coauthored in 1972, *The Limits to Growth*,[55] sold twelve million copies in twenty-nine languages and launched Meadows onto the global stage as a leading environmental thinker and writer.[56] *Limits* made headlines around the world and began a debate about the extent of Earth's capacity to support human economic expansion—a debate that continues to this day.

Born in Elgin, Illinois, in 1941, Meadows received her doctorate in biophysics from Harvard University in 1968.[57] She then became a research fellow at MIT, working in the department of computer engineering with professor Jay Forrester and studying the application of the relatively new field of systems dynamics to global problems.[58] Inspiration from this landmark research led Meadows and her colleagues to write and publish *Limits*.

For a small book, *Limits* packed a huge wallop. Its writing analyzed "the predicament of mankind," including interrelated economic, social, and political problems. Among the topics covered in the book are poverty amidst prosperity, unchecked urban sprawl, environmental degradation, alienation of youth, insecurity of employment, inflation, rejection of traditional values, and loss of faith in institutions.[59]

One of the primary conclusions in *Limits* was that, "If the present growth trends in world population, industrialization, pollution, food production, and resource depletion continue unchanged, the limits to growth on this planet will be reached sometime within the next one hundred years. The most probable result will be a rather sudden and uncontrollable decline in both population and industrial capacity."[60] Years later, Meadows commented to an interviewer, "From my point of view as a scientist, there was nothing more stupidly obvious than to say that the Earth is finite and growth can't go on forever."[61]

Of course, in 1972, the premise of *Limits* was considered to be an extremely radical point of view. Accordingly, the book provoked a firestorm of ridicule, criticism, and vitriol from the business, political, economic, and even academic establishments.[62] On balance, however, *Limits* also garnered reasonable acclaim and applause from a number of skeptics who were gradually becoming more and more concerned about the ever-increasing human population's negative influences on Planet Earth.[63]

All but lost in the controversy was this optimistic solution to this pending global crisis put forth in *Limits*:

> It is possible to alter these growth trends and to establish a condition of ecological and economic stability that is sustainable far into the future. The state of global equilibrium could be designed so that the basic material needs of each person on earth are satisfied and each person has an equal opportunity to realize his individual human potential. If the world's people decide to strive for this second outcome rather than the first, the sooner they begin working to attain it, the greater will be their chances of success.[64]

After the publication of *Limits*, Meadows spent sixteen years writing a weekly syndicated column called the Global Citizen, which appeared in twenty newspapers, and in which she commented on world events from a systems point of view.[65] Through the years, her writing won many awards, including second place in the 1985 Champion-Tuck national competition for outstanding journalism in the fields of business and economics and the Walter C. Paine Science Education Award in 1990.[66] Meadows was also honored as a Pew Scholar in Conservation and Environment (1991) and a MacArthur Fellow, plus she was nominated for a Pulitzer Prize in 1991.[67] Posthumously, she received the John H. Chafee Excellence in Environmental Affairs Award for 2001, the year she died, presented by the Conservation Law Foundation.[68]

In 1996, Meadows founded the Sustainability Institute (SI) at Cobb Hill in Hartland, Vermont. SI's mission statement emphasized implementing systems of sustainable living for all levels of society at locations throughout the globe. After her death, the SI was renamed the Donella Meadows Institute (DMI) and moved its offices to Norwich, Vermont. DMI's overriding message was really quite simple: "Our Mission: to bring economic, social, and environmental systems into closer harmony with the realities of a finite planet and a globally powerful human race by using the disciplines of systems thinking, system dynamics, and collaborative learning that were pioneered by our founder, Donella Meadows."[69] Since its founding, the DMI has been at the forefront of worldwide sustainability thinking and training.[70]

In the latter years of her life, Meadows truly practiced what she preached—adhering to her personal mantra, which was couched in microbiologist and author Rene Dubos's famous quotation, "Think Globally, Act Locally." Because of her worries about climate change, she restricted her own travel to only those events at which she felt her physical presence would do the most good. She also lived for many years on an organic farm, existing simply, and saving energy. She bought a hybrid gas/electric car as soon as they became available.

If Meadows was alive today, her philosophy and lifestyle would provide the perfect components of a climate-change leader. For one thing, she understood all too well what the future will hold if we continue to cling to our "progress is good at all costs" mentality. She also firmly believed that humans had the potential and the power to do what was right for the long-term future of the world and humanity.

We think these last two quotes—included in author Leon Kolankiewicz's *Donella Meadows—A Tribute*—perfectly summarize who Donella "Dana" Meadows was:

> Personally I don't believe that stuff [about just giving up] at all. I don't see myself or the people around me as fatally flawed. Everyone I know wants [both] polar bears and three-year-olds in our world. We are not helpless and there is nothing wrong with us except the strange belief that we are helpless and there's something wrong with us. All we need to do, for the [polar] bear and for ourselves, is to stop letting that belief paralyze our minds, hearts, and souls.
>
> We have no choice but to conform [to a more sustainable future]. If we don't choose to, the planet will make us. And [in fact] our lives will be better if we do. It isn't sacrifice we're selling, it's a more meaningful, time-filled, love-filled, nature-filled existence.[71]

THE ADVENT OF RADICAL ENVIRONMENTALISM

Riding the tidal wave of environmental activities in the "green seventies" were a host of newly formed nongovernment organizations (NGOs)—most focused on slowing economic progress and stopping associated environmental degradation. Included among these "opposition-first groups" were some of the old guard—like the Sierra Club and the National Audubon Society—plus some often-volatile newcomers. Among these were the Friends of the Earth (FOE)—David Brower's personal offshoot from the Sierra Club—the League of Conservation Voters (LCV), Zero Population Growth (ZPG), Earth First!, and Greenpeace. The abundance of federal environmental legislation introduced in the early seventies also ushered in a phalanx of environmental lawyers—many working primarily to stop development—for new nonprofit organizations such as the Environmental Defense Fund (EDF), the Natural Resource Defense Council (NRDC), and the Sierra Club Legal Defense Fund (SCLDF).

In particular, the seventies saw the advent of radical environmental activism. Before this, the environmental movement was led by the peaceful and passive approaches of purists like John Muir, Aldo Leopold, George Perkins Marsh, and Bob Marshall. But as an offshoot of the social extremism of the sixties, this all changed. The primary focus shifted from the persuasive, yet peaceful written words of the aforementioned heroes to actively forcing issues and responses by rallying the support of the masses for environmental causes. In many cases, approaches involved everything from civil disobedience and associated arrests of demonstrators to surreptitious illegal activities and even outright violence.

EDWARD ABBEY: THE CUTTING EDGE OF THE RADICAL LEFT

In the minds of many radical environmentalists, Edward Abbey's outlandish 1975 novel, *The Monkey Wrench Gang*, set the tone and attitude for how to best get things accomplished.[72] With an emphasis on protesting environmentally damaging activities through the use of force, the term "monkey wrenching" soon defined any sabotage, activism, or law-breaking used to preserve wilderness, natural ecosystems, wild spaces, and rare wildlife. *The Monkey Wrench Gang*'s main protagonist was George Washington Hayduke, who epitomized the frustrations of the typical male environmental avenger who was out to smash up the infrastructure of corporate greed and corruption any way he could. Championing the usually unheard voices of the "little people," Abbey's

Hayduke justified his unorthodox, costly, and highly illegal actions of environmental mayhem with the viewpoint that, if not him, then who—it has to be done by someone.[73]

Known for his anarchistic rhetoric and sanctimonious wit, Abbey was often at the center of the hip environmental movement. His writings ranged from blatantly outrageous to sublimely poignant and powerful. While *The Monkey Wrench Gang* generated such radical environmental groups as Earth First!, his nonfiction *Desert Solitaire*[74] has been favorably compared to Henry David Thoreau's *Walden* and Aldo Leopold's *A Sand County Almanac*.

Desert Solitaire is a beautifully told nonfiction piece about Abbey's year as a solitary ranger in the secluded backcountry of Arches National Park. Now often considered a classic piece of natural-history writing, this book is a "forceful presentation of one side"—the side angry at the lost wildness of America's national parks.[75] Abbey despised the industrial tourism that paved the dirt roads and allowed visitors to drive through—and trash—the parks without connecting with the land. As Edwin Way Teale writes for the *New York Times*, this perspective is a side that needs presenting, "a side too rarely presented. There will always be others to voice the other side, the side of pressure and power and profit."[76] Abbey pens a combative diatribe of how desert wilderness should be experienced, watching desert stars from a trail surrounded by nature, not via an automobile. In many of his chapters, Abbey also strongly expresses his deep-seated beliefs about the faults of modern Western civilization, the unethical gyrations of US politics, and the rapid disintegration of America's environment.[77]

After his death in 1989, Abbey's family and writing cohorts unceremoniously buried him at night in the Arizona desert, wrapped only in a blue sleeping bag, near a granite rock famously inscribed with the words Abbey chose: "No Comment."[78] The gesture was fittingly apropos for a man who deeply believed that life should blend as lightly and inconspicuously into the natural environment as possible.

GREENPEACE TAKES CENTER STAGE

For the most part, environmental radicalism never quite achieved the level of mayhem and destruction depicted in Abbey's *Monkey Wrench Gang*. Most of the new NGOs of the seventies relied on getting the public's attention through protest rallies fueled by media involvement and peaceful civil disobedience such as bulldozer blockades, treetop sit-ins, and congressional conservation voting record-tracking.

The primary exception to low-key environmental activism was the organization known as Greenpeace. Although professed to be nonviolent by its leaders, Greenpeace often employed in-your-face, smashmouth techniques—commonly referred to as the direct-action approach—that might even have made Edward Abbey blush. While Abbey's writing was radical at times, he believed in using legal, political, peaceful means to make change . . . such as voting.[79]

More than any other NGO, Greenpeace emphasized using the media to gain attention to their causes. Often described as the most visible environmental organization that ever existed, Greenpeace has always been controversial, acquiring seagoing vessels for the sole purpose of using them to directly confront and interfere with Russian and Japanese whaling factory ships, for example.

In 1976, Greenpeace became directly immersed in a battle to stop the slaughter of harp-seal pups in Newfoundland. Who can ever forget the public information spot showing an adorable doe-eyed and white-furred harp seal pup one minute and word that they were being bloodily bludgeoned to death for their pelts the next? However, this campaign was one of the most controversial in Greenpeace history and had severe local ramifications. It effectively devastated the livelihoods of many indigenous people who relied on the seal for food, clothing, boots, fuel and lamp oil.[80]

During their thousands of dramatic protests, Greenpeace activists also infiltrated nuclear test sites, shielded whales from harpoons, and blocked oceangoing barges from dumping radioactive waste. The *Rainbow Warrior*, flagship of the Greenpeace fleet, was mined and sunk by the French foreign intelligence services in the port of Auckland, New Zealand, in 1985. The ship had been en route to protest a planned nuclear test in Mururoa in French Polynesia when it sank, also claiming the life of Fernando Pereira, a freelance Dutch photographer.

Greenpeace was founded in 1971 in the backroom of a store in Vancouver, Canada, by the Don't Make a Wave Committee.[81] Their first mission of note involved chartering an old halibut seiner, the *Phyllis Cormack*, renaming it the *Greenpeace*, and plowing through unfriendly seas in the Gulf of Alaska to protest nuclear testing on the tectonically unstable island of Amchitka in Alaska.[82] This led to a face-off with a US Coast Guard cutter and eventually generated enough public support to force the United States to end nuclear testing on Amchitka.[83]

No matter what you may think about Greenpeace or their methods for confronting and stopping highly damaging environmental activities, the fact remains that they are today one of the world's largest and most successful NGOs. Greenpeace now has an international organization with five ships, 2.8 million supporters, twenty-seven national and regional offices, and a presence in fifty-five countries. Greenpeace's stated goal is to use "peaceful protest and creative communication to expose global environmental problems and to promote solutions that are essential to a green and peaceful future."[84]

Today, the international chapters of Greenpeace focus their campaigning on such worldwide issues as deforestation, overfishing, commercial whaling, genetic engineering, as well as antinuclear issues and climate change. To keep their noses as clean as possible, the global organization does not accept funding from governments, corporations, or political parties.[85]

PAUL WATSON AND HIS SEA SHEPHERD SOCIETY

The exact founding structure of Greenpeace has never been quite clear. To this day, it's said that you can go into any bar in Vancouver, British Columbia, and sit down next to someone who will tell you they are one of the founders of Greenpeace.

Certainly one of the most noteworthy—and outlandish—characters to ever make this claim is Canadian Paul Watson. Whether or not he was a

Greenpeace founder, Watson was an active participant to the point that he got himself banned from the organization and then went out and formed his own rabble-rousing outfit, the Sea Shepherd Conservation Society.[86] Watson proceeded to sail his ship, the *Sea Shepherd*, around the world, attacking the whaling fleets of Norway, Japan, and, most notably, Iceland, where he actually scuttled and sank two boats while they were at anchor.[87] Somehow, Watson managed to escape serving any actual prison time.

After watching the documentary *A Fierce Green Fire*, it is difficult not to consider Watson a true hero, especially if you love marine mammals. He certainly did not pull any punches when it came to fighting for exactly what he believed—the life of every sperm whale and harp seal. In fact, he repeatedly put his own health and safety in harm's way to protect these majestic and lovely animals. In the end—largely due to the disruptive efforts of Greenpeace and the Sea Shepherd Society—the International Whaling Commission (IWC) enacted a global moratorium on whaling, with only Japan refusing to sign the pact. Despite this success, the suitability of Watson's ways are certainly a matter for conjecture and debate within the climate-change community.

RADICAL TACTICS: TO USE OR NOT TO USE?

Whether or not radical environmentalism should be a key strategy of the climate-change movement deserves serious consideration and debate. Throughout US history, major social and economic upheavals have come about in two primary ways—either by the occurrence of a major disaster or by national groundswells of public opinion that were massive enough to get the attention of Congress. Examples of the former would be revamping of building codes in cities that have suffered massive earthquakes (for example, San Francisco and Anchorage) and shoreline protection and zoning in places hit by powerful hurricanes (for example, Hurricanes Andrew and Rita). Latter examples would be the protest marches in Selma and Birmingham, which hastened passage of the Civil Rights Act and the first Earth Day, which played a large role in driving the passage of federal environmental regulations in the 1970s.

Sometimes these two methods of fomenting change dovetailed, typically involving a massive public outcry after the occurrence of a major disaster. The widespread anger after the highly avoidable death and destruction caused by Hurricane Katrina is one of the best recent examples of this situation.

The downsides of this radical approach are twofold. First, there is the very real possibility that radicalism may prove to be a turnoff to the masses and, as such, result in a loss of credibility for the climate-change movement. As we remember it, the bulk of the American public—rightly or wrongly—considered Greenpeace to be a bunch of displaced hippies who didn't have anything better to do with their time than sail around the world causing trouble for commercial ventures.

The other problem is that civil disobedience often involves breaking the law—either as a result of simple sit-in protests or something more volatile like blockading fossil-fuel extraction activities. Civil disobedience typically means arrests, trials, prison sentences, negative press, and possible loss of voting rights, which stirs up resentment and opposition—instead of support—for the cause.

THE OPEC OIL EMBARGO: FOSSIL FUELS TAKE A DIRECT HIT

In October 1973, the US fossil-fuel industry took a catastrophic hit when the members of the Organization of Petroleum Exporting Countries (OPEC) proclaimed an oil embargo. By the end of the embargo less than six months later, in March 1974, the price of oil had risen from three dollars per barrel to nearly twelve dollars per barrel. The result was many short-term and long-term effects on global politics and the global economy.[88]

The OPEC oil crisis essentially applied the brakes to what had been, up to that time, an ongoing flow of environmental regulation and natural resource protection. But—as with just about everything else that happens in the political world—when people's pocketbooks began to be affected, things changed quickly. Escalating energy prices put a damper on the public's enthusiasm for environmental regulation and heightened corporate displeasure with new energy standards like those for sulfur dioxide emissions. In response to the backlash, President Nixon tried to curtail the burgeoning environmental regulation that was taking place, but Congress thwarted most of his efforts to do so.

GERALD FORD TRIES TO TURN BACK THE CLOCK

In 1974, when Nixon was forced to resign because of the Watergate Scandal, Gerald R. Ford took office and did little to help the country's annoyance with environmental regulations, saying, "I pursue the goal of clean air and pure water, but I also pursue the objective of maximum jobs and continued economic progress. Unemployment is as real and sickening a blight as any pollutant that threatens the nation."[89]

During Ford's brief term, the nation was in the midst of debilitating inflation—with thirty-year fixed mortgage rates skyrocketing—and he had no choice but to concentrate on battling the nation's economic issues, putting environmental issues on the back burner.[90] Never known as a strong proponent of environmental oversight, Ford specifically relaxed regulations on both auto emissions and coal strip mining—two activities that have heavily contributed to our current climate-change dilemma.[91] According to Russell Train, EPA administrator at the time, Ford was "fundamentally bored by environmental issues and the EPA's continuing opposition to his executive office vetoes seemed to be only a minor irritant to him."[92]

THE FIRST WAVE OF CLIMATE-CHANGE HEROES: BROECKER, MANABE, AND WETHERALD

During the three years that Gerald Ford lived in the White House—remarkably as the only person to serve as both vice president and president without being elected to either post—some significant events were taking place in the climate-change arena. In 1975, Dr. Wallace Broecker published a scientific paper entitled "Climate Change: Are We on the Brink of a Pronounced Global Warming?" With this publication, Broecker became the first American scientist to use the term global warming.[93]

In Broecker's opinion, Earth's human occupants have a moral obligation to get rid of the waste we create in the process of living on the planet.[94] Just as we designed methods to collect and dispose of our sewage and garbage, we now need to come up with a way to trap and then permanently store our greenhouse gas emissions—notably carbon dioxide. Because of the dynamic impact of Dr. Broecker's findings and his bold statements on the scientific world, we have included him as one of our contemporary climate-change heroes. We will talk more about Dr. Broecker in chapter 23.

During the 1970s, another change was the increasing use of computer models to predict climate change's effect on Earth's ambient air temperature.[95] Climatologist Syukuro "Suki" Manabe was born in Japan and received his doctorate from the University of Tokyo in 1958, while Richard T. "Dick" Wetherald graduated from the University of Michigan with a degree in meteorology. Working together, Manabe and Wetherald used computer-modeling simulations to verify that radiant heat was being absorbed by carbon dioxide in the earth's lower atmosphere.

CFCs AND THE OZONE HOLE: DIRECT PARALLEL WITH THE CLIMATE-CHANGE CRISIS?

In 1976, the National Academy of Sciences (NAS) released a study that provides an intriguing parallel to the climate-change dilemma we're facing today. The NAS reported that chlorofluorocarbon (CFC) gases were damaging Earth's ozone layer.

The ozone layer is a belt of naturally occurring ozone gas in the stratosphere, the layer above the troposphere (which is our lowest atmospheric level). Positioned at 9.3 to 18.6 miles above our planet's surface, the ozone layer shields life on Earth from the harmful effects of ultraviolet B (UVB) radiation emitted by the sun.[96]

In particular, the NAS study showed that the release of pollutants containing the chemicals chlorine and bromine were breaking down the ozone layer and exposing humans to the carcinogenic (cancer-causing) effects of too much UVB.[97] Overexposure to UVB rays was affecting other life on Earth as well. Most notably, UVB rays were inhibiting the reproductive cycle of phytoplankton, the single-celled organisms (such as algae) that comprise the base of our planet's food chain.[98] At the same time, scientific researchers were also documenting changes in the reproductive rates of young frogs, salamanders, fish, shrimp, and crabs—all attributable to being exposed to excess UVB.[99]

According to the NAS study, CFCs—found mainly in spray aerosols used primarily in developed nations—were the main culprits in the breakdown of the ozone layer.[100] In fact, 90 percent of the CFCs were being emitted by industrialized countries in the Northern Hemisphere. The primary producer of CFCs for use as refrigerants and aerosols—under the patented name Freon—was the powerhouse chemical company, DuPont.[101]

When CFCs reach the stratosphere, they break apart into their primary components, including carbon. The free carbon then attacks the oxygen atoms in ozone, in the process ripping apart the ozone layer.[102] According to the EPA, one atom of chlorine can destroy more than 100,000 ozone molecules. The net effect is that ozone in the atmosphere is being destroyed faster than it is being naturally created.[103]

The primary effect of ozone depletion that most of us are familiar with is the ozone hole over Antarctica that has occurred every spring in the Southern Hemisphere since the early 1980s. In 1984, British Antarctic Survey scientists Joseph Farman, Brian Gardiner, and Jonathan Shanklin discovered this recurring springtime Antarctic ozone hole and blamed it on the use of CFCs—although this cause was still to be proven.[104]

Susan Solomon, an atmospheric chemist with MIT, is widely recognized as a leader in the field of atmospheric science. She has explored the reasons why the ozone hole occurs in Antarctica, and she also obtained some of the first chemical measurements that helped to establish CFCs as its cause.[105] We talk more about Dr. Solomon's work in chapter 23.

Rather than being an actual hole, the ozone depletion area contains extremely low levels of ozone—less than 60 percent of normal during the worst years.[106] Although not as severe as over Antarctica, areas of declining ozone layer were also being recorded over latitudes that included North America, Europe, Asia, Africa, and South America—

making ozone depletion a serious global concern and providing another pointed parallel with our current climate-change situation.[107]

Finally, the world decided to take action to combat atmospheric ozone depletion. First came the Vienna Convention in 1985, which formalized international cooperation on the issue. Next was the signing of the 1987 Montreal Protocol on Substances that Deplete the Ozone Layer—now commonly known as the Montreal Protocol (TMP)—an international treaty that entered into force on January 1, 1989.[108] Finally, in 1992, based on the latest scientific assessments that showed a worsening of the ozone depletion, the proponents of TMP met again and agreed to end production of CFCs beginning in 1996.[109] As a result of the CFC reduction measures mandated by TMP, levels of total inorganic chlorine in the atmosphere peaked in 1997 and 1998 and have been falling ever since.[110]

Climate projections indicate that the ozone layer will return to 1980 levels between 2050 and 2070.[111] Due to its widespread adoption and implementation, TMP has been hailed as an example of exceptional international cooperation, with Kofi Annan, former secretary-general of the United Nations, quoted as saying that "perhaps the single most successful international agreement to date has been the Montreal Protocol."[112]

In summary, effective burden sharing and proposals for solutions to mitigate regional conflicts of interest were among the primary success factors for solving the ozone-depletion crisis. Unfortunately, as we will discuss later, global regulation based on the Kyoto Protocol has failed to accomplish the same results for climate change.[113] To meet the ozone-depletion challenge, global regulation was already being put into place before the establishment of a scientific consensus. Also—as contrasted with climate change[114]—both lay people and public opinion were more aware of and convinced about the possible risks of ozone depletion.

Climate-change activists should consider the way the world dealt with the ozone crisis as a prototype. First, we determined that we had a worldwide problem with our atmosphere and that the primary source was CFCs, manufactured by the chemical company DuPont. Next we convinced other industrialized nations that the problem was real and had to be addressed before we reached the point of no return and subjected ourselves to an unlivable planet.

Following this, we demanded DuPont stop making CFCs and find an alternative that would not harm the ozone layer (which—in the end—they willingly did). Finally, we convened worldwide assemblies in Vienna and Montreal that culminated with the phase-out of ozone-depleting substances. The result has been that now the ozone layer is healing itself. That really sounds like the sure-fire formula for climate-change success to us. While we have a few more mega-conglomerate companies to deal with (i.e., all those aligned with Big Oil), at least we've already proven that it is a doable process.

THE HAZMAT BATTLES BEGIN: LOIS MARIE GIBBS AND LOVE CANAL

Despite the slowdown in environmental actions precipitated by the oil crisis, discovery and management of hazardous waste sites made great strides in the last part of the 1970s under the jurisdiction of the EPA. First, in 1976, came the Resource Conservation and Recovery Act (RCRA), which empowered the EPA to regulate hazardous waste sites from "cradle to grave."[115] Infamously described as "ticking time bombs," hazardous waste sites were thrust into the general public's consciousness when multiple locations were discovered throughout the nation.[116]

The most well-known case began in the spring of 1978, when a twenty-seven-year-old housewife, with no scientific training or activist experience, stood up and took on the powerful world of conglomerated chemical companies. When Lois Marie Gibbs—a very unlikely environmental hero—discovered that her child was attending an elementary school built next to a 22,000-ton toxic waste dump with high levels of toxic chemical residues, she went ballistic.[117] The city hired Calspan Corporation—an independent engineering design and testing firm—to investigate the area and found dioxins and PCBs—chemicals that, with long-term exposure, impair the endocrine, nervous, reproductive, and immune systems, and may be carcinogenic.[118] Gibbs immediately set about alerting her neighbors and organizing the Love Canal Homeowners Association (LCHA) in Niagara Falls, New York.[119]

For a while, Gibbs listened to the bold-faced bluster of the Hooker Chemical Company (now the Occidental Petroleum Company—OPC) until she couldn't take it anymore.[120] OPC's team of representatives and hired consultants—along with their allies in the government—repeatedly lied to cover up the facts that their leaking chemicals were causing miscarriages, birth defects, and cancers in local families.[121]

Gibbs didn't give up and, in the end, scored a significant victory against the corporate interests. By detaining (some may call it "holding hostage") two US Environmental Protection Agency (EPA) representatives, Lois demanded that the White House relocate all families by Wednesday, May 21, 1980, at noon, or else the EPA employees would not be released.[122] President Jimmy Carter agreed to relocate nine hundred Love Canal families to safer homes outside of a thirty-six-block area around the toxic waste dump.[123] Additionally, Occidental Chemical Corporation settled the lawsuit brought against them by the residents of Love Canal for a sum of $20 million—not nearly enough to compensate for the lives they had damaged permanently.[124]

Most importantly, Lois Gibbs influenced legislation against future injustices. RCRA was followed in 1980 by the Comprehensive Environmental Response, Compensation, and Liability Act (CERCLA). CERCLA was better known as the Superfund Act because of the $1.6 billion dollar trust fund it set up.[125] And because of her unwavering efforts to finally secure victory at Love Canal, Lois Gibbs was appropriately dubbed the "Mother of the Superfund Act."[126]

CERCLA enabled the EPA to aggressively pursue the cleanup of abandoned hazardous waste sites.[127] In the process, the agency was able to sue the entities that were responsible and collect reimbursements to pay for cleaning up the toxic messes.[128]

CERCLA also opened the door to a new influx of private consulting firms that specialized in analyzing the contents of the toxic-waste sites—everything from leaking gas-storage tanks to entire despoiled mountainsides and contaminated rivers. Hiring armies of specially trained geologists, petroleum engineers, and hydrologists, these new businesses raked in money by developing and then employing a host of new toxic-waste remediation techniques. It was quite a business model. For at least twenty-five years, these hazmat consulting firms could barely keep pace with the number of toxic-waste sites all over the country.

At Love Canal, once all of the affected families had been moved and were settled into their new homes, Lois Gibbs found that her life had changed forever. While she was fighting her two-year battle, she had been contacted by other families across the country, families who were dealing with similar toxic-waste no-man's-lands. Determined to keep fighting, Gibbs moved to Washington, DC, and created the Center for Heath, Environment, and Justice (CHEJ), aimed at helping people set up grassroots organizations in their own neighborhoods.[129] Today, the CHEJ has helped establish 15,000 grassroots groups and, as executive director, Gibbs regularly speaks to communities about toxic chemicals and children's unique vulnerability to environmental exposures.[130]

For her diligent efforts on behalf of the environmentally oppressed, Gibbs has been featured in hundreds of newspaper and magazine articles and featured on many television and radio shows—including *60 Minutes*; *20/20*; *Oprah*; *Good Morning, America*; and *Today*. In 1982, CBS produced a two-hour primetime movie called *Lois Gibbs: The Love Canal Story*.[131] Gibbs has also received many awards for environmental accomplishments and was nominated for a Nobel Peace Prize in 2003.[132]

Driven by a strong sense of justice, Lois Gibbs provides one of the strongest prototypes for what a single individual can accomplish in the arena of environmental regulation. She stuck to her cause and didn't back down, even in the face of severe resistance to her claims. The climate-change movement needs hundreds of people with the moxie of Lois Gibbs to bring our goals to fruition.

NUCLEAR ENERGY: THE RENEWABLE DEBATE

Although nuclear energy is generally considered to be clean energy, its inclusion in the renewable energy list is a subject of major debate. Some of our heroes, such as James Hansen, feel that it must be included in an effective renewable-energy portfolio.[133] Hansen argues that nuclear power is an important piece of this portfolio, essential to rapidly meet our expanding need for clean energy.[134] According to Hansen, nuclear is "the largest source of carbon-free energy and historically the clean-energy source capable of fastest scale-up."[135] We need to decarbonize as quickly as possible, and therefore we should not exclude any source of energy that is carbon-free, Hansen asserts.

Other heroes, such as Mark Jacobson, assert that the costs of nuclear energy outweigh the benefits compared to wind, water, and solar energy.[136] Jacobson's major point is that we can meet all of our energy needs with wind, water, and solar, so we have no need to subject ourselves to the risks associated with nuclear. (We will talk more about Jacobson and his ideas in chapter 23.) Moreover, Jacobson contends that "the significant lag time between planning and operation of a nuclear power plant relative to a wind, solar, or geothermal plant" prohibits nuclear from being a useful energy source in the time scale that we need it.[137]

Both of these opinions have their merit, but we tend to side with Jacobson. One additional point is that nuclear fission power does not seem to fall within the definition of renewable energy. Solar, wind, and water are renewable, in that Earth has an unlimited supply.[138] Energy from nuclear fission comes from the splitting of uranium-235 into smaller, radioactive isotopes (such as cesium-137 and strontium-90), a process that releases energy.[139] The process requires a very small amount of uranium-235 to generate a large amount of energy, but obtaining U-235 requires mining, and there is not an infinite supply. Uranium-235 is the most ideal isotope for nuclear energy,[140] but it comprises only 0.7 percent of Earth's crust.[141]

All of the above notwithstanding, the horrors associated with nuclear power when it does go wrong are noteworthy, despite the fact that many fail-safes and extensive protections are always put in place on nuclear reactors.

THREE MILE ISLAND AND THE FATE OF US NUCLEAR ENERGY

In an ironic twist of fate, two nuclear accident events—one fictional and one very real—shook the country in 1979. On March 19, *The China*

Syndrome, a Hollywood disaster movie about the near meltdown of a nuclear power plant was released in theaters. Then, just over a week later on March 28, came the real event, the near meltdown of the Three Mile Island nuclear plant on the Susquehanna River just south of Harrisburg, Pennsylvania.

The Three Mile Island accident was a partial nuclear meltdown in one of the plant's two nuclear reactors. It was the worst accident in US commercial nuclear power plant history, with the accident being rated as a five on the seven-point International Nuclear Event Scale.[142]

The Three Mile Island accident crystallized antinuclear safety concerns among activists and the general public. It resulted in the passage of new regulations for the nuclear industry and has also been described as the primary reason for the decline of a new reactor construction program that was already underway in the 1970s. Three Mile Island, coupled with the nuclear power plant disasters at Chernobyl (1986) and Fukushima Daiichi (2011) have put new nuclear construction far back on the burner in the United States.

Despite this, the United States remains the world's largest producer of nuclear power, accounting for more than 30 percent of worldwide nuclear generation of electricity. In 2014, our nation's one hundred nuclear reactors produced 798 billion kilowatt hours, which accounted for 19 percent of our total electrical output. However, lower fossil-fuel prices, especially for natural gas, have put the future economic viability of some existing reactors in doubt, as well as the future of proposed new projects.[143]

THE ENVIRONMENTAL MOVEMENT LOSES ITS EDGE

Near the end of the 1970s, a variety of new problems befell the US environmental movement. First, the movement was perceived—for the most part, justifiably so—as catering to wealthy white Americans. The idea of environmental justice—or lack thereof—was clearly at play here. Clean water and air, abundant wildlife, and plenty of open spaces and wilderness areas began to be viewed as amenities only for the establishment rich. The members of environmental and conservation organizations primarily belonged to upper socioeconomic classes—people who could afford to travel and experience the values of our nation's natural resources that were now protected by the environmental regulations passed during the first half of the decade.

At the same time, engineers, chemists, physicists, mathematicians, and other people working in the hard-science fields—where single, definitive answers could usually be found—started questioning the biologists, ecologists, and the other soft-science folks—where there was usually not just one right answer, but many possible outcomes. Many hard scientists began to go back to the old ideas of thinking that humans could figure out how to deal with anything Mother Nature might throw their way. Science would always find a way to prevail, they believed.

A GEORGIA PEANUT FARMER ASSUMES THE PRESIDENCY

Because of the general loss of motivation in the environmental community when Jimmy Carter beat incumbent Gerald Ford in the 1976 presidential election, Carter inherited a very difficult situation. A Georgia peanut farmer, Carter emphasized that he was a Washington outsider, working against the dismal financial baggage and Watergate legacy that was attached to Ford. In the long run, this scenario did not serve Carter well, as his clear lack of Washington insider knowledge and strength weakened his decision-making powers and led to his defeat as a one-term president in 1980.

One of the most intelligent men to ever serve as US president, Carter did more for the protection of public land than any other president since Teddy Roosevelt. The crowning achievement of Carter's administration was his passage of the Alaska National Interest Lands Conservation Act of 1980. This legislation protected 104 million acres of America's last frontier and represented the single largest conservation initiative in US history. Carter pulled off this bold action by combining the 1906 Antiquities Act with several sections of the Federal Land Policy Management Act.[144] In so doing, he created a splendid tapestry of national parks and monuments, wild and scenic rivers, wildlife refuges, national forests, and wilderness areas. We certainly owe President Carter our deep gratitude for ensuring the preservation of a legacy of unrivaled scenic beauty and untold natural resources for all future generations of Americans to visit and enjoy.[145]

President Carter also took actions to limit natural-resource waste and reduce energy consumption. He set fuel-efficiency standards for automobiles and reduced the national speed limit to fifty-five miles per hour to improve gas mileage. He established a new cabinet department, the Department of Energy (DOE), and tasked it with researching and implementing alternative renewable energy sources. In fact, the DOE was set up to be the world's authority on research into fossil fuels, biofuels, nuclear fuels, and solar and wind power. Establishment of the DOE also included funding for the Solar Energy Research Institute (SERI)—now the National Renewable Energy Laboratory (NREL)—in Golden, Colorado.

From a personal perspective, we lived in Golden, Colorado, from 1976 to 1986, when solar power in the United States first became a really big deal. Just about everyone in our neighborhood jumped at the chance to get a free color television set in return for putting solar panels on their roofs. Most of us also basked in the glow of having an international solar research center—SERI—right in our own backyards.

Ever since we left Colorado and moved to New England, we have wondered what happened to SERI. We had not heard anything about it until we started doing research for this book. In the next chapter, we'll look at what happened when Jimmy Carter left office, which will explain why we stopped hearing about SERI. Ronald Reagan, the man who took Jimmy Carter's solar panels off the roof of the White House because he thought they were "just a joke," severely slashed SERI's budget during his administration.[146]

[Note: The former SERI is now renamed the National Renewable Energy Laboratory (NREL), with the mission statement, "From breakthroughs in fundamental science to new clean technologies to integrated energy systems that power our lives, NREL researchers are transforming the way the nation and the world use energy."[147] Despite this ambitious objective, from our research perspective, NREL doesn't seem to be getting much climate-change play on either the national or international stages.]

Chapter 17

THE REAGAN YEARS

Big Trouble for the Environment (1980–1987)

THE CONSERVATIVE BACKLASH

The last component of the environmental backlash that characterized the end of the seventies was the election of former movie actor and staunch conservative Republican Ronald Reagan to the US presidency in November 1980. The 1980s would be a time of transition, with environmentalists struggling to maintain their gains while alternative elements—notably the Christian right and the Moral Majority—took a more active role. According to Benjamin Kline in his book *First Along the River*, "Although many people still sympathized with the need to protect the environment, there was a backlash against the perceived liberal agenda of the 1970s. Government and public support for the environmental issues waned in the glare of the emerging conservative beliefs."[1]

As Joseph Romm writes, "Reagan gutted Carter's entire multi-billion dollar clean energy and energy efficiency effort. He opposed and then rolled back fuel economy standards. Reagan turned all such common sense strategies into 'liberal' policies that must be opposed by any true conservative."[2] Essentially the entire current Republican agenda—no tax increases, smaller federal government, less environmental regulation, bigger military, and mixing church (Christian religious beliefs) with state—became established precedents when Reagan took office. It didn't seem to matter that our nation was founded on escaping religious persecution in Europe and establishing a new government based on freedom and equality for all. Ironically the Christian right faction of the Republican Party—most notably wealthy white men—stood behind the American flag while overtly stating that all non-Christians were destined to die in hell.

To the staunch Republicans of today, Reagan and his policies are still the gold standard that all future administrations should strive to emulate. Meanwhile, conservationists view Reagan as an intolerant, antienvironmentalist who has led us down the dark and gloomy path to the current precipice of climate peril. They would much have preferred that he remain in Hollywood.

GUTTING THE REGULATORY SYSTEM

Reagan justified dismantling environmental legislation and cutting back regulatory agency funding by saying that environmental issues should be subordinate to material living standards and financial security. In a statement that drives the conservative Republican belief system to this day, he once famously said that, "Government is not the solution to our problem, government is the problem."[3]

After Reagan took office, he decimated the research and development budgets and eliminated tax breaks for anything related to renewable energy at Carter's Department of Energy (DOE).[4] "The Department of Energy has a multibillion-dollar budget, in excess of $10 billion," Reagan said during an election debate, justifying his opposition to Carter's renewable energy policies. "It hasn't produced a quart of oil or a lump of coal or anything else in the line of energy."[5]

Reagan was also the man who said, infamously, that "trees cause more pollution than automobiles do" and that if "you've seen one tree you've seen them all." Clearly he was not a president with much concern for the environment.[6]

WATT AND GORSUCH: FOXES GUARDING THE HENHOUSE

Supported by the economic industrial conglomerate, Reagan appointed notoriously antienvironment people to run his cabinet departments, including James Watt as Secretary of the Interior and Anne Gorsuch

Burford, as EPA Director. Perceived as being hostile to environmentalism, Watt was regularly criticized by conservationists for endorsing development of federal lands for foresting, mining, ranching, and other commercial interests.[7] Greg Wetstone, one of Reagan's chief environment counsels, argued that Watt was one of the two most "intensely controversial and blatantly anti-environmental political appointees" in American history.[8] "James Watt had all the political skills and public relations sense of a boa constrictor," said Jim DiPeso, policy director at Republicans for Environmental Protection (REP).[9] Hardly a good fit as the nation's foremost public land manager, which was Watt's primary task.

Since I (Budd) worked for the National Park Service in Watt's Department of the Interior, I can personally vouch for the high level of discomfort he provided for environmentalists during his tenure. Conservation groups accused Watt of every type of blasphemy, from reducing funding for environmental programs to restructuring the Interior Department in order to decrease federal regulatory power, while also striving to eliminate the Land and Water Conservation Fund and directing the National Park Service to draft rules that would de-authorize congressionally authorized national parks. On the energy front, Watt eased regulations on oil exploration and mining extraction, while recommending federal leases in wilderness areas, national seashores, and national recreation areas for exploring and developing new oil and gas resources.

Watt also resisted accepting donations of private land to be used for conservation. Instead, he recommended opening 80 million acres (125,000 square miles) of undeveloped land in the United States for drilling and mining by 2000.[10] During his time as interior secretary, the total area leased for coal mining quintupled.[11] He also bragged about leasing "a billion acres" of coastal waters for oil and gas development—although only a small portion of that total area would ever be drilled.[12] Especially inappropriate for an interior secretary, Watt's modus operandi was, "We will mine more, drill more, and cut more timber."[13]

Watt became so reviled in the minds of the American public that in 1982 a "Dump Watt" petition demanding his ouster garnered almost one million signatures. He eventually resigned his cabinet position in 1983—to the great relief of environmentalists and conservationists across the Nation.[14]

Anne Gorsuch Burford was no better than Watt. As described in Benjamin Kline's *First Along the River*, she "seemed more determined to dismantle the EPA than to administer it."[15] She slashed the EPA's budget by $200 million and cut 23 percent of the staff.[16] She also relaxed Clean Air Act regulations and rigid ocean pollution allowances and approved the spraying of restricted-use pesticides.[17] Environmentalists argued that her policies were created to favor polluters, not to lead our nation's foremost environmental organization. Burford also resigned in 1983, citing her issues with the congressional controversy over administration policies.[18]

THE MORAL MAJORITY AND THE CHRISTIAN RIGHT

Also entering the sociopolitical mix at this time—to a greater extent than ever before—were the views of the Christian right, led by men like televangelist Jerry Falwell, who helped found the political organization the Moral Majority. Representing the extreme right wing of the Republican Party, the Moral Majority argued that environmental views were at odds with the lessons of the Bible and, as such, were anti-Christian, un-American, and socialist. When the economic downturns in the United States—driven by the oil crises of 1973 and 1979—were factored in, it was clear that the heyday of the US environmental movement had come to an end.

SOLAR AND WIND TAKE FLIGHT

Despite the environmental setbacks that came with the Reagan era, the early 1980s saw several significant and positive renewable energy and conservation related events. In 1981, the Solar One project in the Mojave Desert just east of Barstow, California, was completed.[19] Designed by a team of scientists from the Sandia National Laboratories, Southern California Edison, the Los Angeles Department of Water and Power, and the California Energy Commission, Solar One was a solar-thermal project that provided the first test of a large-scale solar power plant.[20]

The plant's method of collecting energy was based on concentrating the sun's energy onto a common focal point to produce the heat needed to run a steam turbine generator. Using hundreds of large mirror assemblies called heliostats, the facility tracked the sun. The solar energy was then reflected onto a tower with a black receiver that absorbed the heat. In the final step, high-temperature heat transfer fluid was used to carry the solar-generated energy to a boiler on the ground that produced steam that then spun a series of turbines much like a traditional power plant.[21]

The first half of the Reagan decade of decadence hit another positive note when the US wind energy industry blew into existence with the construction of 17,000 turbines, primarily in California. Danish companies (such as Kuriant, Vestas, Nordtank, and Bonus) built most of the world's first wind turbines, including those installed in the United States.

The first wind farm in US history consisted of twenty thirty-kilowatt turbines and was constructed on Crotched Mountain in New Hampshire in 1980.[22] The project was shut down after only a year, however, due to equipment failures and other problems.[23] Meanwhile, throughout the 1980s, the DOE funding for wind power research and development continued to experience a decline, reaching a low point in 1989.

THE ROCKY MOUNTAIN INSTITUTE IS FOUNDED

Several promising new environmentally relevant NGOs were also started in the 1980s. In 1982, experimental physicist Amory Lovins founded the Rocky Mountain Institute (RMI) as an independent, nonpartisan, nonprofit facility. The RMI's mission, "to drive the efficient and restorative use of resources,"[24] involves researching, publishing, consulting, and lecturing in the general field of sustainability—with a special emphasis on profitable innovations for energy and resource efficiency.[25] Since the 1980s, they have helped individuals, communities, and businesses prevent pollution, save energy, preserve natural resources, and bring communities together, an initiative they now call Abundance by Design.[26] RMI also devotes time and resources toward the formation of a new economy via a focus on "natural capitalism." For over thirty years, RMI has emphasized the overlap of business and environmental interests—with a strong focus on the economic value of natural resources as a driver for smart business decisions.[27] Amory Lovins wrote *Natural Capitalism* with Paul Hawken and L. Hunter Lovins in 1999; this "blueprint for a new economy"[28] is featured prominently in our sections on Paul Hawken (chapter 18) and Amory Lovins (chapter 23).

SUSTAINABLE DESIGN: WRI, OCEAN ARKS INTERNATIONAL, AND GUS SPETH

The World Resources Institute (WRI), a nongovernmental global research organization, was also founded in 1982. WRI works to use sustainable natural resource management to create greater equality and prosperity throughout the world. Today, WRI employs more than 450 scientists who look for ways that will simultaneously protect the Earth while improving people's lives.[29]

James Gustave "Gus" Speth, another of our climate-change heroes, created WRI with funding from the John D. and Catherine T. MacArthur Foundation. Hailed by many as "one of the nation's most influential environmentalists"[30] and now leading the way in system-changing activism, Speth went from Washington's "ultimate insider" to being arrested for activism in front of the White House.[31] We talk more about Gus Speth in chapter 24.

Yet another significant conservation organization was founded in 1982 when Dr. John Todd and Nancy Jack Todd established Ocean Arks International. Ocean Arks has worked since then to develop ecological technologies to solve human-created challenges, focusing on water purification. By studying how freshwater ecosystems work, Ocean Arks created what were known as "Living Machines"—now called eco-machines. These eco-machines use the principles of the natural world to clean chemicals and waste from water, allowing for sustainable development.[32]

PAUL CRUTZEN AND THE CONCEPT OF NUCLEAR WINTER

The Cold War had been existing uneasily between the United States and the Soviet Union for more than twenty-five years, when Dutch atmospheric scientist Paul J. Crutzen, together with paleoecologist John Birks, introduced the concept of nuclear winter to the world in their 1982 paper "The Atmosphere After a Nuclear War: Twilight at Noon."[33] In the paper, Crutzen and Birks theorized about the potential climatic effects of the large amounts of sooty smoke from the bomb-generated wildfires—in the forests and in urban and industrial centers and oil storage facilities—reaching the middle and higher troposphere. They believed that the black smoke's absorption of sunlight could lead to an abnormally strong cooling of Earth's surface, coupled with a heating of the atmosphere at higher elevations. Happening concurrently, these two events could create atypical meteorological and climatic conditions that could jeopardize agricultural production for a large percentage of the human population on our planet.

In chapter 23 we will discuss Paul Crutzen in greater detail.

THOMAS LOVEJOY AND DEBT-FOR-NATURE SWAPS

In 1984, Dr. Thomas E. Lovejoy, director of the World Wildlife Fund, came up with the innovative concept of the debt-for-nature (DFN) swaps.[34] Through the program, a creditor will forgive a nation's debt in exchange for setting aside ecologically vulnerable land for protection. Large international nongovernmental organizations such as the World Wildlife Fund, the Nature Conservancy, and Conservation International were among the first to purchase debt from commercial banks for a discounted rate and return it to the indebted country in exchange for land conservation promises.[35] Since then, swaps of international debt for conservation projects have been initiated in developing countries like Bolivia, Ecuador, Madagascar, Jamaica, Costa Rica, the Philippines, and Zambia. Starting in 1987, the process has resulted in making more than one hundred forty million dollars available for conservation work.[36]

The debt-for-nature swaps offers the world two enormous potential long-term benefits. First, the developing countries who take advantage of the swaps ease their debt burden. Second, most of the exchanges involve tropical forests, and since the land acquired through DFN swaps is permanently protected these forests are able to act as stable carbon sinks that aid in stabilizing worldwide CO_2 levels.[37] The financing mechanism for DFN swaps is usually an agreement between the lender (such as the US government), the national government of the debtor country, and a conservation organization that helps to organize the project.[38]

We will talk more about Dr. Lovejoy in chapter 23.

CHERNOBYL NUCLEAR DISASTER

On April 26, 1986, the worst nuclear disaster in the history of the world occurred when a reactor at the Chernobyl Nuclear Power Plant in the Ukraine SSR exploded and released clouds of radioactive particles into the atmosphere.[39] The fallout spread across much of the western USSR and even into Northern Europe and the United Kingdom. Although only thirty-one people died directly in the catastrophe, the long-term effects included abnormally high cancer rates in surrounding areas.[40] Today, the city of Pripyat—formerly with a population of almost 50,000 people—is primarily a ghost town, visited only by tourists who are curious to witness the aftermath of nuclear devastation.[41]

YES . . . WE CAN POLLUTE OUR OCEANS!

Two acts passed in the latter part of the 1980s went a long way toward defeating the crazy, but widely held, notion that our oceans are so vast they can't be polluted. This idea had made people think that we don't need to worry about what we dump in the oceans or do to them.

The Marine Plastic Pollution Research and Control Act of 1987 applied to ship-generated garbage and was aimed at reducing the amount of plastics ships dump into the oceans.[42] The next year, in 1988, the Marine Protection, Research, and Sanctuaries Act (MPRSA), also called the Ocean Dumping Act, was passed, prohibiting all municipal sewage sludge and industrial waste dumping into the oceans.[43]

Since many people—especially climate-change deniers—still seem to hold a similar opinion about the atmosphere—believing that we cannot contaminate it—perhaps it is time for Congress to pass some laws that specifically prohibit the direct emissions of all greenhouse gasses (GHG). Instead of being directly emitted from smokestacks, the GHG would either have to first be treated to reduce their noxious elements and mitigate their impacts or they would have to be captured and disposed of where they are being generated.

If passed, such legislation would be analogous to what the US shipping industry now has to do to comply with the laws against ocean dumping. The industry is required to provide on-shore facilities for containing and disposing of their waste streams—most notably plastics and sewage—things they used to dump directly into the open ocean.[44]

CHICO MENDES: MARTYR OF THE BRAZILIAN RAINFORESTS

On December 22, 1988, a tragic event occurred deep in the Brazilian countryside that had a stark effect on worldwide natural resource conservation in general and today's climate-change situation in particular. On that day, Francisco Mendes Filho, known more commonly as "Chico Mendes"—a Brazilian rubber tapper (a *seringueiro*) and land-rights leader—became world famous when he was gunned down outside his own home by the son of a local rancher.[45] During the shooting, two Brazilian policemen, who had been assigned to protect Mendes from death threats, sat playing dominoes at the kitchen table inside his home.[46] Mendes's assassination happened because the local ranchers viewed Mendes as their enemy—someone who was trying to stop them from expanding their wealthy lifestyles through wholesale clear-cutting of the Brazilian rainforests.[47]

Chico Mendes's life in the Brazilian rainforest had been fairly typical. He went to work as a *seringueiro* when he was only eight years old and did not attend school.[48] The rubber plantation owners did not want their workers to be able to read and write because they feared an uprising. Mendes's father taught him latex tapping and subsistence hunting, but he also taught young Chico to read—providing a critical component of the activist role he was to assume. Chico's education enabled him to write letters to Brazil's president describing the subhuman conditions at the rubber plantations and to teach others in his community to read as well.[49]

Mendes had a strong sense of justice, and he knew that the plantation owners were not treating their workers well, and they were not treating the land well either. Although Mendes and his colleagues were a tiny, marginalized minority, their efforts to change their situation brought them notoriety in parts of Brazil's Amazon during the 1980s. Mendes helped organize the local rubber tappers into a union and developed a technique called an *empate*—which amounted to blockading rubber tree tracts from ranchers and farmers who wanted to clear the land.[50]

Mendes also pioneered the world's first tropical forest conservation initiative that was advanced by the people who lived there. Chico and the organized tappers proposed the idea for extractive reserves—protected forest areas that were inhabited and managed by local communities.[51] The inhabitants could hunt and harvest within the reserves, but developers and plantation owners were prohibited from using the land. In 1987, the Environmental Defense Fund (EDF) and the National Wildlife Federation (NWF) flew Mendes to Washington, DC, in an attempt to convince the World Bank and the US Congress to support the creation of these extractive reserves.[52] Fittingly—less than two years after his death—the Brazilian government established their first extractive reserve, known as the Chico Mendes Extractive Reserve (CMER), in the area where Mendes had lived.[53]

The CMER looks nothing like the conditions that Chico Mendes and his family grew up in as tappers. The reserve buzzes with motorbikes and cars, whereas in Chico's day the *seringueiros* had to walk eight to eleven miles each day to tap the trees. There is electricity, schools, jobs, and even an eco-lodge; many tappers have gone on to university or are employed as forest guides.[54] The locals have extensive knowledge about sustainable harvesting, and care for the forest accordingly. Using the CMER as a model, sixty-eight other extractive reserves—covering more than 136,000 sq km—have been established in the Brazilian Amazon.[55]

Mendes's death proved to be a turning point in the war to save the Amazon rainforest. Today, 40 percent of the rainforest—a total of 58 million acres—is set aside for protection. The Brazilian Ministry of the Environment manages the Chico Mendes Institute for Conservation of Biodiversity (*Instituto Chico Mendes de Conservação da Biodiversidade*), named in his honor.

At first I thought I was fighting to save rubber trees, then I thought I was fighting to save the Amazon rainforest. Now I realize I am fighting for humanity.

—Chico Mendes[56]

Chapter 18

THE CLIMATE-CHANGE DEBATE TAKES OFF (1988–2000)

GLOBAL WARMING AND THE FOUNDING OF THE IPCC

Beginning in the 1980s, the annual average global temperature started to rise and climate scientists realized we were experiencing a worldwide warming trend. For the previous few decades, people had worried that we might enter a new ice age. However, as the media and the general public noticed the warmer temperatures, this theory fell out of favor.[1]

Near the end of the decade, the average global temperatures began to increase even more, and the global warming theory began to gain ground fast.[2] More environmental NGOs were founded, with the goal of preventing further global warming. The media showed striking visual images of smokestacks, side by side with melting ice caps and severe flooding.

This renewed concern in the environment led to the formation of the Intergovernmental Panel on Climate Change (IPCC) in 1988, which was created to collect and process information about climate change. The more than 2,500 scientific and technical experts of the IPCC used existing climate models and literature to predict the future impacts of the greenhouse effect.[3]

Created by the United Nations Environmental Programme (UNEP) and the World Meteorological Organization (WMO), IPCC scientists come from a wide variety of fields—climatology, ecology, economics, medicine, and oceanography—to name a few.[4]

The IPCC creates regular reports on climate change–related data, which give a snapshot of the progress (or lack of progress) brought about by the 1992 United Nations Framework Convention on Climate Change (UNFCCC). UNFCCC is—as we will discuss in more detail later—the main international treaty on that subject, with the goal of "stabilizing greenhouse gas concentrations in the atmosphere at a level that would prevent dangerous anthropogenic (human-induced) interference with the climate system."[5]

To create their assessment reports, the IPCC does not conduct original research or monitor climate phenomena itself; rather they survey the published literature—both peer-reviewed and non-peer-reviewed—and compile their reports from this data.[6] Thousands of scientists and other experts—working as volunteers—contribute to writing and reviewing the IPCC assessment reports. Each report contains a "Summary for Policymakers," which is evaluated by delegates from the participating governments—typically representing more than 120 countries.[7] The IPCC's first assessment report, issued in 1990, concluded that Earth's temperature had risen during the past century and that human use of fossil fuels was likely adding to this rise.[8]

In 2003, the Heartland Institute's Nongovernmental International Panel on Climate Change (NIPCC) was created as an opposing organization, solely to cast doubts on the findings and reports of the climate scientists whose work is reviewed by the IPCC.[9]

To be as candid as possible here, the Heartland Institute that founded the NIPCC is a nonscientific, doubt-mongering entity supported by the fossil-fuel industry and the Koch Brothers—their cohorts in the not-so-fine art of lying. As Naomi Oreskes writes in *Merchants of Doubt*, "The Heartland Institute is known among climate scientists for persistent questioning of climate scientists, for its promotion of 'experts' who have done little, if any, peer-reviewed climate research, and for its sponsorship of a conference in New York City in 2008 alleging that the scientific community's work on global warming is a fake. But Heartland's activities are far more extensive and reach back to the 1990s when they . . . were working with Philip Morris."[10] Philip Morris paid $50,000 to the Heartland Institute to support its pro-tobacco activities, and Heartland met with members of Congress on behalf of the tobacco industry.[11] Greenpeace compiled the annual IRS Form 990 filings by the Koch family foundations, and found that the Koch brothers had given over $88 million to groups denying climate since 1997; Heartland Institute received $100,000 from them between 1987 and 2011.[12] In an interview with Rob Hopkins, writer for the Transition Network, Oreskes stated (referring to the "experts" at Heartland Institute, among the other groups that were merchandising doubt), "These people were scientists in a sense that they had PhDs in science, they had published scientific research, they had prominent positions of power and influence in the American scientific community. But they were not experts about climate change. One of the things that we say in the book is that this is part of the reason they were able to fool so many people. They drew on their scientific credibility to make claims that the people and the press found credible . . . in most cases, the press never pointed out that these people were actually all-purpose contrarians, that they really didn't have expertise on climate change or tobacco and that they really were exploiting their scientific credentials in a way that was quite misleading, in a way that was merchandising doubt."[13] Any documents produced by the Heartland Institute and/or the NIPCC have no credibility—scientific or otherwise—beyond casting doubt.

STEPHEN SCHNEIDER: CLIMATE-CHANGE GURU

One of the primary scientists on the IPCC when it first formed in 1988 was a man who had made the first predictions about global warming, back in 1976. Dr. Stephen Schneider worked at the National Center for Atmospheric Research (NCAR) in Boulder, Colorado from 1973 to 1996, where he cofounded the Climate Project, and was an expert advisor to every president from Nixon to Obama. Over his career, he authored and coauthored numerous books and articles and founded and edited the journal *Climatic Change*. He received many awards for his work on climate change, including sharing the 2007 Nobel Peace Prize with his colleagues on the IPCC and with former vice president Al Gore.[14]

In his writing, Dr. Schneider emphasized that humans were responsible for global warming and talked about its extensive effects, such as the rising ocean temperatures and the increasing strength and frequency of hurricanes, along with the damage to the ozone layer through greenhouse-gas emissions.[15]

According to Princeton professor Michael Oppenheimer, "No one, and I mean no one, had a broader and deeper understanding of the climate issue than Stephen. . . . More than anyone else, he helped shape the way the public and experts thought about this problem—from the basic physics of the problem, to the impact of human beings on nature's ecosystems, to developing policy."[16]

Unfortunately, Schneider died far too early, at the age of sixty-five, from an apparent heart attack while on a flight landing in London in 2010.[17] The climate-change community could certainly use his bold leadership right now.

NASA'S JAMES HANSEN: YES, CLIMATE CHANGE IS FOR REAL! AND THE DENIERS RESPOND

In 1988, another esteemed climatologist stepped up to make the world aware of the dangers of global warming and climate change. Dr. James Hansen, scientist at the National Aeronautics and Space Administration (NASA), testified before Congress that global warming was simultaneously melting the polar ice caps and causing extreme droughts throughout the world.[18] As a climate scientist who has made a lasting impression on the world about the potentially dangerous effects posed by climate change, we will talk more about Dr. Hansen in chapter 23.

Largely to refute Dr. Hansen, fossil-fuel companies and other US industries formed the Global Climate Coalition (GCC) in 1989 with the purpose of convincing politicians and the general public that climate-change science was too uncertain to justify action.[19] While the coalition argued against climate change, however, its own scientific experts were telling them that the science actually proved that humans were responsible for climate change.[20] The GCC dissolved in 2002 due to membership loss after an IPCC assessment report provided overwhelming evidence showing that global warming was indeed occurring.

ACID RAIN: ANOTHER AIR-QUALITY CRISIS

As a wildlife biologist, wetland scientist, and recreational fisherman living and working in New England in the late 1980s, I (Budd) remember another fossil-fuel-related ecological scare. Acid rain is a term for precipitation that contains high levels of nitric and sulfuric acids.[21] While rotting vegetation and erupting volcanoes can release some chemicals that naturally produce acid rain, most of it falls because of human activities. Burning of fossil fuels by coal-burning power plants, factories, and automobiles is the primary producer of acid rain.[22]

When fossil fuels are burned, they release sulfur dioxide (SO_2) and nitrogen oxides (NO_x) into the atmosphere. Once released, SO_2 and NO_x gases react with water, oxygen, and other substances to form mild solutions of sulfuric and nitric acid. Winds can then spread these acidic solutions through the atmosphere—often for hundreds of miles.[23] As a result, pollutants from power plants in heavily industrialized areas in New Jersey, Ohio, or Michigan can—and often did—impact forests, rivers, or lakes in rural parts of New York, New Hampshire, Maine, and Canada.[24]

So what happened to stop this atmospheric contamination? The EPA's Acid Rain Program (ARP) introduced new regulations, requiring power plants to reduce their SO_2 and NO_x emissions. Through 2010, the total amount of SO_2 allowed to be emitted was gradually decreased, with the final 2010 SO_2 cap set at 8.95 million tons—about half the level of emissions in 1980.[25] This reduction was accomplished at about half the expected cost.[26]

Under the ARP, NO_x reductions were achieved through a more traditional, rate-based regulatory process—meaning that the rates are adjusted to inflation minus any emissions reductions. Since the program began in 1995, the ARP has also achieved significant NO_x reductions in emissions from coal-fired power plants.[27]

The ARP was our first national cap-and-trade program, setting limits on emissions and providing financial incentives to reduce pollution. Today, this system allows regulated entities to choose the most cost-effective approach for reducing their emissions. By so doing, the ARP has proven to be a highly effective way for achieving a trifecta of benefits: reducing emissions, meeting environmental goals, and improving human health.[28]

The success of the federal ARP gives us both a protocol and a precedent for the nationwide reduction of our GHG emissions. To solve the climate-change crisis, perhaps we just need to take a closer look at how we successfully handled the reduction in NO_x and SO_2 emissions that were causing acid rain.

EXXON VALDEZ AND THE DESECRATION OF ALASKA'S PRINCE WILLIAM SOUND

Just after midnight on March 24, 1989, the *Exxon Valdez*, a 987-foot tank vessel, hit Bligh Reef in Prince William Sound, Alaska, spilling eleven million gallons of crude oil into the sound. Within two months of the accident, the slick had spread over three thousand square miles and onto more than 350 miles of beaches in the breathtakingly beautiful marine treasure.[29] Eventually, the oil had covered 1,300 miles of coastline and 11,000 square miles of ocean.[30] According to the World Wildlife Fund, immediate effects included the deaths of 250,000 seabirds, 4,000 sea otters, 250 bald eagles, and more than 20 orcas (killer whales).[31] Prince William Sound's herring population still has not recovered; prior to the accident there were 150,000 tons of herring caught, after the spill there are 10,000 tons, causing a collapse of the herring fishery.[32]

After the spill, President H. W. Bush ordered the National Response Team (NRT), headed up by EPA administrator William K. Reilly and secretary of transportation Samuel K. Skinner, to prepare a detailed report summarizing what happened. The resulting NRT Report—published in May of 1989—concluded that, "The lack of necessary preparedness for oil spills in Prince William Sound and the inadequate response actions that resulted mandate improvements in the way the nation plans for and reacts to oil spills of national significance."[33]

Unfortunately, the recommendations given by the *Exxon Valdez* NRT were not followed up, as became clear more than twenty years later after the *Deepwater Horizon* Disaster, which we discuss in detail in chapter 21.

FAMED HOLLYWOOD PRODUCER LAUNCHES HIS ENVIRONMENTAL MEDIA ASSOCIATION

Baby boomers have a special place in their hearts for the entertaining, cutting-edge 1970s sitcoms of innovative Hollywood producer Norman Lear. Shows like *All in the Family*, *The Jeffersons*, and *Sanford and Son* made us laugh, while also reflecting on how we were living our lives. Today, through an organization that he and his wife, Lyn, cofounded in 1989, Lear is rallying support for talking about climate change in the mainstream media.

Lear created his Environmental Media Association (EMA) with the firm belief in the entertainment industry's power to influence millions of people. EMA works with the entertainment industry to educate people about environmental issues, inspiring them to take action. Each year, EMA gives awards to entertainers of all kinds who have featured creative and influential environmental messages.[34]

We talk more about Mr. Lear's background and environmental work in chapter 26.

THE CLEAN AIR ACT: REGULATORY AMBROSIA FOR CLIMATE CHANGE?

From a federal regulatory standpoint, one piece of legislation—the Clean Air Act (CAA)—stands paramount to resolving the climate-change crisis, at least here in the United States. When Congress originally passed the CAA in 1970, it gave the EPA the responsibility of protecting the American people whenever scientific studies showed that new air pollutants threaten our health or environment.[35] In 1990, the CAA was revised and signed into law by President George H. W. Bush. On the EPA's website, the 1990 CAA amendment is characterized as, "Building on Congressional proposals advanced during the 1980s, the President proposed legislation designed to curb three major threats to the nation's environment and to the health of millions of Americans: acid rain, urban air pollution, and toxic air emissions. The proposal also called for establishing a national permits program to make the law more workable, and an improved enforcement program to help ensure better compliance with the Act. . . . [This legislative action] also added provisions requiring the phaseout of ozone-depleting chemicals, roughly according to the schedule outlined in international negotiations (Revised Montreal Protocol)."[36] The strong bipartisan support these CAA bills received in both the House and the Senate clearly demonstrated that clean air and less pollution were goals shared by Republicans and Democrats alike.

According to the Union of Concerned Scientists, over its forty-five-year history the CAA has:

- Cut ground ozone, a dangerous component of smog, by more than 25 percent since 1980;
- Reduced mercury emissions by 45 percent since 1990;
- Reduced the main pollutants that contribute to acid rain—sulfur dioxide and nitrogen dioxide—by 71 percent and 46 percent, respectively, since 1980;
- Phased out the production and use of chemicals that contribute to the hole in the ozone layer;
- Reduced the lead content in gasoline, which has cut lead air pollution by 92 percent since 1980.[37]

However, the most essential part of the CAA, as it applies to our current climate-change situation, is that the EPA is required to regulate the emission of pollutants that "endanger public health and welfare."[38] In 2007, the US Supreme Court, in a landmark decision (*Massachusetts v. EPA*), ruled that global-warming emissions—caused by GHG—are air pollutants and should be subject to EPA regulation under the CAA.[39]

Then, in 2009, the EPA released scientific findings that concluded that global-warming emissions present a danger to public health—now known as the *endangerment finding*. As further reported by the Union of Concerned Scientists,

Citing extensive scientific research, the EPA found that global-warming pollution is connected with:

- Hotter, longer heat waves that threaten the health of the sick, poor, and elderly;
- Increases in ground-level ozone pollution, linked to asthma and other respiratory ailments;
- Extreme weather events that can lead to deaths, injuries, and stress-related illnesses.[40]

Based on this information, it seems obvious to us that the EPA—operating under the CAA—has much of the regulatory authority it needs to immediately control the emission of various sources of GHG. The National Wildlife Federation (NWF) summarized the situation by saying simply,

> It is time for our nation's polluters to finally be held accountable for their harmful emissions that contribute to climate change. ... In passing the CAA, Congress clearly intended it to serve as a living document, in order to ensure that EPA has the tools it needs to respond to new air pollution threats. The science is clear: global-warming pollution poses significant threats to public health and welfare, and EPA is obligated under the law to limit sources of this pollution and address the impacts of climate change.[41]

So—if much of the necessary regulatory authority is in place and has been sufficiently vetted, why isn't this happening? As with anything involving bureaucracy and purported economic impacts, the answer has been severely complicated by political infighting.

In another report, the Union of Concerned Scientists made an even stronger statement:

> Unfortunately, polluters and their allies in Congress are using every opportunity to prevent the EPA from protecting our health by reducing global-warming emissions. Numerous members of Congress in both the Senate and the House of Representatives have announced their

> intention to introduce legislation that would block or delay the agency from reducing global-warming emissions under the Clean Air Act. Some members of Congress even tried to attach bills attacking the EPA to legislation that must pass, such as federal spending and budget bills.
>
> These attacks on the Clean Air Act pose a grave threat to EPA's responsibility to protect our health and environment from the impacts of climate change. Some proposed legislation would delay the EPA from setting standards to limit global-warming emissions for several years, while other bills would indefinitely block the EPA from taking any action on this issue whatsoever. Some proposals would even prohibit the EPA from doing any research or analysis on climate science in its efforts to implement the endangerment finding.[42]

The NWF explained the conservative backlash to federal regulation of climate change and challenged the EPA to stand up for Americans:

> Polluter lobbyists continue to cry foul at any mention of EPA fulfilling its obligation under the CAA with respect to global-warming pollution. This is simply the latest in a string of red herrings that industry has raised time and again to avoid complying with laws that are essential for protecting public health. From seatbelts to catalytic converters to unleaded gasoline, industry falsely claimed that new standards would have devastating economic impacts. History has shown that these requirements have not adversely affected our economy—to the contrary, they have had substantial benefits in saving lives, improving public health, and advancing cleaner technology. . . . It is long past time for EPA to move forward and require the emission reductions necessary to protect America from the urgent threat of climate change.[43]

CLIMATE RECORDS VERIFY GLOBAL WARMING

In the 1990s, some scientists began to believe that global-warming models had overestimated how fast our climate was warming over the past hundred years.[44] This protestation caused the IPCC to review their initial data on global warming, but, in the final analysis, they stood by their past findings that the world's annual average temperature was definitely rising—and at an alarming rate.

As we now know, the ten warmest years on record have all occurred since 2003. This information is certainly sobering and definitely makes it difficult to deny that our world is warming at a much faster rate than any other time in history.[45]

1992 RIO EARTH SUMMIT

The United Nations Conference on Environment and Development (UNCED)—more commonly known as the Rio Earth Summit—took place in Rio de Janeiro, Brazil, from June 2–14, 1992. This conference—attended by government officials from 178 countries and between 20,000 and 30,000 individuals from governments, NGOs, and the media—covered a broad array of topics. Key discussion points emphasized finding solutions for such hefty worldwide subjects as poverty, war, and the expanding gap between industrialized, developed (i.e., wealthier) nations and developing (i.e., poorer) countries.[46]

The central focus of the summit was the question of how to solve these global economic and environmental problems through the introduction of sustainable development. Sustainable development emphasizes that economic advancement (the first leg) and social progress (the second leg) depend critically on the preservation of the natural resource base (the third leg). These areas of focus work together as the three legs of a tripod to prevent environmental degradation.

THE UN FRAMEWORK CONVENTION ON CLIMATE CHANGE (UNFCCC)

Participating countries at the Rio Earth Summit put forth a United Nations Framework Convention on Climate Change (UNFCCC), with the key objective of "Stabilizing greenhouse gas emissions in the atmosphere at a level that would prevent dangerous anthropogenic interferences with the climate system."[47] As part of this treaty, participating developed countries pledged to reduce their emissions to 1990 levels by the year 2000.[48]

The UNFCCC was a remarkable achievement for its time. In 1994, when the UNFCCC fully took effect, there was far less scientific evidence about climate change than there is now.[49]

Unfathomably, the United States blocked calls for serious action on climate change under the UNFCCC—favoring only voluntary emission controls and shifting the emphasis to strictly non-legally binding resolutions.

THE CLINTON PRESIDENCY: HIGH HOPES DASHED

In November 1992, the United States elected President William Jefferson "Bill" Clinton—the first Democrat to become president in twelve years—and most environmentalists breathed a long, deep sigh of relief. Personally, since I (Budd) was still making a living as an environmental consultant, I indeed felt like a sweet spring wind had just swept across the American landscape. Things surely had to get better now that a left-leaning Democrat was back in the White House.

As president, Clinton had many similarities with Jimmy Carter, his Democratic predecessor from the late 1970s. He had a strong southern heritage, a gubernatorial background, status as a Washington outsider, and a brilliant mind. In fact, Clinton had a skill that Carter didn't possess: he was a great communicator—the Democratic equivalent of Ronald Reagan—with the charisma and savvy to be one of America's greatest presidents. But Clinton fell victim to the same weakness that has plagued so many other great—as well as not so great—male world leaders.

Unfortunately, one thing Clinton didn't share with Carter was his high moral ground. It would have been interesting to see exactly how much Clinton could have accomplished in the environmental arena if he had not turned so much of his attention to a White House intern named Monica Lewinsky.

From a climate-change perspective, when Clinton first took office he made an outspoken commitment to reduce CO_2 and other GHG emissions. He proclaimed that climate change was a global strategic threat that required bold leadership. In his first Earth Day address, Clinton announced that he would sign the Biodiversity Treaty and also promised to reduce GHG emissions to 1990 levels by the year 2000[50]—both actions that had been embarrassingly rejected by President H. W. Bush at the 1992 Rio Earth Summit. Unfortunately, when Clinton left office, GHG emissions were nowhere near the 1990 levels that he had promised.

In general, many environmental activists during the Clinton administration began to be known as "Lite Greens"—they wanted to protect the environment, but not when it would cause any downgrading of their own personal quality of life.[51] While definitely possessing the knowledge base and passion for increasing environmental protection, Clinton immersed himself in the belief that the economy had to come first, above all else.[52] This was based on his perception that an affluent, acquisitive society was what the American people wanted.

Despite his strong economic views, Clinton did manage to accomplish some significant natural-resource gains, especially during his last years in office. For one thing, he used the Antiquities Act of 1906 to declare more than 3 million acres of federal land as national monuments, thereby making them off-limits to development.[53] These areas included Utah's Grand Staircase-Escalante, California's Pinnacles, and Arizona's Grand-Canyon-Parashant. He also used his executive power to declare one third of our national forestland—58 million acres in thirty-nine states—off-limits to road building, logging, and oil and gas exploration.[54]

Clinton also took on a variety of commercially complex issues, including restoring the hydrology of the Everglades, restricting flights over the Grand Canyon, banning snowmobiling in national parks, and blocking repeated congressional attempts to open the Arctic National Wildlife Refuge to oil and gas drilling.[55] He also made significant progress with so-called *brown issues* (those related to toxic substances). He cleaned up 515 Superfund sites—more than three times as many as the previous two administrations, established stringent new standards for reducing sulfur levels in gasoline, approved new clean air standards for soot and smog, and doubled the number of chemicals that industry is obligated to report to communities through right-to-know laws.[56] His administration also took significant actions aimed at protecting our waterways and water quality. These actions included launching the Clean Water Action Plan, strengthening the Safe Drinking Water Act, and permanently preventing new oil leasing in national marine sanctuaries.[57]

But some of Clinton's most significant environmental accomplishments came not in the form of what he achieved but what he fended off. As author Paul Wapner describes, "[Clinton] faced an aggressive and hostile Congress that worked consistently to dismantle fundamental environmental laws such as the Endangered Species Act and to frustrate the ability of agencies such as the EPA to carry out their regulatory work. Clinton consistently resisted these attacks by vetoing numerous anti-environmental bills, including the package of legislation that was part of the 1995 Congressional leadership's 'Contract with America.'"[58]

Soon after taking office, Clinton shifted the bulk of his administration's environmental watchdog duties onto his vice president, Al Gore. A staunch conservationist, Gore authored the 1992 bestseller *Earth in the Balance*,[59] which called for mandating much tougher environmental laws and regulations. After leaving office, Gore also became one of our climate-change heroes—most notably because of his 2006 documentary film and later book, *An Inconvenient Truth*.[60] We will talk much more about Mr. Gore—and his substantial influence on the climate-change arena—later in this chapter, as well as in chapter 25.

NAFTA: A BAD ENVIRONMENTAL IDEA

The year 1993 featured a major experiment in worldwide economics when President Clinton signed the North American Free Trade Agreement (NAFTA)—the first such agreement between developed countries (the United States and Canada) and a developing country (Mexico).[61] Since NAFTA went into effect on January 1, 1994, it has systematically eliminated most tariff and nontariff barriers to free trade and investment between the three signatory countries.[62]

The pros and cons of NAFTA are still being debated. On the plus side, the landmark trade agreement certainly made it easier for Americans to purchase Canadian and Mexican goods. Overall, NAFTA significantly increased trade among the three countries with—as might be expected—Mexico being the primary economic beneficiary.

While on the surface, NAFTA certainly appeared to have all its environmental bases covered, the emphasis on maximizing trade between developed and developing countries set a dangerous precedent. As Naomi Klein said in an interview with Oliver Tickell for the *Ecologist*, "The idea was there was only one way to run the world—free markets, free trade, privatization, deregulation, low taxes, investor rights, the cult of consumerism, the cheapest possible everything. But tackling climate change demands the reverse—collective solutions, more regulation, restrained consumption, carbon taxes, and so on."[63]

In its analysis of the twentieth anniversary of NAFTA, the Sierra Club writes that "NAFTA has reduced the ability of governments to respond to environmental issues and has empowered multinational corporations to challenge important environmental policies."[64]

In 2016, as we write this book, a variety of new free trade agreements involving the developed countries and developing countries—most notably through the Trans-Pacific Partnership Trade Deal—are in the works. One iconic phrase comes to mind to summarize this materialism-driven policy: "Those who don't remember history are doomed to repeat it!"

ROBERT BULLARD AND THE ENVIRONMENTAL JUSTICE MOVEMENT

One other notable environmental and social achievement of Clinton's administration is directly related—by global extension—to the developed nations versus developing nations controversy that is a significant component of today's climate-change debate. By executive order in 1994, Clinton decreed that "each Federal agency shall make achieving environmental justice part of its mission."[65]

The roots of the environmental justice movement that Clinton referenced in his decree can be traced back to Warren County, North Carolina, in 1982. With a predominantly African-American population, this mostly poor rural county was selected as a site for a polychlorinated biphenyl (PCB) landfill, disposing of the toxic carcinogens. More than five hundred people were arrested when the community marched in protest.[66]

While their efforts to stop the landfill failed, the demonstrators succeeded in bringing the issue of environmental racism to the forefront of the American public. They emphasized that environmental organizations were usually run by rich, white people advocating for the protection of pristine natural resources while ignoring the conditions of the poor and minority populations of the nation.

The prevailing attitude was that natural resources were more important than the ethnic minority populations of the United States. As a result, many of our nation's most vile waste products—including radioactive materials—were being deposited in areas predominantly occupied by poor minority homeowners. The driving theory behind this disgraceful practice was that there would be less chance of organized opposition in these locations, since the residents were less likely to be aware of what was happening to their communities.[67]

In 1990, Robert D. Bullard—who is widely regarded as the "Father of Environmental Justice"—released his book *Dumping in Dixie: Race, Class, and Environmental Quality*.[68] Suddenly the US government was put on notice that their dirty little secret was out of the bag and their deplorable habit of using minority neighborhoods to dump their toxic trash would no longer be tolerated.[69]This book, considered to be the first on environmental justice issues, took the Civil Rights movement to a new level, as it "constituted a clarion call for environmental justice" claims Eddy Carter, Professor of Philosophy at Prairie View A&M University.[70]

We will talk more about Dr. Bullard's accomplishments in chapter 28.

ONSET OF THE ANTHROPOCENE AGE

The IPCC's second assessment report, issued in 1995, stated conclusively that human activities are having a discernible effect on Earth's climate.[71] This was the first time an organization had definitively stated that humans are responsible for climate change, thus officially recognizing the onset of the Anthropocene Age, at least in the minds of some contemporary climate scientists.

The professional organization in charge of defining Earth's time scale is the International Union of Geological Sciences (IUGS). In the IUGS's opinion, we are officially still in the Holocene epoch, which began after the last major ice age—more than 10,000 years ago. However, some geoscientists are arguing that Holocene is outdated. They are advocating for a switch to *Anthropocene*—from *anthropo*, for human. Their rationale is that human activities—including polluting the oceans, altering the atmosphere, and causing mass extinctions of plant and animal species—have had lasting impacts on the planet.[72]

Ever since atmospheric chemist and Nobel laureate Paul Crutzen—whom we discussed in the last chapter—popularized it in 2000, *Anthropocene* has become an environmental buzzword. In 2013, the term picked up velocity in the scientific community, appearing in nearly two hundred peer-reviewed articles and as the title of a new academic journal.[73]

PAUL HAWKEN, NATURAL CAPITALISM, AND PROJECT DRAWDOWN

In 1999, Paul Hawken coauthored a book called *Natural Capitalism: Creating the Next Industrial Revolution*,[74] with Amory and Hunter Lovins. *Natural Capitalism*—which has been translated into twenty-six languages—popularized the idea that Earth's natural resources should be considered natural capital since ecosystems provide humans with benefits like clean water and waste decomposition. This idea was one of the first concrete steps toward factoring the planet's natural resources into cost-benefit analyses for proposed development projects.[75]

Hawken is currently involved in an ambitious venture that he calls "Project Drawdown". In 2017, Project Drawdown will release a book detailing scores of proposed climate-change solutions—from light bulb technology to tropical forest carbon sinks and alternative strategies for livestock grazing and crop production.[76]

For each potential solution, Hawken and his team will look at the costs and benefits, analyzing what would be required given economic factors and our current technology.[77] Because Hawken is playing such a significant role in assessing the potential benefits of proposed climate-change solutions, we have included him as a current climate-change hero, and we will talk more about him in chapter 24.

Serving on Project Drawdown's board with Hawken is Janine Benyus—biologist, educator, author of six books, and a winner of *Time* magazine's Hero of the Environment award. Benyus cofounded the Biomimicry Institute, which uses designs from nature to create global sustainability solutions. We will also talk more about Ms. Benyus's career in chapter 23.

THE 1997 KYOTO PROTOCOL

The most commonly referenced document in climate-change history is the Kyoto Protocol of 1997.[78] To understand, we need to go back to the 1992 Rio Earth Summit. At Rio, the United Nations Framework Convention for Climate Change (UNFCCC) was one of the primary documents—an international treaty—drawn up. Although all of the nations attending the Earth Summit—195 countries—eventually ratified the treaty,[79] the UNFCCC was a lightweight, non-legally binding document that lacked the judicial and administrative clout to get the job done in addressing climate change.[80]

Under the UNFCCC, regular sessions were to be held to review progress in addressing climate change. These sessions were called Conference of Parties (COP), with COP-1 being held in Berlin in 1995.[81] After the Rio Earth Summit, the UNFCCC signatory countries—especially the US, Russia, and the European Union—spent five years arguing and wrangling about how best to deal with climate change. The 1997 Summit in Kyoto, Japan (COP-3) was the culmination of these contentious debates and provided the perfect opportunity for finally implementing an effective international agreement on controlling GHG emissions[82]—or so it seemed.

The Kyoto Protocol added teeth to the UNFCCC by requiring signatory nations that were classified as developed nations (i.e., wealthier countries) to set GHG emission reduction targets specifying both volumes and dates. Meanwhile, countries designated as developing nations—including such highly populated nations as China, India, and Mexico—were not required to set emission-reduction targets and essentially were given a free pass.[83] In fact, under the Kyoto Protocol the developed countries were supposed to provide both technical and financial support to the developing countries, with a goal of boosting their economic standing. These issues proved very detrimental to the advancement and overall effectiveness of the Kyoto Protocol.

While the Kyoto Protocol was adopted in December 1997,[84] problems associated with ratification and bringing the document into force were just beginning. US vice president Al Gore, representing the Clinton administration, agreed to sign the treaty—not surprising given his strong beliefs in climate change. President Clinton hailed the Kyoto Protocol as "environmentally sound" and "economically strong" and said that it "reflected a commitment from our generation to act in the interest of future generations."[85] However, Washington, DC—specifically Capitol Hill—did not agree with these positive assessments of the treaty, and the Senate voted not to ratify it by a demoralizing and disgraceful vote of 95 to 0.[86]

A hotly debated issue at Kyoto was the relative responsibility of developed countries versus developing countries in footing the bill for solving the climate-change crisis.[87] The Clinton administration believed that since climate change was a global issue all of the world's countries should share equally in resolving the problem. Of course, this opinion did not go over well with the developing countries, which believed that the climate-change problem was overwhelmingly being caused by the rich countries of the world and—furthermore—that the poorer countries were suffering the worst consequences.[88]

Dr. Mark Mwandosya of Tanzania, chairman of the Developing Countries Caucus, expressed this view when he said, "Very many of us are struggling to attain a decent standard of living for our peoples and yet we are constantly told that we must share in the effort to reduce emissions so that industrialized countries can continue to enjoy the benefits of their wasteful lifestyle."[89]

Due in large part to the lack of support from the United States, the Kyoto Protocol languished in limbo and international negotiations on climate change essentially came to a halt for the next seven years.[90] Then, in late 2004, Russia unexpectedly ratified the treaty—as part of their desire to join the World Trade Organization (WTO). According to Nick Paton Walsh writing for the *Guardian*, Russian president Vladimir Putin said, "The fact that the European Union has met us halfway at the negotiations on membership in the WTO cannot but influence Moscow's positive attitude towards ratification of the Kyoto Protocol."[91] Russia's signing of the treaty allowed it to finally enter into full force on February 16, 2005.[92] The United States remained outside of the accord, however. Under President George W. Bush—who called the Kyoto Protocol "fatally flawed" and claimed that its implementation would gravely damage the US economy—the United States disengaged from the treaty in 2001 and has never ratified it.[93]

In retrospect, the Kyoto Protocol was considered to be a major failure and supreme disappointment, especially considering the extremely high hopes at the beginning of the conference. The majority of climate scientists say that the targets set in the Kyoto Protocol are "merely scratching the surface of the problem," and that "in order to avoid the worst consequences of global warming, emissions cuts on the order of 60% across the board are needed."[94]

ENVIRONMENTAL CONCERNS AND GLOBAL WARMING MAKE A COMEBACK

The late 1990s were marked by a renewed concern about environmental decay—both nationally and globally. In August 1998, the *New York Times* reported that "Alaska is thawing, and much of northern Russia and Canada with it, and many scientists are saying that the warming of these cold regions is one of the most telling signals that the planet's climate is changing."[95] Using laser instruments, climate scientists were finding evidence that many of Alaska's hundreds of glaciers were shrinking dramatically.[96] They had been warning for years that these northern climes would be the first to show the physical effects of global warming, and now they were being proven correct.

Over the twentieth century, the world's temperature had risen by about 0.5 C (1.0 F) but, according to scientists at the University of Alaska, Alaska, Siberia, and northwestern Canada experienced a temperature rise of 2.8 C (5.0 F) over the past thirty years.[97] The debate was now not so much about whether global temperatures were increasing but to what extent human-caused global warming was the culprit.[98]

The progressive loss of the arctic and subarctic ice masses meant that the permafrost—ground that had remained permanently frozen year round—was thawing and turning to water for the first time in millennia. This, in turn, caused thousands of acres of taiga forest—scattered clusters of small, ancient trees that had successfully survived atop the frozen ground—to sink into the suddenly mushy soil.[99] Because many of these drowned trees—along with adjacent utility poles—rose from the ground at all sorts of odd angles, they were called "drunken forests."[100]

FAITH-BASED GROUPS JUMP ON THE CLIMATE-CHANGE BANDWAGON

In 1998, at Grace Episcopal Cathedral in San Francisco, California, Reverend Sally Bingham founded the Interfaith Power & Light (IPL)—a staunchly environmental, faith-based organization. IPL now has affiliates in forty states, involving a total of 18,000 congregations.[101] IPL's mission is to be "faithful stewards of Creation by responding to global warming through the promotion of energy conservation, energy efficiency, and renewable energy."[102] They work to protect Earth's ecosystems and support sustainable energy.

In her capacity as president of IPL, Reverend Bingham educates people on the connection between religious faith and caring for the environment. She sees global warming as a moral issue that humanity needs to face.[103] We will talk more about Reverend Bingham's remarkable work in chapter 28.

DAWN OF THE INTERNET AND THE GLOBAL COMMUNICATION AGE

The late 1990s saw the blossoming of a new weapon in the arsenal of environmental activists. The full-on advent of the Internet opened the world of cyberspace to information dissemination and recruitment of new members. Suddenly almost every corner of the planet—no matter how remote—was accessible to open communication with the rest of the world. Plus, nearly every piece of information ever produced could now be transferred to other locations in less than a blink of eye. The onset of the information age had begun and our world would never be the same again.

While he didn't really invent the Internet—as many have jokingly claimed—Clinton's vice president, Al Gore, has consistently been one of the world's most resourceful politicians in using the Internet for spreading information about climate change and other significant environmental issues. In a 1997 letter to the US State Department, which was included in their report on the environment and foreign policy, Gore provided this poignantly insightful assessment of global environmental conditions:

> "Environmental problems such as global climate change, ozone depletion, ocean and air pollution, and resource degradation—compounded by an expanding world population—respect no borders and threaten the health, prosperity, and jobs of all Americans. All the missiles and artillery in our arsenal will not be able to protect our people from rising sea levels, poisoned air, or foods laced with pesticides. Our efforts to promote democracy, free trade, and stability in the world will fall short unless people have a livable environment.
>
> We have an enormous stake in the management of the world's resources. Demand for timber in Japan means trees fall in the United States. Greenhouse gas emissions anywhere in the world threaten coastal communities in Florida. A nuclear accident in the Ukraine kills for generations. . . . Our children's future is inextricably linked to our ability to manage the earth's air, water, and wildlife today."[104]

GEOENGINEERING: CREATING A GLOBAL SUN SHIELD

March 1999 saw the culmination of a study on an unusual event that had the odd potential for connecting the ever-burgeoning pollution of Earth with a solution for climate change and global warming. Over a six week period, scientists discovered and began studying a huge haze of environmental pollutants hovering over the Indian Ocean.[105] Since this filthy cloud was thick enough to reduce the amount of the sun's radiation reaching Earth's surface, some researchers hailed it as a way to stop, and even diminish, warming of Earth's lower atmosphere.[106] The idea grew into an early geoengineering scheme for using giant firehoses to inject sulfur dioxide particles into our planet's upper atmosphere, thereby blocking full sunlight from reaching large areas of land.

Fortunately, clearer heads prevailed on this insane climate-change solution, realizing that we can't pollute our way out of global warming.[107] Scientists opposing the idea were quick to point out that the sulfates comprising this artificial so-called solar shield would drop from the atmosphere in weeks, while the GHG will remain there for centuries.[108] This event, however, exemplified how an anxious public was willing to grasp at anything hopeful in the increasing complexity of the environmental movement.[109]

Chapter 19

A SERIES OF UNFORTUNATE EVENTS (2001–2005)

BIZARRE ELECTION OF GEORGE W. BUSH AND DICK CHENEY

One of the most bizarre events in the history of US presidential elections occurred in the fall of 2000 when the governor of Texas, George W. Bush ("W.") defeated sitting vice president Al Gore. It wasn't so much who won this election as how it occurred. On election night—as we all watched the nail-biting returns with eager anticipation—the television networks twice called the winner, only to renege on their projections both times.

Watching this scenario unfold throughout the seemingly unending election night was unbelievably excruciating. The next day, as we all sat waiting breathlessly for the result to finally be announced, we heard the devastating news. The long months of campaign ads, debates, barbs, and innuendos had come down to this: the deciding vote for the next president of the United States rested solely with the state of Florida.[1] Plus, the vote was simply too close to call, triggering a mandatory recount.

What ensued was a phantasmagorical six weeks of accusations, recounts, legal challenges, and finally US Supreme Court appeals before "W." was declared the winner in Florida by only 537 votes—the closest presidential election in US history.[2] While Gore had won the overall popular vote, he lost the electoral vote by a razor-thin margin.

Without question, the battle against the negative effects of climate change would certainly be much more advanced—and perhaps be on the way to being resolved—if Gore had won this election. For the legions of environmentalists who mistrusted and even despised Bush's strong probusiness and antigovernment policies, the sham of the 2000 presidential election was a most bitter pill to swallow.

SEPTEMBER 11TH, THEN INVASIONS OF AFGHANISTAN AND IRAQ

In US history, two days will always live in supreme infamy—December 7, 1941 (attack on Pearl Harbor) and September 11, 2001 (terrorist attack on the Twin Towers). All Americans who were alive on either one or both of those days will forever remember where they were and what they were doing when they heard the horrific news reports. And—in the case of New York's World Trade Center—many actually watched the heartbreak of the collapsing twin towers happening on live television.

We were able to successfully end the fascist scourge that threatened to overtake the world during World War II, but the jury is still out on our success following the destruction of the World Trade Center on that unfathomable September morning. Fortunately, as we write this in 2016, we have managed to avoid another terrorist attack of this magnitude on US soil. But the fact is that our response to the 9/11 attacks will forever negatively influence the way the United States is perceived by the rest of the world, and not just by Muslim countries. Our ill-advised decision to invade both Iraq and Afghanistan may haunt us forever.

ENVIRONMENTAL CONCERNS YIELD TO TERRORIST FEARS

After 9/11, with the attention of the United States and most of the rest of the developed world so focused on fighting terrorism, there was limited time and incentive left for even thinking about long-term environmental issues like climate change. Americans were preoccupied with imminent threats to personal safety and security and distracted by our wars in Iraq and Afghanistan.

In fact, we believe that this is the main reason why so little has been accomplished to counteract the negative impacts of climate change since the turn of the millennium. The mindset of most Americans seems to be that the enemy in plain sight is the one to fear most, while the enemy hidden in the bushes can be ignored until later. In other words, the threat of worldwide terrorism is so pervasive that it overwhelms all other problems—even something that poses the potential for the greatest environmental disaster in human history.

Besides, as far as most Americans were concerned, the environment was doing just fine. The general attitude seemed to be that we had done enough to protect Mother Nature and now we needed to concentrate on protecting ourselves. Furthermore, there was a pervasive belief that the environmental regulatory process was far too cumbersome and needed to be significantly reduced in terms of its complexity.

In the meantime, while we occupied our thoughts with the War on Terror, the US environment was not really that healthy after all. Forty percent of Americans were breathing unhealthy air, 25 percent lived within four miles of a toxic waste dump, 70 percent of our commercial fisheries were overexploited, and—worldwide—twenty-three billion tons of CO_2 were being released every year. Plus the world's population had tripled—to more than six billion people—in just the past fifty years.[3]

The sad truth is that it may take some sort of environmental disaster to stimulate a substantial worldwide or even national initiative on controlling and reversing climate change. Unfortunately, by the time such an event happens it may already be too late to save Earth from the massive ecological and social changes that are certainly forthcoming if we refuse to act. Meanwhile, the third (2001) and fourth (2005) IPCC assessment reports provided even more definitive information that the anthropogenic emissions of greenhouse gasses were the main cause of temperature rise in the second half of the twentieth century.

UNDOING THE ENVIRONMENTAL REGULATORY PROCESS AND PROMOTING FOSSIL FUELS

As the privileged son of a former president and a devoted Texas oilman, President George W. Bush's policies were taking the country in the absolutely wrong direction for combating climate change. Plus Richard

"Dick" Cheney—Bush's vice president—was given more authority and responsibility that any other vice president in US history. Many Americans (certainly most environmentalists) would soon come to believe that Cheney was actually running the Bush White House.[4]

While the majority of US citizens were focused on the War on Terror, Bush and Cheney concentrated on increasing the US production of fossil fuels throughout the world, including in such incredibly fragile ecosystems as Alaska's Arctic National Wildlife Refuge. They also shaped the environmental policies to align with their views—appointing key regulatory people who shared their pro-development[5] beliefs while deregulating any programs that might impede energy development and economic progress.[6]

During Bush's first three years in office, there was a 75 percent decline—compared to the last three years of the Clinton administration—in the number of federal lawsuits filed against companies for violating national environmental laws.[7] Civil citations for polluters also dropped 57 percent after 2001, while federal prosecutions fell 17 percent. Most appalling according to EPA records, the pace of cleanups of Superfund hazardous waste sites—which had increased dramatically during the Clinton administration—decreased by 59 percent under Bush.[8]

Even worse, the Bush administration refused to seek renewal of the Superfund cleanup tax on polluting industries. This led to the effective bankruptcy of the federal Superfund, resulting in a reported thirty-four unfunded Superfund hazardous waste sites in nineteen states in 2004.[9] This was taking the Republican policy of "no new taxes" to extremes. Allowing the lack of federal funding to place the lives of American families in long-term jeopardy of contracting cancer and other deadly health problems associated with environmental toxins is simply unforgivable.

Finally, specific to climate change, there was a 52 percent reduction in the number of EPA inspections at refineries after 2001, plus a 68 percent reduction in the number of notices of violations refineries received during this same period.[10]

KILLING THE CLIMATE-CHANGE BUZZ

All this, of course, meant any interest for addressing—or even thinking about—climate change or global warming was simply shoved aside. To illustrate the overriding lack of concern about climate change on a nationwide basis, Bush's 2001 announcement that the United States was totally withdrawing from all involvement with the 1997 Kyoto Protocol drew scant attention from the American public.

Bush emphasized that he believed that dealing with climate change would place an undue and unnecessary burden on the American economy. We believe, however, that—based on the total lack of government action of any type on climate change throughout the eight years of his administration—"W." must have breathed a huge sigh of relief when he realized he had dual excuses—the War on Terror and the economy—for not having to tackle the subject.

In fact, the carryover of Bush's general disinterest extended into the first four years of the Obama administration. Just as the word *taxes* sparked a knee-jerk negative reaction from every Republican legislator—as well as the majority of the American public—the terms climate change, global warming, and sea-level rise all became verboten on Capitol Hill. Senators and Congressmen on both sides of the aisle refused—some even seeming fearful of having the issues negatively affect their next elections—to bring these subjects up for discussion, and even less to take actions or propose initiatives to get something done about them.

THE CLEAR SKIES BILL: GOOD, BAD, OR UGLY?

In 2003, the Bush administration proposed a very controversial piece of regulatory legislation. Hailed as the "legislative centerpiece of President Bush's environmental policy,"[11] the Clear Skies Bill was intended to create a mandatory program that would dramatically reduce power-plant emissions of sulfur dioxide (SO_2), nitrogen oxides (NO_x), and mercury by setting a national cap on each pollutant.[12] This bill, however, clearly demonstrated the potential foibles of implementing faulty air quality emission control strategies.[13]

To be fair, the Clear Skies Bill had its array of both detractors and supporters. The *Washington Post* wrote that, "The good things about this bill: A cap-and-trade system could reduce emissions faster and more cheaply than the cumbersome, lawsuit-prone regulatory system. When used to impose meaningful standards, emissions trading could broaden the significant achievements of the air quality regulatory regime in this country, at a lower cost to electricity consumers."[14] Emissions trading is part of a system of economic incentives by which utilities receive permits to produce a specified amount of carbon dioxide and other greenhouse gases, which they then may trade with other utilities.

As described in an editorial of the March 7, 2005, *New York Times*, the Clear Skies Bill came "attractively dressed as a grand bargain under which a market-based system of pollution control would replace the cumbersome regulatory certainty for industry, and cleaner air for everybody."[15] Of course, whenever the Bush/Cheney White House called something an environmental phenomenon, the antennae of the conservation community immediately went up and started searching for the catch, and the Clear Skies Bill certainly did not disappoint. The aforementioned *New York Times* article declared, "Clear Skies is a bad bill, which in the name of streamlining current (federal) law would offer considerably more relief to the industries that pollute the air than to the citizens who breathe it."[16]

The Sierra Club also criticized Bush's plan, claiming that it would actually result in more air pollution than was currently allowed under federal laws.[17] According to Tom Valtin, writer for Sierra Club, the Clear Skies proposal that Bush introduced made "all but the most flagrant polluters virtually immune from government legal action."[18] Furthermore, the Bush administration moved to eviscerate the New Source Review (NSR) provision, which was designed to force the oldest and dirtiest power plants to reduce their pollution if they expanded.[19]

Writing in the December 2004 issue of *Washington Monthly* magazine, David Whitman elaborated on more specific criticism: "In a novel twist, environmentalists have also asserted that Clear Skies is actually weaker than the existing Clean Air Act—and would allow millions of tons of added pollution and inflict tens of thousands of needless deaths during the next decade."[20] In fact, the Clear Skies Bill was so thoroughly criticized for attempting to gut the Clean Air Act that John Kerry featured it in his 2004 campaign list of "The Bush/Cheney Top 10 Environmental Insults."[21]

From a climate-change perspective, the Clear Skies Bill proves that any touted solutions must be well thought out and scientifically strong. The last thing we need is to put forth an idea that becomes a source of ridicule and derision. That's why solutions with an economic component must receive extra scrutiny. Finding solutions that also involve making money are most certainly worthy goals, but not at the sacrifice of public health or the pollution of our common goods, such as air. The Clear Skies Bill died a rather ignoble death in March 2005 when the Senate Environment and Public Works Committee decreed that it would weaken the central pillars of the nation's environmental framework.[22] As a result, the proposed legislation for air regulation was never enacted.[23]

EUROPE'S DEVASTATING HEAT WAVE: 70,000 DEAD

In 2003, Europe experienced the hottest summer on record since at least 1540, with the heat wave killing an estimated 70,000 people.[24] With almost 15,000 deaths and extensive crop damage throughout the entire country, France was hit especially hard. The height of the catastrophe occurred during August—a month when many Europeans are on holiday. In fact, many bodies were not claimed for many weeks because relatives were away on vacation. The staggering amount of unclaimed bodies overwhelmed undertakers—mortuaries were so full that corpses had to be kept in a warehouse outside the city.[25] Amazingly, from our perception, this astounding heat wave and death toll received virtually no play in the United States—again sadly emphasizing the US media's approach that all too often seems to be "if it doesn't happen on US soil, it doesn't count."

RUSSIA RATIFIES KYOTO PROTOCOL, LEAVING THE UNITED STATES OUT IN THE COLD

As we discussed in the last chapter, Russia finally decided to ratify the Kyoto Protocol on October 21, 2004, putting the treaty into effect without the United States having signed on.[26] The United States was responsible for nearly 34 percent of the 1990 carbon emissions; the Russian Federation accounted for just over 16 percent. In order for the Kyoto Protocol to enter into force, one of these two major contributors had to ratify to meet the required 55 percent contributions from the developed nations.[27]

Unfortunately, this wasn't primarily because Russia wanted to stop climate change. According to journalist Nick Paton Walsh writing on the *Guardian*'s website, "Mr. Putin prevaricated over the bill, saying that Russia would only sign it if it was in the national interest and suggesting it would need modifying. His key adviser on the issue, Andrei Illarionov, made Russia's vital ratification of the pact seem unlikely when he described it as an 'economic Auschwitz,' insisting it would cripple Russian economic development. However, Mr. Putin publicly announced he would ratify the treaty after a meeting with EU officials in May [2004], on the same day as the EU dropped its objections to Russia joining the World Trade Organisation.

THE HORRORS OF HURRICANE KATRINA BROADCAST ON LIVE TV

On August 29, 2005, Hurricane Katrina struck the United States and ravaged the city of New Orleans and much of the Gulf Coast.[28] When the storm made landfall, it stretched some four hundred miles across and was rated a Category 3 on the Saffir-Simpson Hurricane Scale—meaning that it was being driven by sustained winds that varied from 100 to 140 miles per hour.[29]

While its storm system caused a lot of direct damage, Katrina is best known for what happened in its aftermath. Levee breaches led to catastrophic flooding and stranded thousands of people without food or water.[30] Many of these people perished needlessly because the emergency services, including local, state, and even the Federal Emergency Management Agency (FEMA) were unable to rescue or—unbelievably in many cases—even reach them. By most accounts, Katrina represented one of the worst-case scenarios of a total lack of emergency preparedness in the history of the United States.[31]

Who can ever forget watching on network television as people clung to the tops of trees, houses, and buildings, pleading for water, food, or help from the contaminated water swirling beneath their feet? Or the thousands of people packed together in the deteriorating conditions of the Superdome, fighting the stinking heat and begging for survival rations that were inexplicably not forthcoming.

Then, on top of watching this appalling human crisis unfold before our eyes, we listened to government leaders like President George W. Bush say things like, "Brownie, you're doing a heck of a job," to ill-trained FEMA director Michael D. Brown, as chaos and carnage churned in the streets all around them. In total, hundreds of thousands of people in Louisiana, Mississippi, and Alabama were displaced from their homes, and experts estimated that Katrina caused more than $100 billion in damage.[32]

According to Justin Worland, writing for *Time* magazine on August 27, 2015, almost ten years after the event, "During Katrina, storm surge pushed ashore the water, causing levees to fail. The storm was recorded as a hurricane of category 1 or category 2 strength when it hit New Orleans—relatively weak for such a devastating hurricane. But the storm surge reached as high as 12 feet in some places, creating flood conditions across the city and wreaking havoc with the city's levee system."[33]

STRONGER STORM SURGES ASSOCIATED WITH SEA-LEVEL RISE

While it may be difficult to tie hurricane intensity directly to climate change, what is vividly certain is the sea-level rise that is associated with global warming. Ongoing sea-level rise will continue to exacerbate the severity of storm surges that are responsible for the bulk of the death and destruction associated with hurricanes in the Atlantic Ocean and the Gulf of Mexico, as well as with cyclones and typhoons in the Pacific

Ocean.[34] To understand storm surges, think of them as tidal waves (tsunamis) caused by high winds blowing the surface water instead of huge walls of water created by underwater earthquakes.

While Katrina's extensive death and destruction were a combination of weather intensity and human-created blunders (such as poorly constructed levees), the event itself was the bellwether of an increasing number of huge storms that occurred around the world from 2005 through 2015. Most notable among these events were ten monstrous Pacific tropical cyclones—including Haiyan, Loke, Rick, Celia, and Marie—along with Superstorm Sandy, which caused unprecedented death and damage along the Northeast coast of the United States in the fall of 2012.

THE HALLIBURTON LOOPHOLE: FRACKING GONE WILD

The year 2005 also brought the most insidious regulatory free pass—now infamously known as the Halliburton loophole—in the history of the US environmental movement. The Halliburton loophole was part of the Federal Energy Policy Act of 2005 (FEPA) signed into law by President George W. Bush.

Named after the fraud-accused[35] corporation Dick Cheney had led before becoming vice president, the Halliburton loophole explicitly exempted hydraulic fracturing—aka fracking—operations from key regulatory provisions of the Safe Drinking Water Act of 1974 and the Clean Water Act of 1972.[36] These exemptions from some of America's most fundamental environmental protection laws provided the oil and gas industry with the immunity it needed to implement a highly polluting and dangerous drilling process on a grand national scale.[37]

The Halliburton loophole effectively turned the nightmare of fracking loose on the American landscape—quite possibly, never to be contained again. Anyone looking down from an airliner flying at 32,000 feet may see the ravages of this regulatory exclusion. In many places, fracking roads and pads stretch as far as the eye can see, obliterating thousands upon thousands of acres of previously unadulterated landscape. And—for the most part—this all has taken place without federal environmental impact documents being prepared or public hearings being held.

But the Halliburton loophole wasn't the only fracking enabler in the 2005 FEPA. The other glaring regulatory misstep was placing the Federal Energy Regulatory Commission (FERC, formerly the Federal Power Commission where Budd started his career as an environmental compliance specialist in 1973) in charge of fracking oversight and compliance under the National Environmental Policy Act (NEPA). This was truly akin to putting the fox in charge of the henhouse. The Federal Energy Policy Act granted FERC sweeping authority to supersede state and local decision-making when it came to choosing the location of infrastructure—including a morass of new pipelines—to service the fracking industry.[38]

How could this happen in the nation with supposedly the world's richest tradition of environmental assessment, regulatory review, and impact compliance? I (Budd) spent thirty-five years as a professional environmental scientist and certified professional wetlands scientist writing compliance documents and testifying at contentious public hearings. The essence of this federal environmental regulatory review process was to make sure the tenets of the CWA and NEPA were given due diligence for proposed projects ranging from new highways, shopping malls, schools, and hospitals to municipal, state, and national parks.

However, the fracking industry, which, by all accounts, runs some of the most dangerous and polluting operations on earth, seems to get a free pass from federal regulations. What in the world is going on? The answer is quite simple: In our haste to free ourselves from dependence on foreign—mainly Middle Eastern—oil supplies, some powers that be decided that the answer lay in finding and developing supplies of "cleaner-burning natural gas" within our own nation's boundaries.

Initially, the oil and gas industry raved about the things that fracking would accomplish, including providing a mother lode of "less polluting natural gas" while doing away with our dependence on foreign oil and putting the United States on a fast track to becoming the world's largest fossil-fuel producer.[39] Because our foreign oil dependency has been a thorn in the side of US diplomats since before the Arab oil embargo in

1973, Congress—on both sides of the aisle—leapt on this newfangled fracking as a once in a generation worldwide business opportunity. Almost everyone on Capitol Hill agreed that any environmental damage resulting from fracking would be immaterial to solving our growing energy crisis for the foreseeable future. As a result, the 2005 FEPA was written, verified, and approved posthaste.

In light of the enthusiasm for fracking that so easily permeated and controlled both the House and the Senate, is the process really so bad? For starters, the access roads and drilling pads needed for the operation destroyed entire swaths of formerly untrammeled landscapes. As we mentioned previously, areas where fracking has already taken place look like hundreds upon hundreds of rows of barren squares, all tethered together with access roads, and with no room in between. Whatever natural habitat that might have existed before fracking occurred is all but obliterated. The fate of the wildlife that used to live in these areas is anybody's guess. Remember, the 2005 FEPA—for the most part—removed the need for federal or state Environmental Impact Statements (EISs) so not only were impacts not avoided, minimized, or mitigated, they typically were never identified or studied in the first place.

And what happens after the infrastructure is in place and the fracking process begins? The process involves blasting millions of gallons of a cocktail of toxic chemicals underground at enormous pressures to break apart subterranean rock and free natural gas so that it can be captured and processed at the surface.[40]

As Wenonah Hauter, executive director of Food and Water Watch, writes, "[As of August 2015] more than 270,000 wells have been fracked in 25 states throughout the nation. More than 10 million Americans live within a mile of a fracking site. This means that 10 million Americans—and in reality many more—have been placed directly in harm's way. Hundreds of peer-reviewed studies have connected fracking to serious human health effects, including cancer, asthma, and birth defects."[41]

It is especially vexing to think about how fracking companies shield the contents of the toxic water and chemical solutions that they blast down into the ground. Fracking-related contamination of groundwater drinking wells is now known to pose a serious threat to public health and safety. But, astonishingly, even doctors who are dealing with fracking-related health complaints are unable to access information on the types of toxins to which their patients may have been exposed.[42]

Chapter 20

CLIMATE CHANGE BEGINS A COMEBACK (2005–2008)

JAMES BALOG AND *CHASING ICE*

Back in the late 80s and early 90s, acclaimed environmental photographer James Balog was a climate-change skeptic.[1] But then he learned about the tangible physical evidence of global warming that was preserved in the ice cores of Greenland and Antarctica.[2] As Balog stated in an interview with *ThinkProgress*'s Stephen Lacey, "That was really the smoking gun showing how far outside normal, natural variation the world has become. And that's when I started to really get the message that this was something consequential and serious and needed to be dealt with."[3] In 2006, Balog received a challenging assignment from *National Geographic* to document Earth's changing glaciers—these pictures became the June 2007 cover story for *National Geographic*, "The Big Thaw."[4] In very short order, his mind was completely and irrevocably changed. Now he is one of climate change's most gifted, articulate, and passionate activists.

As described in the trailer for Balog's documentary *Chasing Ice*: "Through [Balog's] Extreme Ice Survey, he discovers undeniable evidence of our changing planet. During this climate change classic, Balog deploys revolutionary time-lapse cameras to capture a multi-year record of the world's shrinking glaciers. His hauntingly beautiful videos compress years into seconds and capture ancient mountains of ice in motion as they disappear at a breathtaking rate."[5]

On a personal note, our first viewing of *Chasing Ice* left us mesmerized, exhausted, and enraged. At the conclusion of Mr. Balog's questions and answers, the entire movie auditorium audience of five hundred people instinctively rose as one in a show of support for the film's dramatic message and the man who had risked so much to put it together. It is impossible to watch *Chasing Ice* and come away still doubting that climate change is happening. In fact, it's difficult to watch this epically beautiful, yet supremely sad documentary and not be moved to tears.

For his dramatic efforts to capture never before seen visual evidence of climate change, we have named James Balog one of our climate-change heroes. We will talk more about him in chapter 26.

GRASSROOTS LEADERSHIP AT THE MEGACITY LEVEL

The year 2005 saw the establishment of the world's first grassroots organization for coordinating climate-change action throughout the world's largest cities. The C40 Cities Climate Leadership Group (C40) advances the climate action agendas of the world's megacities to achieve meaningful reductions of GHG emissions and climate risks.[6]

Currently, C40 is a global network of more than eighty engaged municipalities—representing more than 600 million people throughout the world[7]—who are committed to reducing their GHG emissions and increasing climate resilience. These actions are accomplished on the local level through the implementation of measurable, replicable, and sustainable climate-related policies, which address the issue of climate change on a global basis.[8]

In 2006, C40 first expanded through a partnership with President Clinton's Climate Initiative (CCI). Former New York mayor and climate-change activist Michael Bloomberg served as president of the C40 board from 2010 to 2013 and is still fully involved with the operation of the organization. Steering committee members are located in the cities of Tokyo, Hong Kong, Johannesburg, Buenos Aires, Houston, Jakarta, London, Berlin, Los Angeles, Rio de Janeiro, Copenhagen, and Seoul.[9]

The critical importance of C40 is that it demonstrates the potential for dealing with climate change from the bottom-up, instead of the more traditional top-down method. Based on the overwhelming opinion of climate scientists and activists that we have interviewed, the solutions to climate change will be found in initiating activities at the grassroots level, then letting the motivated populace bring change at the national level.

For details on Michael Bloomberg's other grassroots climate-change initiatives, see chapter 28.

MICHAEL MANN AND THE HOCKEY STICK CONTROVERSY

Several important climate-change-related events occurred in 2006, starting with the "Hockey Stick Controversy." In 1998, climate scientist Michael Mann and his colleagues created a graph of several hundred years of average global temperatures.

The Hockey Stick Controversy arose after the scientists had extended the graph to show almost a millennium's worth (from 1000 CE to 1900 CE) of steadily and slowly declining global temperatures, followed by radical increases in temperatures beginning in the 1900s and extending into the early 2000s. The shape of this graph looked like a hockey stick with a long and relatively straight handle ending in a sharply angled, short blade pointing up.[10]

As environmental journalist David Appell describes it,

> To construct the hockey-stick plot, [Michael E.] Mann [of the University of Virginia], Raymond S. Bradley of the University of Massachusetts Amherst and Malcolm K. Hughes of the University of Arizona analyzed paleoclimatic data sets such as those from tree rings, ice cores and coral, joining historical data with thermometer readings from the recent past. In 1998 they obtained a "reconstruction" of Northern Hemisphere temperatures going back 600 years; by the next year they had extended their analysis to the past 1,000 years. In 2003 Mann and Philip D. Jones of the University of East Anglia in England used a different method to extend results back 2,000 years.[11]

Climatologist Jerry Mahlman coined the term hockey stick to succinctly describe the pattern that the graph showed.[12] Soon afterward, the

"Hockey Stick Graph" was elevated to iconic status for climate believers and a target for climate deniers.[13]

Climate-change skeptics often cite events such as the Little Ice Age and the Medieval Warm Period as evidence to refute the shape and scientific relevance of the Hockey Stick Graph. Mann's response to these skeptics is quite succinct:

> From an intellectual point of view, these contrarians are pathetic, because there's no scientific validity to their arguments whatsoever. . . . But they're very skilled at deducing what sorts of disingenuous arguments and untruths are likely to be believable to the public that doesn't know better.[14]

We talk more about Michael Mann's work in the climate-change arena in chapter 23.

AL GORE'S INCONVENIENT TRUTH

The story behind how Al Gore's *An Inconvenient Truth*[15] was made is rather complicated. What ended up as an award-winning documentary film started out as a slideshow on global warming that the former US vice president presented—by his own count—at least a thousand times. The original slideshow was a labor of love created by a man who had poured a large part of his life into something he passionately believed. In fact, Mr. Gore openly laments the fact that his dire warnings about climate change and global warming never seemed to be getting through to the general public—no matter how many presentations he gave.[16]

Environmental writer Andrew Revkin describes what happened:

> *An Inconvenient Truth* came about after Laurie David, a prominent Hollywood environmentalist, saw Mr. Gore give a short version of his [slide] presentation [in 2004] at an event held just before the premiere of the climate disaster movie *The Day After Tomorrow*. Ms. David said she was stunned by the power of Mr. Gore's talk and helped organize presentations in New York and Los Angeles for people involved in the news

media, environmental groups, business and entertainment. By the time she had done the Los Angeles event, "I realized we had to make a movie out of it," she said.[17]

Laurie David recruited a team of filmmakers and investors and then convinced Mr. Gore, to let a film crew follow him around. "In the film, directed by Davis Guggenheim," Revkin writes, "Mr. Gore comes across as a professorial guide who uses science, humor, his own life lessons, depictions of perilous climate-driven events and even cartoons to make his case."[18]

An Inconvenient Truth—the documentary film—premiered at the Sundance Film Festival in 2006 and had gala openings in both New York City and Los Angeles. The film was a critical and a box-office success; it won two Academy Awards—for Best Documentary Feature and Best Original Song. Plus, it grossed more than $50 million worldwide, making it eleventh highest grossing documentary film in US history to date.[19]

All in all, we believe Mr. Gore must have been quite pleased with the film's results. After hundreds of what he himself considered to be futile efforts, *An Inconvenient Truth* was suddenly being praised for raising public awareness of climate change at the international level. Furthermore, schools all over the world started including it in their science curricula.[20]

Since its release, *An Inconvenient Truth* has had a pronounced dual effect—sharpening the understanding and sense of foreboding in believers while further polarizing and alienating deniers. We talk much more about Mr. Gore—as one of our climate-change heroes—in chapter 25.

MELTING OF THE GREENLAND AND ANTARCTIC ICE SHEETS

The next serious climate-change event that took place in 2006 requires a bit of background explanation—we need to understand ice sheets, specifically how they are formed, how they move, and how they are different than the more widely known and discussed glaciers. First, an ice sheet forms when annual snowfall exceeds annual snowmelt over a specific portion of Earth's landmass.[21] As you can imagine, when this happens over long periods of time—say, thousands of years—an outrageously large solid block of ice results. Then, when this ice block becomes heavy enough, the whole frozen mass begins to flow.[22] Sounds a lot like the glacier that we all studied in school, right?

Well, it is—sort of. The big difference between a glacier and an ice sheet is that a glacier generally flows in only one direction. An ice sheet, on the other hand, flows outward in all directions from its center. The result is that ice sheets are huge—much larger than your average glacier. In fact, by definition, an ice sheet has to cover more than 20,000 square miles.[23] That's an area more than twice the size of the state of Vermont.

Our Earth has just two major ice sheets—one on Greenland and the other on Antarctica.[24] The Greenland Ice Sheet covers approximately 850,000 square miles—that's almost 300,000 square miles more than Alaska, our largest state. The Antarctic Ice Sheet covers an estimated 5.4 million square miles and is divided into three sections: the East Antarctic Ice Sheet, the West Antarctic Ice Sheet, and the Antarctic Peninsula.[25] Taken together, the Greenland and Antarctic Ice Sheets contain about 99 percent of the world's freshwater ice.[26] If the Greenland Ice Sheet melted completely away, the world's sea levels would rise roughly twenty-three feet. If the entire Antarctic Ice Sheet were to liquidate, sea levels would rise an unfathomable 187 feet.[27]

Now, herein lies the rub: Satellite radar data from 2006 showed that the flow velocities of large ice streams underlying southern Greenland had doubled in the past five years—something most experts had thought was impossible. This gave rise to concerns that maybe speculations about lubrication of the base of an ice sheet were correct. Perhaps the flows under the ice sheet could indeed be powerful enough to cause massive sections of ice to break off into the ocean—causing catastrophic increases in worldwide sea levels. Fortunately, the Greenland ice streams soon slowed down again, showing that this lubrication effect was only temporary. Indeed, a long-range study reported that these particular streams were discharging ice into the sea no faster—on average—than what was occurring a decade earlier.[28]

But glaciologists were not completely reassured. As physicist and author Spencer Weart writes:

> Considering how ice streams around Antarctica had also been observed to accelerate and slow down suddenly, it seemed that these systems were more sensitive to perturbations than the scientific community at large assumed. Moreover, a new satellite was transmitting disturbing data. It measured gravitational force so sensitively that it could detect changes in the mass of an ice sheet from year to year. Both Greenland and West Antarctica were in fact still losing substantial amounts of ice into the oceans. Observers were dismayed to see mass around the margins of Greenland dwindling at a rate that doubled in less than a decade.[29]

All the scientific data pointed to alarmingly steady worldwide increases in air temperature that could only be classified as global warming. The climate scientists all knew it, although many chose to hold to their beliefs in the scientific method, stating only that anthropogenic activities appear to be causing global temperatures to rise. They continued to couch their remarks in terms of *ninety-five percent confidence intervals* and *likely statistical trending*. Trained scientists, they judiciously avoided using words like *definite* and *certainty* to describe their findings. This was, of course, exactly what climate skeptics needed to hear to keep their "merchants of doubt" train chugging aggressively along the wrong tracks.

BILL McKIBBEN'S STEP IT UP CAMPAIGN AND THE BIRTH OF 350.ORG

In 2007, renowned climate-change activist Bill McKibben started a nationwide environmental campaign that he called "Step It Up."[30] Step It Up's primary goal was convincing the US Congress to take action on climate change. On April 14, 2007, McKibben organized hundreds of rallies in cities and towns all across America. The battle cry of Step It Up was "Curb Carbon Emissions by 80 Percent by the Year 2050."[31] The Step It Up Campaign spread like wildfire and quickly earned the unified support of a wide variety of environmental, student, and religious groups.[32]

Step It Up was the result of McKibben's realization that spreading the facts about climate change was not enough. For nearly a decade, he had been writing about the loud climate-change alarm being sounded by scientists, hoping that "reason would prevail" and "governments would take care of the problem."[33] McKibben started to realize that "power, not reason, was ascendant,"[34] and if environmentalists wanted change, they were going to have to assemble more than a mass of scientific knowledge—and it had to be more powerful than Big Oil (and its associates). As McKibben writes in *Oil and Honey*, "Environmentalists clearly weren't going to outspend the fossil fuel industry, so we'd need to find other currencies: the currencies of movement. Instead of money, passion; instead of money, numbers; instead of money, creativity."[35]

McKibben and a group of dedicated undergraduates from Middlebury College, in Vermont, where he teaches, formed several protests against fossil fuels, and found supporters in the legions of people who were "deeply concerned about global warming but felt powerless in its face."[36] The result of the Step It Up campaigns was that both Barack Obama and Hillary Clinton, running for the democratic nomination, changed their platforms on climate change. Their new goals reflected Step It Up's aforementioned mission: cut carbon dioxide emissions by 80 percent before 2050.[37]

But McKibben and his team of students knew this was not enough. By 2050, irreversible damages would already be in place. As he writes in *Oil and Honey*, "We'd need to work faster, on a larger scale. NASA's James Hansen, the planet's premier climate scientist, provided us with a number: in January 2008, [Hansen's] team published a paper showing that if the concentration of carbon dioxide in the atmosphere rose above 350 parts per million, we couldn't have a planet 'similar to the one on which civilization developed and to which life on earth is adapted.'"[38] Based on this, McKibben founded 350.org, knowing that the initiatives we needed to move us toward a life-sustaining planet would require a massive, organized, global movement to override the money and power of the fossil-fuel industry. Taking their name from Hansen's dire warning number, McKibben's group of intrepid undergraduates reasoned that Arabic numerals crossed linguistic barriers—making the group accessible on a global scale.[39]

Because of all he has done in devoting a significant portion of his life to climate-change awareness and response, Bill McKibben is featured as one of our premiere climate-change heroes. We will talk more about Mr. McKibben in chapter 24.

GREEN FOR ALL COMES ALIVE

A very significant event that occurred in 2007 was the formation of Green for All, a grassroots organization with the dual goals of promoting sustainable development and environmental justice. Van Jones, founded Green for All to contend with the unprecedented dual danger of the world's economic crisis and the ecological threats of climate change.[40] The solution, in his eyes, is to create a "'green-collar economy'—one that will create good, productive jobs while restoring the health of our planet's living systems."[41] Jones defines a green-collar job as one that supports family, offers long-term career potential, and contributes directly to enhancing or preserving environmental quality.[42]

Mr. Jones suggested that picking up a caulking gun and a clipboard will help lift a big chunk of the economically-stranded people in rural and urban communities.[43] Several workers with caulking guns can help weatherize and retrofit homes for energy efficiency. An energy auditor can find energy-saving opportunities for homeowners and renters with some training and a clipboard.[44] According to Jones, this is a way to catch young black people who are in "economic free fall" due to the loss of blue collar manufacturing jobs.[45] In a column by Thomas Friedman in the *New York Times*, Jones described the potential for green jobs in America:

> "You can't take a building you want to weatherize, put it on a ship to China and then have them do it and send it back," said Mr. Jones. "So we are going to have to put people to work in this country—weatherizing millions of buildings, putting up solar panels, constructing wind farms. Those green-collar jobs can provide a pathway out of poverty for someone who has not gone to college."[46]

This Green for All campaign proved enormously successful. Since its inception in 2007, the organization has raised over $12 million to support its staff and initiatives, has aided in economic recovery by creating over one million environmentally conscious jobs and engaging 175,000 people in the promotion of climate-change solutions.[47] Bolstered by the support of Sustainable South Bronx's Majora Carter, Mr. Jones noted that the key was first to get young people involved on the ground floor of renewable energy businesses. From there, they'll then have the opportunities to become managers in five years and owners in ten years.[48] He concluded poignantly: "The green economy has the power to deliver new sources of work, wealth and health to low-income people—while honoring the Earth. If you can do that, you just wiped out a whole bunch of problems. We can make what is good for poor black kids [also] good for the polar bears and good for the country."[49]

The overriding significance of Green for All in the climate-change arena is two-fold. First, it's another demonstration of how starting at the grassroots level can mobilize people and eventually bring about solutions at state, regional, and then national levels. Secondly, it clearly demonstrates the problems of environmental justice but—instead of just exposing the inequities—presents ways to solve the dilemma for the good of all.

In chapter 28, we will talk more about Mr. Jones's landmark work in dovetailing environmental justice with sustainable design for the creation of a cleaner, healthier climate for everyone.

HANSEN'S UNSAFE CO_2 READING IS SURPASSED

Another major proof of climate change came along in 2008 when the Keeling Project—half a century after initiating measurements at Mauna Loa in Hawaii—conclusively showed that CO_2 levels had risen significantly, increasing from 315 parts per million (ppm) in 1958 to 380 ppm in 2008.[50] Suddenly, the relevance of 350.org's message hit home. Not only had we passed James Hansen's unsafe proclamation on global CO_2 concentrations, we were way beyond it. Also, according to climate scientists, we were headed for multifaceted worldwide environmental disasters—such as inundation of major coastal cities and mega-droughts covering large portions of entire countries—that would leave many parts of Earth unsuitable for human inhabitants. Furthermore, many climate scientists stated that even if all GHG emissions were halted immediately, global warming would still continue for many millennia.[51]

TIM DeCHRISTOPHER'S BOLD ACTIONS LEAD TO FOUNDING OF PEACEFUL UPRISING

On December 19, 2008, University of Utah economics student Tim DeChristopher pulled one of the greatest acts of civil disobedience in the history of the US environmental movement. On that date, DeChristopher attended a Bureau of Land Management (BLM) oil and gas lease auction in Salt Lake City, Utah. Designated "Bidder 70," DeChristopher had no money, but that didn't stop him from submitting the winning bid on thirteen federal land parcels in the vicinity of Arches and Canyonlands National Parks.[52] His bold action ended up saving 22,000 acres of land—worth $1.7 million—from oil-company exploration and development.[53] Weeks after his illegal bid, a federal judge, Ricardo Urbina, essentially agreed with DeChristopher, blocking the sale of all of the land leases, due—in part—to their potential air pollution implications.[54] Of course, DeChristopher ended up spending nearly two years in prison for pulling off this stunt.[55]

Not one to be easily cowed, DeChristopher took things a step further while awaiting trial for his actions. He started a grassroots organization known as "Peaceful Uprising," whose mission emphasizes nonviolent actions aimed at maintaining a sustainable future for the United States and the world.[56] Doing this established DeChristopher's reputation as a charismatic and ingenious climate-justice leader.[57] We will talk more about DeChristopher in chapter 28.

Chapter 21

ONSET OF A SOCIAL REVOLUTION OR JUST BUSINESS AS USUAL? (2008–2016)

PRESIDENT BARACK OBAMA: ENVIRONMENTAL GODSEND?

The 2008 US National Election brought many precedent-setting changes to Capitol Hill when Barack Obama was elected president. In addition to being the first African American to serve in the White House—an astonishing turn of events given our country's muddled history of race relations—Mr. Obama was a progressive environmentalist in every sense of the word. Even before he began campaigning for presidency, he embraced an environmental ethic, stating in October 2007 that "We are not acting as good stewards of God's earth when our bottom line puts the size of our profits before the future of our planet."[1] As president-elect, he immediately jumped onto the 350.org bandwagon, creating the Obama-Biden comprehensive New Energy for America plan, which pledged that 25 percent of energy consumption in the United States would come from renewable sources by 2025 and that greenhouse gasses (GHG) would be reduced by 80 percent by 2050.[2]

President Obama's success in the environmental arena—while not living up to his ambitious campaign promises—was certainly notable in the early years of his presidency. He demonstrated his astuteness in environmental matters—especially the battle against climate change—by stating in his 2010 State of the Union address,

> I know there have been questions about whether we can afford (energy efficiency) changes in a tough economy. I know that there are those who

> disagree with the overwhelming scientific evidence on climate change. But here's the thing—even if you doubt the evidence—providing incentives for energy-efficiency and clean energy are the right thing to do for our future because the nation that leads the clean energy economy will be the nation that leads the global economy. And America must be that nation.[3]

In general, the American public either undervalued or just flat out failed to recognize the supreme environmental efforts and successes being made by the Obama administration. Environmentalists, however, were taking notice. In fact, Carl Pope, national executive director of the Sierra Club, said, "This is by far the best first year on the environment of any president in history, including Teddy Roosevelt. Most presidents have done their best environmental work late in their term. This is a very, very strong opening."[4]

This was high praise indeed for Obama's diligent efforts on the environmental front. In fact, if the Republican-controlled Congress had not been repeatedly sticking to Mitch McConnell's often-stated "single most important thing we want to achieve is for President Obama to be a one-term president,"[5], and thereby opposing anything that did not support Republican legislative goals, we believe that Obama would be celebrated as the most accomplished environmental president to ever serve in the White House.

ICONIC ARCTIC DENIZEN LISTED AS A THREATENED SPECIES

In 2008, the effects of climate change made news headlines when the US Fish and Wildlife Service officially listed the polar bear—Earth's greatest living icon of the frozen barrens of the Far North—as a "threatened species." One of our most revered and fearsome predators—the alpha animal of its territory—was suddenly in danger of disappearing from the wild forever.

The cause of the polar bear's decline was simple. Since the great bear lives in a frozen landscape, with no land beneath the ice for thousands of miles, it depends on this ice for support and hunting platforms.[6] With the warming seas in the Arctic regions, the bear's critical habitat—ice floes—were rapidly becoming fewer and fewer, even vanishing altogether in some places during the summer months. And once the sea ice disappeared altogether, so did the polar bear populations in those areas.[7]

ANTARCTICA'S WILKINS ICE SHELF COLLAPSES

As further evidence of global warming, in March 2008, part of the Wilkins Ice Shelf on the Antarctic Peninsula suddenly collapsed, losing more than 160 square miles.[8] Even though the Southern Hemisphere was entering its winter at this point, which brought extreme frigid temperatures to the region, the ice shelf continued to break up.[9] Then, in April 2009, an ice bridge connecting the Wilkins Ice Shelf to Charcot Island disintegrated completely, leaving the remainder of the ice shelf vulnerable to further collapse.[10]

CYCLONE NARGIS RAVAGES MYANMAR WITH 138,000 FATALITIES

On Friday May 2, 2008, Cyclone Nargis caused the worst natural disaster in the recorded history of Myanmar (formerly the nation of Burma), bringing more evidence of the devastating effects of climate change.[11] When the cyclone roared ashore, it sent a storm surge twenty-five miles up the densely populated Irrawaddy Delta, causing catastrophic destruction and at least 138,000 fatalities. Damage was estimated at over $10 billion, which made it the most damaging cyclone ever recorded in this region.[12]

WALL STREET BANK IMPLOSION UPSTAGES CLIMATE-CHANGE CONCERNS

Wall Street's infamous mortgage banking implosion of 2008 had nothing to do with climate change. However, the resulting global financial crisis—

considered by many economists to have been the worst since the Great Depression of the 1930s—certainly once again took the public's mind off climate change and other environmental concerns.

This crisis—which was essentially created by big US banks playing mortgage roulette with American homeowners—threatened the collapse of many large financial institutions. Saying these banks were too large and important to be allowed to go under—the "too big to fail" logic we heard about so often—the US government bailed out the largest of the banking institutions. Despite this ultimate Big Brother maneuver—performed, of course, at the expense of the American taxpayers—stock markets still tumbled worldwide.

JEREMY GRANTHAM FOUNDS HIS RESEARCH INSTITUTE

On a more positive note, financial wizard Jeremy Grantham founded the Grantham Research Institute on Climate Change and the Environment in 2008. Established within the London School of Economics and Political Science, the institute takes an interdisciplinary approach to climate-change action, "bringing together international expertise on economics, finance, geography, the environment, international development, and political economy."[13] The institute was created to help people throughout the world learn about climate change and research which policies would best deal with climate issues.[14]

We will talk more about Mr. Grantham in chapter 27.

JOE ROMM'S CLIMATE BLOG HITS THE BIG-TIME

The climate section of *ThinkProgress* was started by climate scientist Dr. Joseph J. Romm. Dr. Romm's blog discusses climate and energy and political news related to climate change, and provides responses to the media on climate-change issues. In 2008, *Time* magazine named *ClimateProgress* one of the "Top 15 Green Websites."[15] In 2009, Thomas Friedman, in his column in the *New York Times*, called *ClimateProgress* "the indispensable blog"[16], and in 2010 *Time* included it in a list of the "25 Best Blogs of 2010."[17] The UK's *Guardian* also included *ClimateProgress* in its "Top 50 Twitter Climate Accounts to Follow."[18]

We will talk more about Dr. Romm and his climate-change writing in chapter 24.

THE CLIMATEGATE CONTROVERSY

In November of 2009, an unknown hacker stole and leaked one thousand emails, spanning thirteen years, from one of the world's leading climate-change research centers, the Hadley Climatic Research Unit (CRU) at the University of East Anglia (UEA).[19] Climate skeptics focused on ten to twenty messages between a few scientists, who had made some thoughtless remarks about firing the associate editors of peer-reviewed journals.[20] The skeptics misinterpreted these exchanges to portray these climate scientists as falsifying climate-change data. As Peter Kelemen, geochemistry professor at Columbia University puts it, "alleged problems with a few scientists' behavior do not change the consensus understanding of human-induced, global climate change, which is a robust hypothesis based on well-established observations and inferences."[21] It hardly seems like a coincidence that climate skeptics foisted this email controversy—aka Climategate—on the American people several weeks before COP-15, the UN Climate Change Conference in Copenhagen, Denmark, was scheduled to begin in December 2009. Because of the timing, Michael Mann, one of the accused scientists, said that the release of the emails was nothing more than a smear campaign intended to undermine the proceedings at COP-15.[22]

Climate-change deniers argued that the emails clearly proved that global warming was a scientific hoax and that scientists manipulated climate data and attempted to suppress critics. CRU rejected these notions, by stating that the emails had been taken out of context and only represented an honest exchange of ideas.[23] CRU's assertion is supported by the investigations of six official agencies that cleared the scientists of wrongdoing.[24]

As FactCheck journalist Jess Henig writes in a *Newsweek* article, "Leading scientists are unequivocally reaffirming the consensus on global warming in the wake of 'Climategate.' . . . The American Association for the Advancement of Science released a statement 'reaffirm[ing] the position of its Board of Directors and the leaders of 18 respected organizations, who concluded based on multiple lines of scientific evidence that global climate change caused by human activities is now underway, and it is a growing threat to society.'"[25]

THE WORLD WELCOMES AN EVANGELICAL CRUSADER FOR CLIMATE CHANGE

In 2009, Katharine Hayhoe, a devout evangelical Christian who is married to a church pastor, revealed that she is a staunch supporter of the science behind anthropogenic climate change and the immediate need for the world to do something about it.

Actually, Hayhoe's announcement was not all that surprising to people who knew her as a professor of atmospheric science and director of the Climate Science Center at Texas Tech University. Instead, evangelicals throughout the nation were the ones who were surprised—many of whom steadfastly believed that climate change was just a liberal hoax.

Defying the evangelical Christian stereotype has not been without its bumps for Katharine. In an interview published on the Christian-based website, the *Plough*, Benjamin Dolson writes that the evangelical community "has both embraced and rejected her message. She has endured criticism and even death threats, especially after an appearance on Bill O'Reilly's show and a public falling out with conservative politician Newt Gingrich over a book chapter that he unceremoniously pulled at the last minute."[26]

The book Dolson referred to is Professor Hayhoe's conservative Christian book entitled *A Climate for Change: Global Warming Facts for Faith-Based Decisions*.[27] The book won widespread praise from a rather unusual combination of religious leaders, fact-based scientists, and environmental activists. The book's supporters included the past president of the American Association for the Advancement of Science (AAAS) and the president of the National Wildlife Federation (NWF).[28]

Hayhoe's book certainly engendered some skepticism from conservative Christians such as Gingrich, but it also garnered many accolades. Jonathan Merritt, faith and culture writer and national spokesperson for the Southern Baptist Environment and Climate Initiative, wrote,

> In the midst of one of the most heated debates of my lifetime, there are some voices that offer calm and respectful insight to people of faith. Balancing passion with civility, Hayhoe and Farley speak with such a voice. No matter what you believe about climate change's causes or effects, *A Climate for Change* will challenge you to think critically about both the issue and your responsibility to respond.[29]

While Dean Hirsch, president of World Vision International declared,

> This is a book all Christians should read. It makes the science, the issues, and even the misunderstandings come alive. Christians need to help in the fight against climate change and its effects. Jesus calls Christians to serve "the least of these"—the poor and oppressed of the developing world—who are most affected by climate change. This book is a compelling call to action for any Christian who cares about the issues our world is facing today. And that should be all of us.[30]

After the publication of her book, director James Cameron recruited Professor Hayhoe to narrate and appear in portions of his climate-change documentary, *Years of Living Dangerously*. She has also appeared on talk shows and at scientific conferences across the country.

According to a national poll by the Yale Program on Climate Change Communication, evangelicals who believe that global warming is happening are now at 51 percent,[31] a significant increase from the 34 percent found in a 2008 poll by the Pew Research Center.[32] The impact of outspoken evangelical scientists such as Katharine Hayhoe has no doubt influenced this considerable progress.

We will talk more about Professor Hayhoe's unique dual sphere of influence in chapter 23.

HIGH HOPES IN COPENHAGEN (COP-15): NOT SO FAST!

In December 2009, representatives from 115 countries met in Copenhagen, Denmark, at the Copenhagen Climate Change Conference (COP-15), with the hope of establishing a definitive, worldwide climate-change agreement.[33] Representing governments, intergovernmental and nongovernmental organizations, faith-based communities, UN agencies, and media, more than 40,000 people applied for accreditation (the official authorization required to attend the proceedings).[34]

The Copenhagen Accord that came out of COP-15 included pledges made by many countries to reduce greenhouse gas emissions by whatever they felt they could do. For the first time, China and other major developing countries were also involved. Unfortunately, the accord didn't include a treaty and the pledges were not binding.[35] There are varying accounts of what happened on the final day of COP-15. This is one of the more intriguing—and even humorous versions—that our research turned up: In what has been described as a surprise meeting between the US delegation—including President Obama and Secretary of State Hillary Clinton—and the delegations from China, India, South Africa, and Brazil, the basic terms of the Copenhagen Accord were hammered out on the final day of the conference.[36] Obama was on his way to meet with Wen Jiabao, the Chinese premier, when he was told that the Indian, South African, and Brazilian leaders were in the room as well. This surprised Obama and Clinton, as they had been told these leaders had already left the conference.[37] According to an article in the *Telegraph* by Philip Sherwell, when Obama was informed of this, he responded, "Good," and then strode in saying, "Are you ready for me?"[38] The leaders of the four nations met for eighty minutes and brokered the Copenhagen Accord, which began unraveling the day after it was made.[39] China's representative at the conference insisted that the 80 percent by 2050 target for industrialized countries be taken out of the deal.[40]

In the end, the limitations of the Copenhagen Accord were glaringly evident. It did not set overall emissions targets or deadlines and it was not a legally binding treaty. While all nations had to submit written plans for reducing CO_2 emissions by January 2010, the Copenhagen Accord did not include the committed and detailed policies many envisioned in the months leading up to COP-15. Obama admitted as much when he said, "People are justified in being disappointed about the outcome in Copenhagen."[41]

DEEPWATER HORIZON: FOSSIL-FUEL HAZARDS ARE ON FULL DISPLAY

On the morning of April 20, 2010, a bubble of high-pressure methane gas was accidentally released, causing an explosion on the *Deepwater Horizon* oil platform's exploratory oil-drilling rig.[42] The resulting fire raged on for more than a day until the rig finally sank into the Gulf of Mexico, some forty-one miles off the Louisiana coast. Eleven of the 126 people who had been working on the rig when the explosion occurred were never found and were declared dead.[43] While climate change was not directly responsible for this tragedy, the explosion of the *Deepwater Horizon* (aka the "BP oil disaster") clearly demonstrated the ongoing hazards of our all-consuming reliance on fossil fuels.

The rig's underwater well continued gushing geysers of unrefined crude oil into the Gulf for nearly ninety days, until its pipe was capped in mid-July 2010.[44] The estimated total discharge of oil was 206 million gallons,[45] making the BP oil disaster the largest marine oil spill in US history. The well was finally declared sealed on September 19, 2010, five months after the explosion.[46]

As you might expect, the natural-resource impacts of the BP oil disaster were overwhelming. Thousands upon thousands of animals—including wading birds, shorebirds, marine mammals, and sea turtles—many of them state or federal protected species, died in the aftermath. Thousands of acres of seagrass beds, coastal beaches, tidal marshes, and mangrove swamps were fouled forever. And this was just the damage that could be seen on the surface.[47]

In actuality, we may never know the full impacts the spilled oil—plus that of the estimated 1.8 million gallons of chemical dispersants used in cleaning it up[48]—had on the ocean floor. Many marine experts believe that the long-term damage to the marine life and overall environmental quality of the Gulf of Mexico may not be known for decades, or even centuries.[49]

Of course, the BP oil disaster had dire consequences for the human environment as well. The coastal fisheries industries were hardest hit, and many long-term operators of shrimp boats and crabbing vessels were put out of business.[50] Plus the numerous tourist businesses all along the Gulf Coast—from Texas to Florida—were flat-out body-slammed for years after the disaster occurred.

As of 2015, BP finally agreed to pay out more than $20 billion dollars of reparation money for natural-resource damages, state and local response costs, and individual compensation to Gulf Coast businesses.[51] But for many families who had been making their living for generations from motels, restaurants, and souvenir shops, it was too little, too late. They had to give up, move elsewhere, and start all over again. In fact, the economic costs to people living all along the Gulf Coast states will likely continue to be felt for decades.[52]

While the impacts to the natural and human environments were inestimable, the biggest travesty in the BP oil disaster had to do with the people who could have—and should have—stopped it all from happening in the first place. We are talking about the petroleum engineers and marine geologists who worked for BP and—most of all—the bureaucrats employed by the former Minerals Management Service (MMS).[53] The MMS was the federal agency at the time that was supposed to be watchdogging and safeguarding the marine drilling and fossil-fuel extraction process.[54]

But within hours of the devastating explosion and blowout, it became painfully clear that there was no one—not any petroleum geologists or marine engineers working for BP, not any of the bureaucrats working for the MMS, not any local or state safety experts—who had the slightest idea what to do to control the undersea spew of gushing crude oil.[55] In fact, the response plan they did have in place was embarrassing in places. For example, BP listed Peter Lutz as its wildlife expert—apparently not realizing that he had passed away years before they submitted their plan; the company also listed seals and walruses—species that do not exist in Gulf waters—as species of concern in the case of an oil spill.[56]

So how could this have been allowed to happen? To understand, we need to first take a look at the federal regulatory process under which the *Deepwater Horizon* well was operating. A categorical exclusion under the National Environmental Policy Act (NEPA) is based on the following regulatory language: "category of actions which do not individually or cumulatively have a significant effect on the human environment, and which have been found to have no such effect in procedures adopted by a Federal agency."[57] Environmental attorney Jaclyn Lopez elaborates in her paper *BP's Well Evaded Environmental Review*, "The main purpose of a categorical exclusion is to reduce delay where the agency has identified categories of activities that are too minor to warrant environmental review."[58] As you can derive from this language, a categorical exclusion includes implicit guarantees that nothing bad is going to result from the action being taken. Hence, there is no need to conduct an environmental impact assessment for a categorical exclusion. Believe it or not, the *Deepwater Horizon* well was being drilled under just such a categorical exclusion when it blew up.[59]

For all of the hundreds of Environmental Impact Statements (EISs) that I (Budd) worked on during my career, the primary emphasis was always on first avoiding, then minimizing potential impacts. Finally, those impacts which couldn't be avoided or further minimized were mitigated by taking compensatory actions. All three of these nested phases of environmental impact assessments—avoidance, minimization, and mitigation—involved conducting detailed analyses under NEPA and producing the best contingency plans for each anticipated level of potential impacts. This tightly structured procedure—technically known as "404(B)(1) analysis"—is always supposed to be the basis of every environmental impact evaluation process.

But because of the categorical exclusion granted under NEPA, construction and installation of one of the world's deepest oil wells was proceeding with no tested and approved environmental emergency plan ready to be put into place in the case of the most basic and potentially disastrous type of underwater oil drilling mishap: an accidental underwater blowout.[60] Furthermore, the safety procedures that were put in place were overridden in the interest of expediency to save $500,000 in operating costs.[61] Exacerbating this safety oversight, the safety mechanisms that were in place such as the blowout preventer that was attached to the underwater wellhead evidently did nothing at all, except provide a false sense of security that a blowout like this couldn't happen—until one actually did.[62] That's what made the head-scratching, arm-waving, jerry-rigging chaos that followed the disaster so sickening to watch and impossible to fathom.

While the professional engineering specialists and safety experts dithered and dallied about what to do, the unmitigated plume of crude oil spread across the ocean bottom and sea surface, finally reaching invaluable wetland estuaries and pristine barrier island beaches—coating everything it touched with contaminating layers of crude.

MERCHANTS OF DOUBT PUTS EVERYTHING INTO PERSPECTIVE

Also in 2010, science historians Naomi Oreskes and Erik Conway released their boldly eviscerating book, *Merchants of Doubt*.[63] Later made into a movie, *Merchants of Doubt* exposed the anti-climate-change lobby for exactly what they are—paid "free-market" naysayers posing as professional scientists to line their own pockets with money from the fossil-fuel industry. Remarkably, the same scientists who stated—for the record—that smoking was not detrimental to human health and that the ozone hole didn't exist were now proclaiming to the world that climate change was a liberal-inspired hoax. Their main ploy involved capitalizing on real climate scientists' reluctance to state their findings with one-hundred-percent certainty—using this to suggest doubts that didn't exist.

We will talk more about Professor Oreskes's work and background in chapter 24.

JOSH FOX PRODUCES HIS DOCUMENTARY FILM *GASLAND*

You will recall that in chapter 19 we discussed a process called hydraulic fracturing, or *fracking*, used for finding and harvesting natural gas. We talked about how fracking had somehow been exempted from federal environmental regulations and how this just didn't make sense, given that the process was both damaging to natural resources and hazardous to people.

In 2010, a fellow with a theater background named Josh Fox decided to do more than just sit by and watch when the fracking industry came to him offering $100,000 for the rights to drill on his land in Pennsylvania's Delaware River Basin. Instead of taking the money, he started doing his own research on the fracking process and ended up with the documentary film *Gasland*.[64]

The film is, as the HBO website on *Gasland* explains it, "Part verité road trip, part exposé, part mystery and part showdown, *Gasland* follows director Fox on a 24-state investigation of the environmental effects of hydraulic fracturing. What he uncovers is mind-boggling: tap water so contaminated it can be set on fire right out of the tap; chronically ill residents with similar symptoms in drilling areas across the country; and huge pools of toxic waste that kill livestock and vegetation."[65]

Fox draws a variety of conclusions in *Gasland*, including the fact that the general public remains largely unaware of the environmental hazards of fracking. This is the primarily because state and local environmental agencies do not have the manpower or money to educate people about fracking. They also typically don't have the capability of investigating and regulating the sprawling natural gas industry.[66]

As another aspect of this, if property owners are fortunate enough to receive compensation from the gas companies for fracking on their land they must sign nondisclosure agreements. These documents then keep them from bringing lawsuits or telling others (neighbors, friends, or family) about any negative experiences they may have had with the process.[67]

FUKUSHIMA DAIICHI POWER PLANT DISASTER: ANOTHER NUCLEAR WAKE-UP CALL

On March 11, 2011, a massive earthquake and the resulting tsunami struck Japan's Fukushima Daiichi Nuclear Power Station and gave the world a frightening wakeup call about the dangers of reliance on nuclear power.[68] Not since Ukraine's Chernobyl disaster twenty-five years earlier, in 1986, had the world sat and watched in fear, wondering what would be the ultimate outcome and extent of this radiological tragedy. The Fukushima nuclear disaster proved to us all that it is impossible to build a nuclear reactor that can't be eventually taken out by the overwhelming power of a natural disaster.

When the tsunami hit the Fukushima Plant, it knocked out the backup power systems that were needed to cool the reactors. This caused the nuclear cores containing the radioactive fuel to melt down, resulting in hydrogen explosions and the release of the radioactive material at three of the plant's nuclear reactors.

Communities up to twenty-five miles away were affected by radioactive contamination, forcing the evacuation of up to 100,000 residents. While there were fortunately no fatalities directly linked to short-term radiation exposure, the earthquake and associated tsunami were a horrible tragedy—resulting in more than 15,000 deaths with another 2,500 people missing.[69]

THE KEYSTONE XL PIPELINE: ICON OF THE CLIMATE-CHANGE BATTLE

In 2011, environmentalists began to worry about the Keystone XL Project, noting that, among other issues, "Alberta Tar Sands Crude" is worse for global warming than regular crude because of all the extra energy—and associated GHG emissions—it takes to extract and refine the gunk. In fact, NASA scientist James Hansen warned that burning all of the oil in these vast tar sands deposits would mean "game over for the climate."[70] Since President Obama could make the final decision about the Keystone XL project, the proposed pipeline brought climate-change activists together to apply pressure on him, reminding him of his promise to address global warming.[71]

In the early 2000s, crude prices were rising rapidly, and fossil-fuel companies started searching for ways to bolster their reserves.[72] In part, this meant investigating the huge swaths of tar sands—totaling almost

142,000 square miles (about the size of the state of Iowa)—that lay under the forests of Alberta, Canada.[73]

These tar sand deposits are located in northeastern Alberta, roughly centered on the boomtown of Fort McMurray. *Esquire* magazine was hardly exaggerating when it called Fort McMurray "the little town that might just destroy the world."[74]

The problems with accessing this underground larder of crude-containing bitumen were numerous—that's why it was still lying buried in the ground. The first challenge came in trying to figure out how to actually extract petroleum products in this remote area of boreal forests and muskeg (peat bogs). The tar sands are thick and sticky, a type of bitumen with the consistency of peanut butter. To get an idea of exactly what is being dealt with here, consider Oxford Dictionary's definition of bitumen: "A black viscous mixture of hydrocarbons obtained naturally or as a residue from petroleum distillation. It is used for road surfacing and roofing."[75]

The end result of tar sand processing is oil similar to that pumped from conventional oil wells. But because the tar sands are such a gooey mix of bitumen, sand, clay, and water, extracting oil from tar sands is much more complicated and difficult. Because the mixture is too thick to pump, miners use open pit mining techniques to get at the tar sand deposits.[76] Workers use hot water and agitation to separate the bitumen from the rest of the mix. It requires two tons of tar sands to make one barrel of oil (42 US gallons) using this method. For bitumen buried too deeply for pit mining, workers use steam injection, solvent injection, and firefloods (injecting and burning oxygen) to thin the oil enough to pump it out of the ground. These methods can use vast amounts of water and energy.

Environmental problems notwithstanding, this is a totally nonsensical approach to solving our energy problem. We use huge amounts of energy to extract and process the tar sands so we can use the oil for energy. Moreover, because the tar sand is so viscous, the extracted bitumen also requires additional processing before it can be refined. Before pipelines are able to transport it, it must be diluted with lighter hydrocarbons (adding even more fossil fuels).

Once the crude oil has been extracted from the tar sands and processed, companies need to figure out how to move it to refineries (most of them located great distances away) where it can be turned into usable fuel. In 2005, TransCanada—a Canadian energy company—planned the Keystone pipeline system project, which would transport the processed product from Alberta to refineries in Illinois and Texas. Most of the system was built without many people noticing.[77]

The final phase of the project involved a pipeline that would stretch down from Hardisty, Alberta, to Steele City, Nebraska—a distance of almost 1,200 miles at a total cost of $8 billion. As proposed, this pipeline would eventually transport 830,000 barrels of crude oil per day to refineries in Port Arthur, Texas.[78] This final portion is the Keystone XL project that has been the subject of so much controversy and debate during the Obama administration.[79]

In 2008, TransCanada applied for a permit from the US State Department to run the Keystone XL pipeline across the US-Canada border and down to Nebraska. Such permits had been granted for the earlier phases of the pipeline project, by the George W. Bush administration.[80] The review process included both a national interest assessment from the State Department and an Environmental Impact Statement (EIS) under the National Environmental Policy Act (NEPA).[81] In 2010, the US EPA, looking through the preliminary EIS, noted serious concerns. For one thing, the pipeline was intended to cut directly through the Sand Hills in Nebraska, an ecologically sensitive area that had been declared a National Natural Landmark in 1984. That area also contained the Ogallala aquifer, the drinking water source for millions of people. Even when the route of the pipeline had been adjusted to avoid the Sand Hills, though, the EPA still felt that Keystone XL would increase greenhouse gases and would unnecessarily increase the risk of a serious oil spill.[82]

The battle over the Keystone XL pipeline went on for years. While environmentalists fought against the pipeline, climate-change deniers accused Obama of, yet again, trying to hold back the energy industry and impede progress.[83]

Then, on November 6, 2015, President Obama finally rejected the Keystone XL project. Secretary of State John Kerry stated that the project "is not in the country's national security interest,"[84] and Obama agreed, saying, "America is now a global leader when it comes to taking serious action to fight climate change, and frankly, approving this project would have undercut that leadership."[85]

THE SOLUTIONS PROJECT: DOES IT HAVE ALL THE ANSWERS?

Founded in 2011 by Stanford university professor Mark Z. Jacobson, banker Marco Krapels, documentary filmmaker Josh Fox, and Hollywood actor Mark Ruffalo, the Solutions Project is a nonprofit organization that has the mission of helping the world move toward using one hundred percent renewable energy.[86] The Solutions Project defines renewable energy as wind power, water (hydroelectric, wave/tidal, and geothermal) power, and solar power—conveniently abbreviated as WWS.[87]

The Solutions Project has looked at the research and believes that America should be able to transition to using one hundred percent renewable energy by the year 2050.[88] They believe that we already have the technology to make this happen and that the real hurdles are primarily social and political.

In their "50 States-50 Plans Initiative," the Solutions Project has developed a plan for each of the United States outlining the best combination of renewable energy types for that state, taking into account its climate and geography.[89] Not only would this plan be environmentally sustainable, but there would be other benefits for society as well. Adopting one hundred percent renewable energy would mean savings in energy costs, savings in health costs, and the creation of long-term jobs. The Solutions Project has created infographics for each of the United States and for 139 countries, showing the theoretically best renewable energy combinations, but also the financial, health, and job-creation benefits.

For more details about two of the founders of this landmark organization—Mark Jacobson and Mark Ruffalo—see chapters 23 and 26, respectively.

IS CLIMATE CHANGE LINKED TO SEVERE WEATHER EVENTS?

Throughout 2011 and 2012, the debate about just how much human activity was affecting worldwide weather events continued to intensify. In general, climate scientists did not disagree that human activity was mostly to blame for the rising global air and ocean temperatures, but they did not all agree as to whether climate change is responsible for extreme weather events[90]

The editors of *Nature* suggested that, "Better [computer] models are needed before exceptional events can be reliably linked to global warming."[91] But the American Meteorological Society reported in their *Bulletin* that "approximately half of the analyses [of various kinds of weather disasters] found some evidence that anthropogenic climate change was a contributing factor to the extreme event examined."[92]

When the online magazine *Yale Environment 360* asked eight climate experts about the effect of climate change on severe weather, here's what they found:

> [The climate experts] responses varied, with some contending that rising temperatures already are creating more tempestuous weather and others saying that more extreme weather may be likely but that not enough data yet exists to discern a trend in that direction. Scientists in both camps said two physical phenomena—warmer air holds more moisture, and higher temperatures exacerbate naturally occurring heat waves—would almost by definition mean more extremes. But some argued that the growing human toll from hurricanes, tornadoes, floods, and heat waves is primarily related to burgeoning human population and the related degradation of the environment.[93]

Kevin Trenberth, a senior scientist in the Climate Analysis Section of the US National Center for Atmospheric Research (NCAR), however, was more definite:

> Yes, undoubtedly. The environment in which all storms form has changed owing to human activities. Global warming has increased temperatures and directly related to that is an increase in the water-

> holding [capacity] of the atmosphere. Over the ocean, where there are no water limitations, observations confirm that the amount of water vapor in the atmosphere has increased by about 4 percent, consistent with a 1 degree [Fahrenheit] warming of sea surface temperatures since about the 1970s. The human component does not change much from year to year and affects all storms.[94]

A SURREAL HEAT WAVE

In the United States, the year 2012 also brought a "surreal March heat wave" and a severe drought to the Midwest.[95] Several states in the corn belt—from Kansas to the Ohio Valley and Nebraska through Iowa and up to southern Wisconsin—suffered from weeks of drought and days of over 100-degree temperatures.[96] Their corn crops were demolished, which precipitated a corn crisis, or more accurately a corn disaster, that year. Plus, it turned out to be the hottest year ever recorded in the contiguous United States. Usually, records are broken by fractions of a degree. However, the average temperature for 2012 was a full degree hotter than in 1998, which had previously been the hottest year on record.[97]

Guy Walton, a meteorologist with the Weather Channel, documented 34,008 daily high temperature records set at weather stations throughout the United States in 2012, as opposed to only 6,664 record lows.[98] In the past, the number of record highs and lows was pretty much in balance, but since the 1970s the ratio has been tilting toward the high-temperature records. The disparity in 2012 was greater than ever before, however. Jake Crouch, a scientist with the National Climate Data Center in Asheville, North Carolina, told the *New York Times*, "The heat was remarkable. . . . It was prolonged. That we beat the record by one degree is quite a big deal."[99]

The temperatures in 2012 were not an anomaly. In fact, 2015 now ranks as the second warmest year, following 2012, in the 120 years (1895–2015) of recorded US weather history; 2015 marks the nineteenth consecutive year of above-average annual temperatures in the contiguous United States.[100]

GREENLAND'S ICE SHEET: DISAPPEARING FAST

Writing for Britain's *Daily Mail* in July 2012, Rob Preece reported that "Scientists have raised fears of rising sea levels after pictures showed Greenland's ice sheet has melted over a larger area than at any time in more than 30 years of satellite observations. Almost the entire ice cover of Greenland has experienced some melting at its surface during this month, according to data collected by three independent satellites."[101]

This information came as a shock. Typically, only about half of the Greenland ice sheet melts during the summer.[102] Dr. Son Nghiem, was analyzing satellite data at NASA's Jet Propulsion Laboratory when he noticed this extreme melting. Not sure whether to believe what he was seeing, Dr. Nghiem rushed to check transmissions from other satellites as well. Afterward, he remarked, "This was so extraordinary that at first I questioned the result—was this real or was it due to a data error?"[103]

In fact, scientists later documented that the melting had spread with remarkable speed—impacting about 40 percent of the ice sheet on July 8, 2012, and about 97 percent only four days later.[104] Quite fittingly, *Chasing Ice*—the Academy Award–winning documentary that we discussed earlier was released a few months later, providing irrefutable visual evidence that the Greenland Ice Sheet was rapidly wasting away.

DIVEST HARVARD FINDS A BEGINNING

In 2012, Chloe Maxmin and a small group of other student leaders at Harvard University created Divest Harvard. The organization's goal was to convince Harvard to stop investing in the fossil-fuel industry—to, rather, divest—get rid of—those income-generating stocks and bonds that supported fossil-fuel use, and put their money instead into industries that would be better for the planet. They hoped their campaign would teach the Harvard community and the public about climate change. Furthermore, in an interview Chloe described how divestment is an entry point in which everyone can participate: "With a movement that is trying to reach everyone, you need something that everyone can relate with, or connect to. Everyone is part of an institution that has something to divest—a tactic that finally matches the scope of the problem."[105] They hoped, too, to spread this idea beyond Harvard and help change the way people everywhere thought about fossil fuels.[106]

Though Divest Harvard did not start the divestment movement, it was one of the earliest adopters in an initiative that has taken off since 2012.[107] The fossil-fuel divestment movement is now global, and, according to the organization Fossil Free, more than 550 institutions and organizations are participating.[108] Along with these institutions, divestment has also been endorsed by many prominent people worldwide, including UN Secretary General Ban Ki-moon, climate change chief Christiana Figueres, and the World Bank's president Jim Yong Kim.[109]

A REPUBLICAN CONGRESSMAN MAKES A CLIMATE-CHANGE SACRIFICE

Robert "Bob" Inglis served as a US Republican congressman from South Carolina from 1993 to 1998 and then again from 2005 to 2009. At first, Congressman Inglis toed the party line and was a steadfast climate-change denier. But in 2004 when his son asked him to do better on environmental issues, Inglis set about studying everything he could find on climate change. His final assessment: Saving his soul—in terms of protecting the quality of life of his five children—was much more important than winning elections.[110]

In 2006, Mr. Inglis traveled to Antarctica twice to see first-hand the melting that he had heard was occurring there. What he learned and saw about climate change during this period eventually led him to propose his Raise Wages, Cut Carbon Act of 2009. According to GovTrack.us, Inglis's bill was designed to tilt the US taxation system away from income, specifically the Social Security payroll tax, to one that taxes carbon outputs.[111] Inglis's rationale was that, while the science on climate change is clear the economics are even clearer: "It is a no-brainer: Change what you tax. . . . Get off of income, get on emissions. You can't find a member of Congress that disagrees with that. Go look for one. The biggest subsidy of all is being able to dump into the trash dump of the sky without paying a tipping fee."[112] Unfortunately, Congressman Inglis's brave attempt at implementing a carbon tax died in Congress.

Plus, after publicly announcing his newfound belief in climate change and the imminent need for finding solutions, Inglis lost his bid for reelection in a 2010 Republican primary. But instead of regretting his choice to follow his heart, Inglis continued to advance his backing of the need for climate-change action by founding the Energy and Enterprise Initiative (E&EI) in 2012.[113]

E&EI is an "educational outreach that lives to demonstrate the power of accountable free enterprise."[114] They believe that climate change can be solved by eliminating subsidies, especially the lack of accountability for emissions—an implicit subsidy. "By creating a level playing field in which all costs are transparently 'in' on all fuels," states E&EI, "[We believe] that the free enterprise system will deliver innovation faster than government regulations could ever imagine."[115]

Inglis appears in both the film *Merchants of Doubt* and in the Showtime series *Years of Living Dangerously* (episodes 3 and 4). For his work on climate change in the face of fierce political opposition from within his own party, Inglis was given the 2015 John F. Kennedy Profile in Courage Award.[116]

We will talk more about Bob Inglis's career in chapter 25.

SUPERSTORM SANDY DEVASTATES THE NORTHEAST

On October 29, 2012, Superstorm Sandy made landfall in New York and New Jersey. It killed more than one hundred people, uprooted thousands, and caused an estimated $65 billion in damages. A 2013 study by NOAA found that "Climate change related increases in sea level have nearly doubled today's annual probability of a Sandy-level flood recurrence as compared to 1950."[117] The fear has become that with the way our CO_2 emissions are trending, the areas hit by Sandy could soon face similar storm surges every year.

When Joe Romm of *ThinkProgress: Climate* talked to Dr. Jennifer

Francis of Rutgers University's Institute of Marine and Coastal Sciences about climate change and Sandy, she said,

> I think the case has strengthened. I've done a bit more research into the linkage with the very warm Arctic following the record 2012 ice loss, and it appears that the heat released from the Arctic Ocean in the fall created a substantial positive anomaly in the upper-level atmospheric heights in the North Atlantic. This likely contributed to the strong ridge and blocking high that existed when Sandy came along, and that ultimately not only steered Sandy westward but also set up the strong pressure gradient between Sandy and the blocking high that caused the enormous expanse of tropical-storm-force winds from Delaware to Nova Scotia.[118]

THE DOOMSDAY CLOCK: TICKING CLOSER TO MIDNIGHT

The "Doomsday Clock" is a symbolic clock face representing a countdown to a possible global catastrophe. Since 1945, members of the Science and Security Board of the *Bulletin of the Atomic Scientists* have maintained the clock's setting.[119] As a double check on their rationale, the *Bulletin*'s scientists are advised by the Governing Board and the Board of Sponsors, which includes fifteen Nobel laureates. The closer the Doomsday Clock is to midnight, the closer these scientists believe the world is to global disaster.[120]

Originally, the Doomsday Clock represented an image of how close we were to global nuclear war. Since 2007, it now includes an assessment of the threat climate change poses to humanity on a worldwide scale.[121] In other words, the atomic scientists and engineers at the *Bulletin* view climate change as a significant threat on par with nuclear war.

Appropriately, the Doomsday Clock was moved forward in 2015—to three minutes to midnight—partly because of "unchecked climate change." As described on the *Bulletin*'s website, "Despite some modestly positive developments in the climate change arena, current efforts are entirely insufficient to prevent a catastrophic warming of Earth."[122]

OIL-BY-RAIL: A VERY BAD IDEA

Several major rail accidents in 2013 finally began to get the attention of reporters, environmental groups, and community organizations. As Jason Mark of *Earth Island Journal* states, they started to realize that "shipping oil by rail is 1) a growing practice that 2) poses a real threat to public safety and 3) is frightfully under-regulated."[123]

In fact—as we discussed earlier—one of the top 2013 environmental stories involved a disastrous oil train explosion in July in the village of Lac Megantic in Quebec, Canada, which killed forty-seven people and wiped out half the town center.[124] Then, in November 2013, a train carrying 2.7 million gallons of crude oil exploded near Aliceville, Alabama.[125] A train collision in Casselton, North Dakota, spilled 400,000 gallons of petroleum a month later. Lastly, an oil train derailed and caught fire in Lynchburg, Virginia, on April 30, 2014, forcing the evacuation of three hundred people.[126]

In the wake of these disasters, federal regulators finally started taking action in 2014. The US Department of Transportation recommended new rules to regulate shipping crude by rail—a move that even Republicans seemed to like. But shipping oil by rail is convenient, and therefore regulating it is controversial. It seems likely that the issue will continue to be discussed for some time.

CLIMATE-CHANGE RALLIES ARE GROWING IN SIZE

Led by people like activist and founder of 350.org Bill McKibben, climate rallies are growing in size and power. In Washington, DC, on February 17, 2013, 35,000 people gathered for the Forward on Climate rally, organized by 350.org, the Sierra Club, the Hip Hop Caucus, and the National Resources Defense Council, among others.[127] The group of activists headed to the White House to demand that President Obama take immediate action to contain climate change. People from more than thirty states had come together at the rally to protest the effects climate change was having on their health and their homes and land. They were joined by students, scientists, indigenous community members, and many others, in this largest climate rally in US history.[128] "For 25 years our government has basically ignored the climate crisis," Bill McKibben said; "now people in large numbers are finally demanding they get to work. We shouldn't have to be here—science should have decided our course long ago. But it takes a movement to stand up to all that money."[129]

Two of the focal points at the rally were convincing the president to oppose the Keystone XL pipeline project and gaining his support for controlling emissions from the nation's dirtiest power plants.[130] The participants hoped he would embrace the message from his 2013 State of the Union address, where he had said, "For the sake of our children and our future, we must do more to combat climate change."[131]

"Twenty years from now on President's Day," executive director of the Sierra Club Michael Brune said, "people will want to know what the president did in the face of rising sea levels, record droughts, and furious storms brought on by climate disruption. . . . President Obama holds in his hand a pen and has the power to deliver on his promise of hope for our children."[132]

KOLBERT'S *THE SIXTH EXTINCTION* POINTS A FINGER

In January 2014, Elizabeth Kolbert, staff writer for the *New Yorker* and Williams College professor, published her book *The Sixth Extinction*,[133] which issued a clarion call for the protection of biodiversity. As background, Kolbert explains that the dinosaurs were killed during the Fifth Extinction—which scientists suspect was caused by an asteroid. But today, we are living through an epoch that many scientists describe as the "Sixth Extinction," and this time, human activity—through our use of fossil fuels, causing climate change—is the culprit. As one scientist put it, "We're now the asteroid."[134]

"We are effectively undoing the beauty and the variety and the richness of the world which has taken tens of millions of years to reach," Kolbert told Terry Gross, from NPR's *Fresh Air*. "We're doing, it's often said, a massive experiment on the planet, and we really don't know what the end point is going to be."[135]

We have provided details on Ms. Kolbert—another climate-change hero—and her career in chapter 24.

CALIFORNIA DROUGHT BURNS ON

In 2014, the drought in California had been going on for four years and showed no signs of ending. This is especially important since nearly half of the country's fruits, nuts, and vegetables come from California, along with most of our dairy products. Without a doubt, the state's ongoing lack of water is affecting us all.[136]

The drought is undoubtedly related to climate change. A report published in a supplement to the *Bulletin of the American Meteorological Society* on September 29, 2014, described how a team led by Stanford climate scientist Noah Diffenbaugh used computer simulations and statistical techniques to study the links between atmospheric pressure, the path of storms, and modern greenhouse gas concentrations.[137] Their results? "Our research finds that extreme atmospheric high pressure in this region—which is strongly linked to unusually low precipitation in California—is much more likely to occur today than prior to the human emission of greenhouse gases that began during the Industrial Revolution in the 1800s," said Diffenbaugh.[138] The drought in California shows us what might happen in more areas of the American West, as it continues to get even hotter and drier. It gives us a really good idea of exactly how vulnerable our agriculture is to climate change.[139]

Although somewhat improved, California's drought is still ongoing in 2016. According to the Weather Channel's Tom Moore, "As of the start of the water year (Sept. 29, 2015), a whopping 92 percent of the state was at least in a severe drought and 46 percent was in exceptional drought status, according to the U.S. Drought Monitor. As of late April 2016, 74 percent of the state was in at least a severe drought with 21 percent in an exceptional drought. These numbers are an improvement, but the state remains in a precarious situation."[140]

WORLD WILDLIFE DISAPPEARING AT AN ALARMING RATE

In September 2014, the World Wildlife Fund (WWF) announced that, between 1970 and 2010, there has been a 52 percent reduction in the overall populations of birds, reptiles, amphibians, and fish. The WWF's Jon Hoekstra was clear: "There is a lot of data in this report and it can seem very overwhelming and complex. . . . What's not complicated are the clear trends we're seeing—39 percent of terrestrial wildlife gone, 39 percent of marine wildlife gone, 76 percent of freshwater wildlife gone—just in the past 40 years."[141]

Human populations have more than doubled in the time wildlife populations have been halved. Our numbers have grown from 3.7 billion in 1970 to more than 7.4 billion in 2016.[142] Humans are multiplying more quickly than Earth can handle, and there isn't as much room or as many resources left for the other inhabitants.[143]

This situation is really a shame. By applying just a little forethought to any land development project, the loss of wildlife habitat can be significantly minimized. Instead of immediately clearing and leveling the land, we must begin by determining where the most valuable habitats are on a piece of property. Then we can design the development to avoid disturbing the best wildlife habitats—which typically also harbor the best vegetative plots (e.g., a stand of old growth forest). Many times, preserving the best areas of both natural vegetation and wildlife habitat on a piece of property will actually enhance the overall marketability of the development. The developer ends up with a more money in his/her pocket, while the homebuyers are happier living in a landscape where some of the natural features have been protected.

RENOWNED DIRECTOR COMMITS TO THE CLIMATE-CHANGE BATTLE

In 2014, after two years of trying, Oscar-winning director and life-long environmental activist James Cameron convinced Hollywood to produce *Years of Living Dangerously* (*YLD*), his indictment of the climate-change crisis. With nine episodes, each narrated by a celebrity personality, *YLD* is at once both hammer-fisted and poignant. It's impossible to watch any one of the episodes without jumping up from the couch and shouting, "I'm going to take action on climate change and I'm going to do it right now!"

We will talk more about Cameron's background in environmental activism in chapter 26.

FOX NETWORK LAUNCHES *COSMOS* REMAKE FEATURING NEIL DEGRASSE TYSON

In 1980, PBS released its thirteen-episode series entitled *Cosmos: A Personal Voyage* and struck televised gold. Narrated by astrophysicist and science communicator Carl Sagan, the original *Cosmos* set a milestone for scientific documentaries, examining such enigmas as the origin of life and humankind's place in the universe. Along the way, *Cosmos* became the most popular science series in the United States, and, to date, more than 500 million people in sixty countries have viewed the original *Cosmos*.[144]

In 2014, Seth McFarlane, successful animator and creator of the *Family Guy* television series, convinced the Fox Network to produce a *Cosmos* remake with world-renowned astrophysicist and equally eminent Dr. Neil deGrasse Tyson assuming Dr. Sagan's narrator role. Watching Dr. Tyson take complex scientific principles and spin them down for everyone to understand is indeed something to behold. He is exactly the type of person who will be most effective at explaining the complexities of climate change to the American public and then getting them to understand the importance of taking immediate action.

In chapter 26, we have provided details about Dr. Tyson's incredibly varied and accomplished background.

PEOPLE'S CLIMATE MARCH IN NYC: ANOTHER WORLD RECORD

The organizers were not expecting the response they received for the September 21, 2014, People's Climate March in New York City. The march was held just two days before a landmark summit on climate change convened in New York City. The goal was to call attention and demand action to end the climate crisis—and it succeeded beyond expectations. Over 1,500 environmental and social justice groups, including Bill McKibben's 350.org, joined together to plan the march, hoping to attract a minimum of 100,000 participants, which would make it the largest climate demonstration in history. In fact, more than four times that many people actually turned out.[145]

Jason Mark, writing for the *Earth Island Journal* wrote, "While the sheer size of the march was clearly important, the diversity of the participants was even more so. There's a persistent and pernicious assumption among political observers that only white, affluent, college-educated people care about the environment and climate change. The New York demonstration—along with other marches in cities and towns worldwide—revealed what a distortion these thoughts are."[146]

Reflecting the ethnic, religious, and age diversity of New York City itself, the march captured the attention of global leaders.[147] In fact, there were marches and demonstrations held at the same time in 162 other countries around the world.[148] In a speech a few days later to the UN General Assembly, Obama said, "Our citizens keep marching. We cannot pretend we do not hear them. We have to answer the call."[149]

UN CLIMATE SUMMIT IN NEW YORK CITY

The climate march was a perfect prelude to the United Nations Climate Summit in New York City. UN Secretary-General Ban Ki-moon invited global leaders and participants from business, finance, civil society, and local communities to come together in the Big Apple on September 23, 2014, for the UN Climate Summit. This Climate Summit was not officially part of the UN Framework Convention on Climate Change (UNFCCC) meetings, but rather aimed to mobilize political leaders to reach a global climate agreement before the Paris Climate Change Conference (COP-21) that was scheduled to take place in December 2015.[150]

Throughout the 2014 Climate Summit, Rajendra Pachauri, chair of the Intergovernmental Panel on Climate Change (IPCC), emphasized three points: "the unequivocal human role in causing climate change; the urgency to accelerate the pace of action; and the availability of tools to cope with the impacts of climate change."[151]

Former US vice president Al Gore emphasized job creation, innovation, economic growth, and prosperity as some of the many reasons to act on climate change. He concluded his remarks by underscoring the significance of expanding the green bond market and saying that "political will is a renewable resource."[152] (Green bonds are those financial bonds specifically meant to support positive environmental ventures.[153])

Speaking for climate refugees, poet Kathy Jetnil-Kijiner of the Marshall Islands read a poem she had written for her infant daughter, in which she promises to keep fighting for the planet.[154]

Actor Leonardo DiCaprio, who has been named a UN Messenger of Peace for his work fighting climate change, also spoke during the Climate Summit. Noting the need to phase out fossil-fuel subsidies, he underscored that renewable energy is good economic policy and called for "courage and honesty" from global leaders.[155] Mr. DiCaprio then led off his conclusion by saying, "This is not a partisan debate; it is a human one. Clean air and water, and a livable climate are inalienable human rights. And solving this crisis is not a question of politics. It is our moral obligation—if, admittedly, a daunting one. We only get one planet. Humankind must become accountable on a massive scale for the wanton destruction of our collective home. Protecting our future on this planet depends on the conscious evolution of our species."[156]

We talk more about Mr. DiCaprio's passionate involvement in the climate-change movement in chapter 26.

NAOMI KLEIN'S BESTSELLER CALLS FOR GLOBAL ECONOMIC AND SOCIAL RESTRUCTURING

In her 2014 book, *This Changes Everything*,[157] Naomi Klein argues that our capitalistic system is broken and failing us fatally. Trade agreements such as NAFTA and the Trans-Pacific Partnership (TPP) willfully erase the climate crisis by allowing trade to trump climate. Ontario, Canada, had a classic example—the province planned to revive its "moribund manufacturing sector" by bringing the solar energy market to the province.[158] Materials and the workforce had to be sourced locally according to the plan created by the province, called the Green Energy and Green Economy Act. The plan resulted in the creation of 31,000 new jobs.[159] In an ironic twist, however, the European Union declared that the buy-local clauses in the Green Energy and Economy Act were a violation of the World Trade Organization rules for trade; and the World Trade Organization agreed.[160] Ontario had to drop its low-carbon, source-local provisions in favor of carbon-intensive overseas sourcing to honor trade agreements. In addition to trade trumping climate, Klein cites the connection between pollution and labor exploitation. The companies that are willing to work laborers to the bone for less than meager wages are equally willing to burn mountains of fossil fuels, yet spend next to nothing on pollution controls.[161]

Klein joins many of our other heroes—Donella Meadows, James Gustave Speth, Bill McKibben, Jeremy Grantham, Naomi Oreskes—in suggesting that we need a post-carbon economy, where our success is not measured by our gross domestic product but by how we care for one another and the planet.[162] She calls it "Growing the caring economy, shrinking the careless one,"[163] and emphasizes that we should expand the sectors that are not driven by increased annual profit, such as the public sector, co-ops, local businesses, and nonprofits, as well as the caregiving professions, which are generally occupied by women and people of color.[164] Furthermore, Klein suggests shorter work hours—which will create more jobs, but will also provide people more time to engage in low-carbon, low-consumption activities such as gardening and cooking.[165] With suggestions such as these, Klein meticulously lays out a plan for massively reducing our GHG emissions in order "to simultaneously reduce gaping inequalities, re-imagine our broken democracies, and rebuild our gutted local economies."[166]

See chapter 24 for more information on Ms. Klein's background and career.

AN IMPORTANT MIDTERM ELECTION YEAR TURNS SOUR FOR CLIMATE CHANGE

The midterm elections of November 2014 were important for environmentalists. Tom Steyer, a businessman and philanthropist founded the political action committee (PAC) NextGen Climate to help get environmentalist politicians elected and defeat climate deniers.[167] Focusing on seven key races, NextGen Climate spent tens of millions of dollars in Maine, New Hampshire, Florida, Colorado, Michigan, and Iowa.

NextGen Climate wasn't able to stir up the enthusiasm they wanted, however. Most of the money raised was Tom Steyer's own, and only three of the seven candidates they backed won their races.[168] In addition, Senator James Inhofe of Oklahoma—a staunch climate-change denier—was elected to chair the powerful Senate Environment and Public Works Committee.[169] This all amounted to a very bad sign for maintaining the growing national momentum of the climate-change movement.

Even though the election results were a setback, there was good news as well. A voter survey released by NextGen Climate just before the election showed that the vast majority of younger voters recognized that climate change is real, and they wanted the federal government to take action to deal with GHG emissions. Stan Greenberg, a prominent political strategist, explained that, "This issue matters for Millennials. It is a defining issue, and leaders that deny or decline to act will pay a serious price politically."[170]

Since Tom Steyer and his NextGen Climate PAC play such a prominent role in pushing for the election of legislators who will be willing to take the reins and initiate real changes to address climate change, we are featuring him as one of our climate-change heroes and we will talk more about him in chapter 27.

WEATHER RECORDS CONTINUE TO HEAT UP

The year 2015 became the hottest year on record, with California, Idaho, Oregon, Utah, and Washington all experiencing their hottest Junes and

Florida recording its hottest spring months.[171] But that was only the beginning for this epic year. During 2015, regions of the United States experienced record-breaking weather events nearly every month. Here are some examples:

- The snowpack in California hit a record low in April, with essentially no snow in the Sierra Nevada Mountains. The previous low was 25 percent of the long-term average snowpack, which happened twice—once in 1977 and again in 2014. The snowpack in 2015 was a staggering 6 percent of the long-term average for that time of year.[172] When I (Mariah) visited California that year, the daily news headlines from the *San Francisco Chronicle* were a wake-up call—suggesting that the drought was California's new state of being.[173] Governor Jerry Brown placed mandatory, statewide water cutbacks for the first time in history, and they continue in 2016, with some temporary water restrictions becoming permanent.[174]
- While the drought withered California, much of the rest of the country was getting drenched with water in May. According to the National Oceanic and Atmospheric Administration (NOAA), "The May precipitation total for the contiguous U.S. was 4.36 inches, 1.45 inches above average. This was the wettest May on record and the all-time wettest month in 121-years of record-keeping."[175] Texas and Oklahoma took the brunt of this extreme weather, with precipitation averages soaring to more than twice the long-term average; twenty-three people died in the subsequent floods.[176]
- With 110.6 inches of snow falling between July 1, 2014, and June 30, 2015, the city of Boston had its all-time snowiest year. The seventy-five-foot high "snow farm" where plows had dumped the snow and ice they cleared from the roads throughout the winter didn't melt until July 14.[177] And, as it melted, it revealed the trash and debris that had been scooped up along with the snow. During Boston's "snowmageddon," I (Mariah) could not see out of the windows in our house for nearly two weeks, and my husband had to cross-country ski to work due to overburdened plows! As an upside, our kids became expert snow-fort makers.

- Becoming the second-earliest tropical or subtropical storm to make landfall in the United States, Tropical Storm Ana hit South Carolina on May 10, 2015. The only tropical storm to make an earlier landfall was a February 1952 storm in Florida.[178]
- The extremes did not abate come fall. In September, one of the deadliest floods in Utah history occurred when twenty people died during extreme flash floods along the Utah and Arizona border.[179] During October, South Carolina experienced catastrophic and deadly flooding as well. Residents said they had never seen such powerful rainstorms, and at least nineteen people were killed.[180]

G7 SUMMIT STRIKES ACCORD

On June 8, 2015, the G7 nations—Canada, France, Germany, Italy, Japan, the UK, and the United States—agreed to cut GHG by eliminating the use of fossil fuels by the end of this century.[181] Many climate-change activists considered this to be a very significant achievement. While the agreement wasn't binding, it did provide a strong sense of both progress and hope.

Under the leadership of German chancellor Angela Merkel, the G7 summit—held at a Bavarian castle—convinced even the most reluctant countries, Japan and Canada, to commit to the need to "decarbonize the global economy."[182] The seven nations also agreed to limit the rise in average global temperatures to a maximum of 2.0 C (3.6 F) higher than preindustrial levels. In addition, the participating countries agreed to provide $100 billion dollars a year for dealing with the effects of climate-change mitigation, especially in developing nations.[183]

May Boeve, executive director of 350.org, in a statement about the summit, said, "The G7 is sending a signal that the world must move away from fossil fuels, and investors should take notice. . . . If you're still holding onto fossil fuel stocks, you're betting on the past. As today's announcement makes clear, the future belongs to renewables."[184]

THE POPE HIMSELF WEIGHS IN ON CLIMATE CHANGE

When it comes to religious authorities on Earth, it's difficult to go beyond the pontiff of the Roman Catholic Church, which has a membership of 1.25 billion worldwide. That's why it was so astounding when the current Pope Francis—now widely known as the most progressive pontiff in the history of Catholicism—came out on June 18, 2015 and strongly supported environmental consciousness, in particular working against climate change.

Jim Yardley and Laurie Goodstein, in their *New York Times* article, wrote,

> Pope Francis on Thursday called for a radical transformation of politics, economics and individual lifestyles to confront environmental degradation and climate change, blending a biting critique of consumerism and irresponsible development with a plea for swift and unified global action. The vision that Francis outlined in an historic 184-page papal encyclical is sweeping in ambition and scope: He describes relentless exploitation and destruction of the environment and says apathy, the reckless pursuit of profits, excessive faith in technology, and political shortsightedness are to blame. The most vulnerable victims, he declares, are the world's poorest people, who are being dislocated and disregarded.[185]

ENVIRONMENTAL VISIONARY CREATES BATTERY BACKUP FOR RENEWABLE ENERGY STORAGE AND EXPANDS MARKET FOR ELECTRIC CARS

In April 2015, Elon Musk, one of the world's wealthiest men and most innovative thinkers, announced the start-up of Tesla Energy. The company will produce rechargeable, lithium-ion batteries for storing solar energy—Powerwall for home use and Powerpack for businesses and utilities. The battery packs will be manufactured in a $5 billion dollar factory—Gigafactory I—in Storey County, Nevada.[186] As the CEO of Tesla Motors, Musk has also pioneered new technology and greater adoption of electric cars. The company's award-winning Model S vehicle gets nearly three hundred miles from less than five minutes of charging—and goes from 0 to 60 in 2.5 seconds.[187]

We will talk more about Mr. Musk's concepts and ideas for dealing with climate change in chapter 27.

OBAMA'S CLEAN POWER PLAN

On August 3, 2015, President Obama and the EPA announced the Clean Power Plan (CPP). For climate-change activists, the CPP represents an important and historic step in reducing carbon pollution from power plants. Under the CPP, power plants will have to follow stricter regulations and each state must make its own plan to cut pollution. The CPP will, within twenty-five years, lower carbon pollution to 32 percent below 2005 levels.[188] Many experts view the CPP regulations as historic, with Michael Gerrard, director of the Sabin Center for Climate Change Law at Columbia Law School, going so far as to call the regulations the most important ever issued by the United States.[189] The objectives of the CPP are threefold, according to David Graham of the *Atlantic*: 1) to decrease emissions immediately in the United States, 2) to set the precedent for other countries, and 3) to add to Obama's legacy as president.[190]

In regards to the first point, the cuts in allowable pollution and emissions will have an immediate effect on the United States and its goals toward climate action.[191] According to the EPA's website, the Obama administration designed the CPP to "strengthen the fast-growing trend toward cleaner and lower-polluting American energy."[192] Each state will have its own goals, tailored to that state's mix of resources and challenges, with tough, but achievable regulations for power plants.[193] Because coal is the most carbon-intensive fuel, one of the primary benefits of the CPP is that it will likely close many existing coal-fired plants and block new ones.[194]

The CPP illustrates the United States' commitment to staying at the forefront of the world's fight against climate change, as well, which illustrates Graham's second point.[195] President Obama announced the CPP just months before the Paris climate conference (COP-21), as a bold example of the type of American action that will be needed to persuade other countries to act as well.[196]

As for his third point, Graham suggests that the CPP is one of the most essential pieces of Obama's legacy, together with the Affordable Care Act.[197] President Obama feels strongly about fighting climate change. During his speech for the UN Climate Summit in New York City on September 23, 2014, Obama said,

> There's one issue that will define the contours of this century more dramatically than any other, and that is the urgent and growing threat of a changing climate. . . . We cannot condemn our children, and their children, to a future that is beyond their capacity to repair. . . . For I believe, in the words of Dr. King, that there is such a thing as being too late. And for the sake of future generations, our generation must move toward a global compact to confront a changing climate while we still can. This challenge demands our ambition. Our children deserve such ambition. And if we act now, if we can look beyond the swarm of current events and some of the economic challenges and political challenges involved, if we place the air that our children will breathe, and the food that they will eat, and the hopes and dreams of all posterity above our own short-term interests, we may not be too late for them. . . . We can act to see that the century ahead is marked not by conflict, but by cooperation; not by human suffering, but by human progress; and that the world we leave to our children, and our children's children, will be cleaner and healthier, and more prosperous and secure.[198]

Expanding on this point at a White House press conference on August 3, 2015, Obama talked about the call for action Pope Francis had made six weeks before, and also highlighted the moral case for reducing emissions. He rebuked those critics who claimed the CPP would hurt minority and poor communities, stating, "If you care about low-income minority communities, start protecting the air they breathe and stop trying to rob them of their health care."[199]

Of course, the CPP then faced serious challenges on Capitol Hill. Senate Majority Leader Mitch McConnell led the opposition against the plan, backed by a large number of Republican politicians, as well as by energy corporations. McConnell is one of the senators from Kentucky, a major coal producing state.[200]

On October 23, 2015, twenty-four states and a coal company—Murray

Energy—sued the EPA over the CPP, claiming the agency is "going far beyond the authority Congress granted to it."[201] They wanted the Court of Appeals for the District of Columbia Circuit to overturn the CPP and asked that the court "immediately stop its implementation" until the legal questions are decided.[202]

As matters stand now, in 2016, the CPP is in limbo. According to the EPA, "On February 9, 2016, the Supreme Court stayed implementation of the Clean Power Plan pending judicial review. The Court's decision was not on the merits of the rule. EPA firmly believes the Clean Power Plan will be upheld when the merits are considered because the rule rests on strong scientific and legal foundations."[203]

COP-21 TAKES PLACE IN PARIS, FRANCE

We had the pleasure of attending the 21st UNFCCC Conference of the Parties on Climate Change (COP-21), held from November 29 through December 11, 2015, at the Le Bourget Conference Center in the suburbs north of Paris, France.[204] We use the word pleasure to describe our experience at COP-21 for two main reasons.

First, it was remarkable to see firsthand the profound resiliency of Parisians in overcoming and going about their typically enthusiastic lifestyles in the wake of the horrific terrorist attacks, which had occurred just over two weeks prior, leaving 130 people dead and scores more injured. While there was certainly plenty of extra security around, if you somehow had not been aware of what happened on the night of November 13, 2015, you would have felt like the glorious City of Lights was functioning as normal.

Second, the mood within the exhibit halls—attached to the main conference center where the high-level discussions were taking place—was very upbeat, exciting, and spirit lifting. The booths from countries all over the world featured a wealth of information, as well as technology, hands-on demonstrations, colorful speakers, educational brochures, and give-away souvenirs. In almost all cases, the exhibitors were positive, believing absolutely that by working together we can solve climate change.

All around the exhibitor's booths swirled a festival of international cuisine, music, and people wearing the clothing of their native countries. We were especially impressed by a gentleman from South America wearing red body paint and not much else, holding a large blue-and-green globe inscribed with the letters SOS.

Unfortunately, the same glowing review cannot be given for the results of the meeting of more than 190 heads of state or their representatives. Yes, a significant milestone occurred when all of the parties in attendance—for the first time in the two-decade history of the COP process—signed an international treaty (the Paris Agreement). But in the minds of most activists, scientists, and attendees, the contents of this agreement simply did not go anywhere near far enough to make a significant dent in solving climate change worldwide. This was vividly driven home with a mass rally and march—led by many prominent climate-change activists—down the Champs-Elysees to the Arc de Triomphe after the agreement was signed on December 12th.

What went wrong at COP-21? For one thing, the Paris Agreement reaffirmed the goal of limiting global temperature increase to "well below" 2.0 C (3.8 F). This is the ultimate in "waffle language," giving unnecessary wiggle room to allow countries to do the minimum. The maximum worldwide temperature increase really needs to be limited to 1.5 C (2.7 F). As almost very climate scientist in the world will tell you, a 2.0 C—or greater—global temperature increase will have dramatic negative impacts on the quality of life of millions of people living in coastal environments all over the world. To be blunt, we simply can't go there!

The Paris Agreement also contained binding commitments by all parties to "make Nationally Determined Contributions (NDCs)" and "to pursue domestic measures aimed at achieving them."[205] Once again we have classic weasel wording. Exactly what does this statement mean, anyway? Does everyone promise to go home and simply try their best to do better with their emission-reduction targets?

Exactly when will each country's specified NDC be accomplished? Then, to what extent will the NDCs be increased each year? These are the things we need to know—immediately. Only then will we be able to get a handle on how the future of solving worldwide climate change must be pursued.

Each country also committed to submitting revised (ideally increased) NDCs every five years. We believe that countries' NDCs must be conscientiously and honestly upgraded annually and then presented each year for review at future COPs.

On the positive side, there were two items in the Paris Agreement that actually go a long way toward providing a solution to a long-standing global problem: First, the signing countries agreed to "reaffirm the binding obligations of developed countries to support the efforts of developing countries."[206] Then, the agreement also promised to "extend the current goal of mobilizing $100 billion a year in support of developing countries by 2020 through 2025, with a new, higher goal to be set for the period after 2025."

We will revisit the approach of the Conference of the Parties in chapter 31.

THE BREAKTHROUGH ENERGY COALITION IS FORMED

On the opening day of COP-21, Bill Gates, founder of Microsoft and arguably the world's wealthiest man, joined Presidents François Hollande and Barack Obama in introducing their newly formed Breakthrough Energy Coalition (BEC). The members of the BEC are twenty-eight of the world's wealthiest private investors—collectively having assets of $350 billion—who have all committed to finding innovative alternative energy sources. The United States, China, and seventeen other countries have also announced plans to double their spending on energy research during the next five years.[207]

For more information on Mr. Gates's involvement in climate change, see chapter 27.

SENATOR PRESENTS HIS 125TH CONSECUTIVE CLIMATE-CHANGE SPEECH

When it comes to climate-change dedication on the national legislative scene, no one can beat Rhode Island senator Sheldon Whitehouse. On

January 27, 2016, for the 125th time since he took office, Senator Whitehouse prepared and presented a passionate plea for Congress to take action against climate change.[208]

Although most of his presentations are in front of a nearly empty chamber, Senator Whitehouse refuses to give up. A sign beside him during all of his presentations reads, "It's Time to Wake Up!"[209] He believes that because he represents the Ocean State, he has a special onus to fight for Rhode Island's current and future populace. The diligence shown by Senator Whitehouse is exactly what we need to really make a difference on climate-change policy at the federal level.

For more information on Senator Whitehouse and his personal climate-change crusade, see chapter 25.

BEST ACTOR FEATURES CLIMATE CHANGE AT THE OSCARS

During the 2016 Academy Awards, Leonardo DiCaprio finally won a well-deserved Oscar for Best Actor, after five previous nominations. Standing on stage after receiving his statuette, DiCaprio confidently and graciously thanked his fellow actors in *The Revenant*, the directors who guided him along the way, and his parents.

Then he concluded his remarks by talking about the way the changing climate had affected the filming of *The Revenant* and how we need to come together to deal with the problem. "Climate change is real," he said. "It is happening right now. It is the most urgent threat facing our entire species, and we need to work together and stop procrastinating."[210]

While Mr. DiCaprio is a professional actor and not a scientist, he is one of our climate-change heroes for exactly this reason. He never misses an opportunity to use his celebrity status to promote environmental causes, especially one that he believes in with all his heart—climate change. It's exactly this type of boldness from well-known people that will move climate change to the forefront of the public's mind and keep it there until real global transformations are underway.

Chapter 22

WHERE DOES THE FUTURE STAND ON CLIMATE CHANGE?

The US environmental movement has gone through several significant phases—most notably in the past fifty years—from the Radical '60s to the Regulatory '70s, the Backlash '80s, the Complacent '90s, the Hopeless 2000s, and now the Introspective 2010s. From the viewpoint of climate change, we now stand on the precipice of the decision-making process as we move forward into what will hopefully be the Solution 2020s.

In our opinion, effective solutions to climate change will result in one of two ways. One way is based on positive, proactive progress, while the other exists as a reactive response to negative events. Achieving the first of these ways will require us to de-emphasize social and religious infighting among cultures and countries (i.e., focus on human values versus human conflicts). International leaders will also need to come together in agreement and all work together to accomplish results that will provide long-term benefits for our planet.

The major environmental crises currently facing earth's human population—global warming, ozone layer depletion, the sixth extinction, overpopulation—all seem overwhelming to most people. This sense of impending, irresolvable doom breeds, in turn, hopelessness and indifference. The prevailing attitude can be summarized as "Why bother doing anything when it won't make any difference?"

Ridding ourselves of the current shroud of hopelessness and planting a worldwide seed of hopefulness will give us the energy and motivation we need to create the massive groundswells of public support we must have to force world leaders to take immediate, tough actions and find realistic solutions. These might include mandating widespread conversions to renewable energy sources, stopping approvals of new fossil-fuel extractions, and imposing heavy fees on existing fossil-fuel generators.

The other scenario is much less desirable—involving a major cataclysmic event that can be directly associated with climate change. But such an incident might, in the long-term, actually be more effective in stimulating quick results. History teaches us that—unfortunately—significant changes in regulatory policies and political mandates often follow catastrophic events, where years of debate had previously failed. Examples include earthquake-proof building codes after a major quake, updated bridge design after the collapse of a major span, and more intelligent land use and zoning maps after disastrous hurricanes or floods.

We firmly believe that the time is ripe for climate-change solutions that are not precipitated by responses to calamitous events but instead by the hopeful and forceful voice of the masses, speaking together their desire to make a difference for all future generations. In part 5, we discuss our suite of solutions to climate change—which are based collectively on our research for this book coupled with our interviews with the scientists, activists, and celebrities that we have designated as our climate-change heroes.

Part Four

ALL HAIL TODAY'S HEROES

Who They Are and What They Have Done

Chapter 23

SCIENTISTS / RESEARCHERS

JAMES HANSEN: PIONEER / COMMUNICATOR

The most well-known climate-change hero comes down to a three-way tie between Bill McKibben, Al Gore and James Hansen. Dr. Hansen is often called "the world's best-known climate scientist,"[1] though he calls himself a "reticent, Midwestern scientist."[2] He studied physics and astronomy under Professor James Van Allen in the space science program at the University of Iowa.[3]

Since the late 1970s, James Hansen—adjunct professor of Earth and Environmental Sciences at Columbia University's Earth Institute—has focused his research full time on Earth's changing climate. Hansen was head of NASA's Goddard Institute for Space Studies and served as Al Gore's science advisor for *An Inconvenient Truth*. His famous climate-change testimony to a congressional committee on June 23, 1988, helped raise broad awareness of the issue—in *This Changes Everything*, Naomi Klein called Hansen's testimony the "birthday" of the climate movement.[4] Dr. Hansen is consistently recognized for speaking truth to power, for identifying ineffectual policies as "greenwash"—the term for when a company spends more money advertising their environmental commitment than actually acting on it—and for outlining the actions that the public must take to protect the future of young people and other species on the planet.[5]

His Indiana Jones–style hat and winning smile are endearing, but his science and opinions on climate change are fierce. He leaves no room for people to doubt the urgent need for climate-change action: "Imagine a giant asteroid on a direct collision course with Earth. That is the equivalent of what we face now [with climate change], yet we dither."[6]

In 1981, Dr. Hansen published a paper in *Science* magazine concluding that the observed warming in the prior century was consistent with the greenhouse effect of increasing carbon dioxide.[7] The paper further emphasized that this warming would exceed the normal up and down fluctuations of random weather patterns by the end of the century—a prediction that turned out to be true. His research indicated that the twenty-first century would see "shifting climate zones, creation of drought-prone regions in North America and Asia, erosion of ice sheets, rising sea levels and opening of the fabled Northwest Passage."[8] Check, check, and double-check—like a very depressing shopping list, we are slowly crossing off each one of these effects as they occur.

As a communicator, Dr. Hansen uses clear, conceptual analogies to help us understand global warming. He talks about how adding carbon dioxide to the air is like throwing another blanket on the bed—energy is coming in, but it can't get out, creating an energy imbalance.[9] The energy imbalance is enormous: "It's equivalent to exploding 400,000 Hiroshima atomic bombs per day 365 days per year—that is how much extra energy Earth is gaining each day."[10] And this extra energy is what has caused our planet's temperatures to increase and the changes in climate we have seen. But most concerning, says Dr. Hansen, are the positive feedback loops, the effects that cause that change to accelerate—melting ice sheets make the planet darker, causing more melting; warming oceans release more carbon dioxide, just as a Coca-Cola does as it gets warm and flat, causing more warming.[11]

The consequences of this will be devastating, as we have been discussing, and Dr. Hansen ticks them off matter-of-factly: "We will have started a process that is out of humanity's control. Ice sheets would continue to disintegrate for centuries. There would be no stable shoreline. The economic consequences are almost unthinkable. Hundreds of New Orleans-like [Hurricane Katrina] devastation [will occur] around the world. What may be more reprehensible . . . is extermination of species."[12] The beloved monarch butterfly, he tells us, could be one of the 20 to 50 percent of all species ticketed for extinction by the end of the century if we continue with business-as-usual fossil-fuel use.[13]

In his book *Storms of My Grandchildren*, Hansen lays bare the facts on climate change, stating bluntly, "The planet is hurtling even more rapidly than previously acknowledged to a climatic point of no return."[14] In a talk at MIT, Hansen told the audience that, while most policy recommendations focus on limiting greenhouse gasses to no more than 2.0 C (3.6 F), this would still be a "disaster scenario."[15] The amount of sea-level rise caused by ice melt from 2.0 C of warming would render most of the world's coastal cities uninhabitable.

For climate-change solutions, Hansen believes that changes like imposing a rising fee, or tax, on carbon emissions will incentivize entrepreneurs to develop no-carbon and low-carbon energy and products. One hundred percent of this fee would be returned to all legal residents, "equal shares on a per capita basis deposited monthly in bank accounts," offsetting any increases in prices the consumer would see, explains Hansen in a 2008 draft letter to President Obama.[16] Hansen thinks this system will work because the dividend and rising fossil-fuel costs will continually remind the public to reduce their emissions.[17] He claims that "fossil fuels have been artificially cheap because their true costs to society, including pollution and climate change, have been ignored."[18]

Research indicates that a near-universal carbon fee on fossil fuels could create three million new jobs for Americans and reduce US carbon emissions by more than half within twenty years.[19] Even better, several top economists support similar measures, and it sounds very much like a conservative climate plan. Jim DiPeso of Republicans for Environmental Protection describes the plan like this: "Transparent. Market-based. Does not enlarge government. Leaves energy decisions to individual choices. Takes a better-safe-than-sorry approach to throttling back oil dependence and keeping heat-trapping gases out of the atmosphere."[20]

THOMAS LOVEJOY: EXPLORER / ORIGINATOR

On a bright sunny day at the end of August, we arrived at George Mason University to speak with Professor Thomas Lovejoy. His assistant, Carmen, told us without prompting that he was a fantastic boss. This immediate praise from someone who has worked for him for decades, let us know a bit about the character of the man we were to meet. What we were not prepared for was the passion, depth, and breadth of Thomas Lovejoy's knowledge.

When we entered his book-lined office, his warm and welcoming demeanor put us right at ease. We had barely seated ourselves before he launched into an excited story about stromatolites—the fossilized remains of blue-green microorganisms. These microorganisms did themselves in by producing too much waste oxygen but, in the process, generated the oxygen levels in the atmosphere that enabled the rise of higher life forms.[21] Lovejoy compared their demise to what humans are doing with carbon dioxide emissions: "In the process of oxygenating the atmosphere, they made themselves pretty scarce, and they hang on here and there today. And we are doing exactly the same thing, except with carbon dioxide. You don't really need to say any more than that."[22]

Nature will always find a way to recover, but it gives no preference to a species that does itself in with excess and lack of systems thinking.[23] Dr. Lovejoy puts it bluntly: "There have been times in the past when there has been dramatic climate change and it would not have been a great time to be living, especially with a population of seven billion. We have had 10,000 years of unusual stability in the climate system. I don't think anybody really knows why that is, but whatever the story there, those are the conditions that nourished the rise of civilization. To essentially kick the props out from under that is pretty dumb."[24]

Dr. Lovejoy received his BS and PhD in biology at Yale University. Credited in 1980 with coining the now commonly used term "biological diversity," Lovejoy directed the World Wildlife Fund's US program from 1973 to 1987 and was one of the first advisers to *Nature*, the popular long-running public television series.[25] In the 1970s, he started one of the world's largest and longest-running experiments, the Biological Dynamics of Forest Fragments Project (BDFFP). His work in Brazil's Amazon, and specifically at BDFFP, over the past fifty years has been called "the most important ecological experiment ever done."[26]

Dr. Lovejoy makes it very clear that we need action: "The 2.0 C target that the negotiators agreed on is actually is too much. It was never chosen as a meaningful number, it was chosen as a doable number—it would be a world without tropical coral reefs, and we are seeing all kinds of change at the moment like in bark beetle-caused die back in the coniferous forests. We are seeing that at 0.9 C, so 2.0 C is too much."[27] Additionally, the more climate change there is, the harder it is to counter and manage. The last time the planet was 2.0 C warmer, the oceans were thirteen to twenty feet (four to six meters) higher.[28] If we do not stop pumping carbon into the atmosphere now, not only will we have to deal with ending fossil-fuel emissions, concurrently we will be struggling with massive intrusion of seawater. It will make today's problems seem like a cakewalk.

Dr. Lovejoy is one of several scientists who advocates strongly for removing the carbon from the atmosphere while at the same time limiting the amount we are adding. Burning old ecosystems (which is what fossil fuels are) adds carbon to the atmosphere, but the destruction and degradation of modern day ecosystems—for example, via cutting down rainforests, poor soil management practices, or filling wetlands—removes a carbon sink as well. The standard approaches, such as limiting the CO_2 from smokestacks, are, of course, important. But Lovejoy believes that ecological restoration could bring global temperature averages down by at least half of a degree if not more. He says that humans need to recognize that the planet works as a linked biological and physical system. To manage it, we need to manage ourselves in a way that actually restores a lot of the destroyed and degraded ecosystems.[29]

To Lovejoy, this means restoring forests, grasslands, coastal wetlands, and agricultural ecosystems—"which have been leaking carbon like crazy."[30] Many of the aforementioned, if actually restored, will accumulate carbon that, in turn, makes for greater soil fertility—a win-win situation. Coastal wetlands, so called "blue carbon"—"blue" due to their being composed of water, and "carbon" for their capacity as a carbon sink—are particularly key. In addition to removing carbon from the atmosphere, restoring coastal wetlands provides added ecological benefits, including protection against storm surge.[31]

While ecological restoration is a dream solution for Dr. Lovejoy, he is clear that it won't be enough. Carbon dioxide needs to be removed from the atmosphere biologically (via planting trees and ecological restoration) but also physically (that is, by doing things that will bring down the levels of pollution without using biology). The first strategy for physical remediation is to reduce emissions drastically. What can't (or won't) be reduced needs to be pulled out of the atmosphere, and turned into something that is essentially inert and stable.[32] The problem with this second physical strategy is that, besides the radical geoengineering strategies that are riddled with unpleasant side effects, we do not have clearly established ways to pull carbon out of the atmosphere—though many are working on it, exemplified by Paul Hawken's drawdown project and some other methods that are mentioned in Janine Benyus's bio in this chapter. Even if we did have good ways to physically precipitate carbon out of the atmosphere, it would not solve the problem, but it may hold off the major damages for a few more years, which we could be essential for re-tooling our energy system.[33]

Dr. Lovejoy has explored biological and physical ideas for climate-change solutions, but he hasn't stopped there. He told us about his ideal economic approaches, as well. His current approach for economic solutions is to put a price on carbon. The simplest way to do that is with a carbon tax that helps our economic system better reflect environmental costs. He recognizes that this can be really difficult to accomplish politically, unless people are really willing to overhaul the existing tax system. He asserts that a carbon tax could be implemented, but suggests taking the tax out of wages—a strategy echoed by Bob Inglis, whose bio is provided in chapter 25. Dr. Lovejoy calls this "taxing productivity."[34]

What gives Thomas Lovejoy hope is the spirit and passion of the upcoming generations. He also credits his optimistic intellectual approach. He says, "Happily, I have this mindset—where every morning I start out and it's a fresh puzzle. Because otherwise you get bogged down in gloom."[35]

MICHAEL MANN: THEORIST / COMMUNICATOR

Professor Michael Mann Mann was one of the scientists responsible for what is known as the "Hockey Stick Graph," which, as we described in chapter 20, dramatically and conclusively shows that today's climate change is not a long-term cyclical event.

Born in Amherst, Massachusetts, in 1965, Michael Mann received his undergraduate degrees in physics and applied math from the University of California at Berkeley, a master's degree in physics from Yale University, and a doctorate in geology and geophysics from Yale University. He currently serves as Distinguished Professor of Atmospheric Science at Penn State University and director of the university's Earth System Science Center (ESSC). The emphasis of his current research is gaining a better understanding of Earth's climate system through the use of theoretical models and observational data.[36]

Throughout his career, Mann has received numerous awards and honors. These include being named by *Scientific American* as one of the fifty leading visionaries in science and technology in 2002 and receiving NOAA's outstanding publication award, also in 2002. Plus—along with other Intergovernmental Panel on Climate Change (IPCC) authors—he contributed research work that led to the award of the 2007 Nobel Peace Prize.[37]

Finally, Mann is the author of more than two hundred peer-reviewed papers and two books—all of which have solidified his reputation as a premiere climate scientist. In fact, the titles of his books—*Dire Predictions: Understanding Climate Change*[38] and *The Hockey Stick and the Climate Wars: Dispatches from the Front Lines*[39]—give you a really good idea of how deeply he is immersed in the climate-change debates. Mann has also cofounded the science website RealClimate.org, which has won several awards.[40]

Mann's innovative research was a key component in building Earth's historical temperature record—something that is essential for understanding the climate-change debate. These critical findings allowed Mann and his co-researchers to produce the now-famous hockey stick graph. Since it was first introduced in 1998, the "hockey stick" has been one of the key tools for gaining the overwhelming scientific consensus that our planet is warming and human activities are responsible.[41]

Of course, Mann's climate pathway has not all been smooth traveling since his hockey stick first hit the streets. In fact, Mann has often been dragged through the mud by the brigade of climate-change deniers. As journalist Joshua Holland describes it, "He was a central figure in the trumped-up 'Climategate' scandal [that we discussed in chapter

21], accused with other scientists of fraud by conservative bloggers and pundits before being vindicated by eight separate independent investigations. He was later the subject of an 'academic witch-hunt' by former Virginia Attorney General Ken Cuccinelli until a circuit court judge ruled that Cuccinelli had provided no 'objective basis' for his crusade."[42]

A fierce defender of scientific discourse, Mann has filed defamation suits against a variety of his antagonists, including right-wing columnist Mark Steyn, the *National Review*, and the Competitive Enterprise Institute (CEI)—a libertarian think tank that the *Washington Post* has labeled "a factory for global warming skepticism."[43]

Mann laments the fact that climate naysayers are still so prevalent. He staunchly believes that if the public had a better grasp of how science works, there would be no climate skepticism. They would understand that climatology is based on the same scientific method used in other fields. But he also understands that much of the negative flashback is being stoked by the fossil-fuel industry and supporters.

Mann wrote on *Huffington Post*, about fossil-fuel companies misleading the public and policymakers as to the risks of their products for decades: "As early as the late 1970s, executives at fossil-fuel companies were well aware that burning oil, gas and coal could cause irreversible and dangerous climate change. Indeed, as early as 1981, Exxon-Mobil was weighing whether or not to develop carbon-intense gas reserves off the coast of Indonesia because of the climate risks associated with the project."[44]

Mann then continued with his indictment, writing that Exxon-Mobil and other fossil-fuel companies chose to suppress what its own scientists knew: "From 1979 to 1983, the American Petroleum Institute operated a scientific task force to study climate change. According to a researcher who worked on the project, it was taken out of scientists' hands and quickly buried—and forgotten—until reporters rediscovered it just last year."[45]

JANINE BENYUS: INNOVATOR / CONNECTOR

Janine Benyus thrills at finding nature's adaptations and applying them to human problems, something she calls "biomimicry." She emphasizes that nature has already done the research and development needed to solve climate change—we just have to heed its teachings. To follow nature's lead, Janine Benyus heads the Biomimicry Guild, the Innovation Consultancy, and the Biomimicry Institute. These organizations help innovators apply ideas, designs, and strategies from biology to the creation of sustainable human systems and products.[46]

From the moment our conversation began, we could sense Janine's deep reverence for every creature of the natural world. She spoke to us from a lakeside cabin in Montana—during the call she spotted a bald eagle and interrupted herself to admire its flight. "Pretty much anything I look at as a climate-change solution—something in the natural world has already invented a pretty elegant way to do it . . . that's what gives me hope actually. We don't have to make this up. We don't have time to make this up. We have to depend on what is proven at this point," explains Benyus.[47] Biology has been perfecting systems for 3.8 billion years of research and design, and Benyus thinks we need to capitalize on these years of R&D: "Quieting human cleverness and asking nature first is the way that people should invent—that is not that hard a switch. It really isn't."[48] In fact, to facilitate this process, Benyus created the website AskNature.[49] The website is an open-source platform featuring the world's most comprehensive catalog of natural solutions to human design challenges and environmental problems.

Through the Biomimicry Institute, and AskNature, Benyus has collected countless examples of biomimicry technology that offer climate-change solutions. Innovators have studied the no-drag structure of shark skin and the splashless entry of a kingfisher into water and used nature's design to improve the efficiency of ships and trains, respectively. Renewable energy engineers take inspiration from the shape of tuna tails and kelp blades.[50]

Benyus's interest in climate change began when she saw the impact of a Free-Air Carbon Dioxide Enrichment (FACE) experiment while she was working for the US Forest Service. In FACE experiments, scientists can simulate the effect of varying levels of carbon dioxide on plants. "The findings were scary—Frankenstein frightening," says Benyus, "The first few years, carbon dioxide really increased growth, but it also changed the carbon to nitrogen ratio in the leaves. It made the leaves SO tasty to pests that leaf insects and aphids of all kinds would pile up on the leaves—a half of an inch, even an inch thick in some places. It was incredible."[51] Over a few years, the increase in growth stopped, in part because the vegetation had been eaten back to a greater than usual extent, but also because the altered carbon to nitrogen ratio changed the biosynthetic pathways of photosynthesis. Plants ceased to photosynthesize a month earlier than normal.[52] "When I saw those emitters [large pipes that release CO_2 into the air] . . . and what happened to that growing forest . . . it was like a nightmare. I have never been able to forget it. Everybody should have to walk through that forest—and see the future," Benyus said.[53]

As temperatures change due to global warming, plant and animal communities will start to migrate to more suitable climates in whatever way they are able. Moving to a new habitat is challenging for a species, not just due to the change of abiotic conditions such as soil, temperature, pH, and salinity, but also due to the breakup of its community. It loses its symbiotic partners, mutualistic partners, and even its predators, making it very hard for species to excel. "This biological pollution [due to climate change] will make chemical pollution seem like a walk in the park. . . . This is what I look at, and why I am adamant that we can't just adapt to climate change yet. We actually have to put our energy toward reversing it . . . pulling down the carbon that is already there and getting the context back to biologically relevant parameters," emphasized Benyus.[54]

Benyus's strategy is threefold: cease fossil-fuel emissions completely, transfer to clean energy sources, and decrease the amount of carbon in the atmosphere. The latter is a special focus for her. According to Benyus, there are two things that biomimicry has been talking about for almost two decades as a way to draw down carbon—biosequestration and using carbon dioxide as a feedstock for products, such as concrete, described below.[55]

Janine Benyus draws a very important distinction between biosequestration and geoengineering: "When people hear carbon drawdown, they assume that I am talking about geoengineering—this is very different."[56] Biosequestration uses ecosystem-inspired agriculture and land use to bury carbon very deep in the soil layers and roots of plants. Explained Benyus, "Our [current] land use—farm, forestry, agriculture—is about twenty-four percent of the carbon dioxide emissions right now

... we burn forests, we cut them, we dig up soils and basically expose that carbon to oxidation or soil erosion. Over the last two hundred years in the United States, we have lost half of our soil carbon. To me, that looks like a bathtub that is half full. We could bring some carbon home, but it is going to take a different kind of agriculture."[57]

Biomimetic agriculture uses the local ecosystem as a model.[58] A diverse prairie significantly enriches the health of the ecosystem, as it has deep-rooted species that hold carbon very deeply in the soil.[59] These plants, in turn, feed mycorrhizal fungi and turn carbon into liquid carbon, which the fungi push out through their roots into deep soil. "A mycorrhizal or fungal rich field holds fifteen times more carbon than sterile, industrial soils," says Benyus.[60]

An ideal biosequestering ecosystem would also include animals—and might look similar to the plains back when bison roamed across them. Back in those days, Native Americans used the "three sisters" approach to agriculture, rotating crops and planning them carefully to enrich the soil naturally instead of using fossil-fuel-based fertilizers. "With this approach, if we are wrong about drawing down carbon into our landscapes, the only mistake we make is better soils, better food, more water health and less erosion ... so it is kind of a no-regrets approach," says Benyus.[61]

Using carbon as a feedstock is another exciting climate-change solution. "You are going to see this meme a lot—that carbon dioxide is precious so let's make something out of it," Benyus postulates.[62] In fact, Elon Musk's XPRIZE Foundation announced a four and a half year competition that will award $20 million to the best idea for turning carbon dioxide from a liability into an asset.[63] Several companies have already put strategies into effect. A green-energy company, Calera, bubbles carbon dioxide through sea water and makes concrete, aggregate for sidewalks, and gypsum for dry wall—removing it from the atmosphere and locking it in human structures.[64] In Asia, technology developed by Asahi Kasei Chemicals, uses carbon dioxide to produce polycarbonates. Polycarbonates are a type of plastic found in many products, from water bottles to lenses for glasses.[65]

For Benyus, "The key flip really is this idea that life on earth sees carbon as a feedstock and a building block. We [humans] are the only ones who do not. We see it as the poison of our time because there is too much of it. But when organisms are in the presence of abundant resources, they find a way to use them. We have got to become the generous species that actually helps to regulate climate. We affected climate, now we have to bring it back. That is quite different than we have thought of ourselves before."[66] It is a monumental task, but one which we have no choice but to overcome.

MARK JACOBSON: INNOVATOR / SOLVER

Our conversation with Mark Jacobson was akin to pinning down a bumblebee. Dr. Jacobson, director of the atmosphere and energy program and professor of civil and environmental engineering at Stanford University, is positively buzzing with frenetic scientific intensity. Plus, he is truly working on something that is as essential to our human existence as pollination. The professor has compiled decades of research on how to transition to one hundred percent renewable energy by 2050.

Professor Jacobson, along with twelve peer-reviewed studies, has proven that this switch is economically and technologically viable. In the process, he has debunked the myths that we lack sufficient storage capacity to provide sustained reliable energy and that fossil fuels are cheaper than renewable energy sources. Not only are renewables one-hundred-percent viable, transitioning to them will create nearly five million new forty-year construction and operation jobs, avoid nearly $590 billion in annual mortality and illness costs, avert nearly 45,000 air pollution deaths annually, and will pay for itself in as little as two years from air pollution and climate-cost savings alone. Climate-change problem? What climate-change problem? Unfortunately, social and political obstacles are blocking this plan from being implemented—more on that later.

Jacobson's fascination with air pollution and climate issues started on a visit to Los Angeles as a teenager. The air pollution in the city hit him and he thought, "Why should people live like this?"[67] This launched him into decades of detailed research on environmental engineering, aerosol microphysics and chemistry, climate microphysics and chemistry, radiation transfer, and weather and climate on global through urban and local scales. He discovered a key insight that cancelled out many fuel options—black carbon from diesel exhaust, burning biofuel, burning biomass, and kerosene burning was the second most important cause of global warming after carbon dioxide, and ahead of methane.[68] He studied renewable energy options extensively as well, and then began comparing different energy solutions to global warming, air pollution, and energy security. His conclusions were that wind, water, and solar (WWS) are the best solutions to global warming, air pollution, and energy security; and that electric vehicles are the best vehicle options.[69]

Scientific American brought Jacobson's research into the limelight in 2009 when they published a paper he wrote with Dr. Mark Delucchi, "A Plan to Power 100 Percent of the Planet with Renewables."[70] This became the backbone research of the Solutions Project (TSP) we discussed in chapter 21, providing a plan for one hundred percent WWS for all fifty of the United States and, so far, 139 countries.

As part of this analysis, Jacobson looked at job creation, health benefits, climate benefits, policies that could be implemented, land area required—and even the number of devices needed for each type of energy—wind turbines, solar panels, and so forth.[71] TSP's website (www.thesolutionsproject.org) has an interactive map so you can see exactly what the plan is for your state and how you can participate. Sarah Shanley Hope, executive director of TSP says, "Three years ago, he [Jacobson] got a lot of 'Oh, that is not possible'" when people examined his proposed plans, "But the crystal clarity of his model—that through the technology today for wind, water, and sun, along with the energy efficiency upgrades required and the efficiency gained by transferring to clean renewable energy—we can actually power not just our electric grid, but by electrifying that grid we can power our transportation systems, our heating and cooling systems, and our industrial uses of energy. It really is a whole system transition that we have the potential, and it is totally possible . . . to achieve over the next thirty-five years."[72]

Jacobson frequently notes the clear technological and economic feasibility of a transition to one hundred percent renewable energy by

2030, but he also points out that it is unlikely due to social and political reasons. The public is on board for the most part—a recent poll showed that 70 percent of people are in favor of wind and solar, while coal and nuclear are way down at the bottom, with gas close to the bottom.[73] The bottleneck is in the politics, suggests Jacobson: "You have mostly extremists in government that are not necessarily representative of what the public wants. Plus you have some people who have some financial interests who will oppose."[74]

A lot has gone on that is beneficial toward a transition to renewable energy. Since Dr. Jacobson wrote the *Scientific American* paper in 2009, there has been a huge drop in the energy cost and electric power prices of renewable energy technologies. "Solar is now in the United States, at the utility scale, cheaper than natural gas. Slightly cheaper; it is almost the same, but slightly cheaper. Wind is now half the cost of natural gas. . . . Since then [2009] we also have more electric vehicles on the road, more choices, longer mileage, and lower cost vehicles available. Heating and cooling is starting to be electrified more. And also more people are talking about it, and more policies put in place to implement clean, renewable energy," says Jacobson.[75]

Though the science about climate change can be dark and the solutions seem far away, Jacobson's work gives us a light at the end of the tunnel.

KATHARINE HAYHOE: UNIFIER / TRAILBLAZER

Professor Katharine Hayhoe is, first and foremost, a Christian. She outlines her beliefs in no uncertain terms in her book *A Climate for Change: Global Warming Facts for Faith-Based Decisions*, which she wrote with her husband, Andrew Farley, a pastor at a Christian church in Texas.[76] She says that she worships the creator of the universe, believes that God spoke the world into existence and sustains it by his power, that Jesus Christ is the way to eternal life, that the Bible is God's word, and that the message of the gospel is of the highest import.[77] On these issues, Katharine Hayhoe's ideas line up squarely with the majority of the Christian community.

In her book, Dr. Hayhoe also enumerates what she does not believe, "We don't believe that life came from nothing or that humans evolved from apes. We don't believe in government running our lives or in destroying the economy to save the Earth." One other really important thing that she does not believe in—climate change. When asked if she believes in climate change, she laughed, smiled mischievously, and exclaimed, "NO! I do not believe in climate change! Book of Hebrews: 'Now Faith is the substance of things hoped for, the evidence of things not seen' [11:1, King James Version], Science is the evidence of what we do see. As a scientist, I crunch the data. I study the world around us. I look at the projections to the future; I do not *believe* that the climate is changing, I *know* that it is; I know that humans are responsible, and I know that we have an important choice to make."[78]

For the faith-based community, which Hayhoe is deeply connected to, this is a critical distinction. When you ask someone if they "believe" in something, you are using faith-based language, in her opinion.[79] By using this language, you are essentially offering people an alternate religion and then asking them if they believe in this alternate religion. More than 80 percent of people in the United States and six out of seven people worldwide have a religion already.[80]

Hayhoe has made many important contributions to the science and study of global climate change. She is an atmospheric scientist who has authored more than sixty peer-reviewed publications and served as an expert reviewer for the Intergovernmental Panel on Climate Change's Fourth Assessment Report. Dr. Hayhoe has worked at Texas Tech since 2005 as an associate professor of political science and director of their Climate Science Center.

Hayhoe is policy agnostic as a scientist, in the sense that any solution is better than no solution. But as a human, not an expert, she thinks that we need to discontinue our current system that takes our tax money and gives it to the energy companies in the form of massive subsidies. Hayhoe suggests we remove the taxes and subsidies from all of the different types of energy and put a price on carbon that will make it cost prohibitive to burn fossil fuels.[81] The price we assign to carbon, she says, should account for the price we are already paying in terms of our health, in terms of crop insurance, in terms of FEMA bailouts, and so forth. "We are paying those costs, make no mistake, but we need to be paying them in a direct way so that you and I, when we make our decisions about what car to buy, or how much energy to use, we can make that decision with the correct price signal," Hayhoe explains.[82]

The problem with this solution, according to Hayhoe, is "in the United States, politics has become so polarized that half the country would probably cut off their right arm rather than give the government any more power than it already has. . . . Climate change and its associated impacts and health issues have become a casualty of the polarization of society that has occurred over the last thirty years. People feel like it is incompatible with their identity and who they are to agree that we have to do these very common set of things [politically] . . . to have clean air, healthy kids, and avoid all of these health costs. We live in this polarized society where people won't even admit something is true because they would have to give up who they are as a person."[83]

Something that decidedly does not work is telling people that if they do not care about climate change they do not have the right values. It is vital to find and understand their value system and then connect it to climate-change issues. As a corollary to this, Hayhoe mentions that climate change is fully understandable without believing in evolution: "As we go back in Earth's history, our satellite, thermometer, natural and written records consistently validate the dramatic and unprecedented nature of the recent increase in heat-trapping gases in the atmosphere. And this recent increase corresponds directly with the dawn and growth of our industrial age less than three hundred years ago."[84] Those who believe in evolution (as we, the authors, do) can still connect with someone who believes in creationism, working together in the fight against climate change.

SUSAN SOLOMON: SOLVER / COMMUNICATOR

Susan Solomon, Ellen Swallow Richards Professor of Atmospheric Chemistry and Climate Science at Massachusetts Institute of Technology, has discovered some depressing news about carbon dioxide emissions: What we emit today will persist for a thousand years. "Even if we stop emitting," Solomon says, "we will have the same temperature for at least one thousand years. This is due to the long lifetime of carbon dioxide and the time scale at which heat is kept in the ocean."[85] This underscores the potential for essentially irreversible climate impacts. This is one way to leave your mark on the world—20 percent of the carbon dioxide you used to drive to work in 2016 will remain in the atmosphere until 3016 CE.[86] Something to think about every time you start your car—unless it is an all-electric Tesla (see bio on Elon Musk in chapter 27).

Dr. Solomon is a leading atmospheric scientist and a recipient of the highest honor granted to scientists in the United States, the Presidential Medal of Science. At MIT, she was the founding director of their Environmental Solutions Initiative from 2014 to 2015. She is a member of the National Academy of Sciences,[87] a scientist at NOAA, and co-chair for the Intergovernmental Panel on Climate Change (IPCC). *Time* magazine recognized Solomon as one of the most influential people in the world in 2008. Two geographic features in Antarctica have been named after her—Solomon's Saddle and Solomon Glacier.

Solomon presents a bleak picture about climate change. In January 2015, Boston, where she lives, had twenty-four inches less snow than January 2014 and was 6.0 F warmer. This in itself is not indicative of climate change, but it does tell you something about what climate change would look like if you had every winter, not just one winter, 6.0 F warmer than the previous. While many New Englanders might welcome that change in the winter, it would not feel so great in the summer. Solomon says that if we continue with the business-as-usual scenario, by 2100 Boston's annual weather will feel more like the Carolinas of today.[88] We do not even dare imagine what that scenario will mean for the Carolinas.

Dr. Solomon describes climate change like overfilling a bathtub—the rate at which we are adding carbon to the atmosphere is much faster than the rate at which the planet can remove it. "We need to decrease our emissions by eighty percent just to stop the carbon due to the bathtub problem," she says.[89] Interestingly, while sea level has risen globally by eight inches in the twenty-first century, the sea levels on the East Coast of North America have risen more than two feet. To return to the bathtub analogy, this is like the slosh in a bathtub, and East Coasters are on the wrong side of the slosh.[90]

Dr. Solomon's early scientific contributions informed our understanding of the ozone hole greatly, helping us avert that catastrophe, as we discussed in chapter 16. She is credited with recognizing the role of polar stratospheric clouds (PSCs) as a surface on which chemical reactions leading to ozone destruction could proceed far more rapidly than previously acknowledged.[91]

Dr. Solomon takes great solace in the climate-change agreement made at the 2015 climate-change talks in Paris: "It really was quite amazing.... For the first time [we dealt] with this issue as a planet. I don't think it could have been any better than how it came out. It was a remarkable piece of diplomacy and international cooperation. And I am pretty hopeful, despite all of the negative messages about the nature of climate change. It is a pretty scary thing, it will be a different planet, but I think we are on a much better trajectory thanks to what happened last December in Paris."[92]

Something else that gives Solomon hope is the fact that investments in renewables in various nations have gone way up. "Penetration of renewables has been happening extremely quickly," she says.[93] "What the world really needs is a lot more clean energy development, deployment, and investment. There are countries that are leading on developing new technologies like the US, Japan, China, Germany, UK, France. It is a very limited subset of countries, but many of them are deeply invested in this process now. They are sending a signal to industry, and that signal is being picked up around the world."

AMORY LOVINS: INNOVATOR / AUTHOR

When my wife, Debby, and I (Budd) interviewed singer John Denver several times in the early 1980s, we had no idea that we were looking at the cusp of what would become today's sustainable design and development movement. At the time, practically nobody knew anything about Denver's environmental activism. They just knew Denver as a "hippie-like" singer and songwriter who produced nearly a dozen number-one hits and was fond of exuberantly shouting "far out" to his legions of fans—of which we were certainly two. (During our thirteen years in Colorado, we never missed one of John's annual concerts at the spectacular outdoor Red Rocks Amphitheatre in Golden.)

As contributing editors to *Colorado Homes and Lifestyles* magazine, Debby and I were always on the lookout for good outdoor-oriented stories that we could feature in each month's issue. So imagine our surprise when we read that our favorite entertainer was also a dedicated conservationist who had just purchased nearly a thousand acres of land on which he could build something he called his Windstar Research Farm. Windstar's objectives were finding and refining methods for living more compatibly and efficiently in tune with the natural world. Ideas like bio-intensive gardening, solar insolation, greenhouse cultivation, and geodesic dome housing were all de rigueur at Windstar. Of course, Debby and I (Budd) thought this was all "very cool" (as well as "far out") and it made great copy for our magazine photo-essays, but we never thought much about it as an actual way of life. It was just something that seemed nice to think about for the fairly distant future.

Some thirty-five years later, all the sustainable living concepts we first heard about at John Denver's Windstar Research Farm are in the forefront of the scramble to find solutions for climate change. Of course, our friend John is now gone—far too soon—but we are fortunate that one of John Denver's long-time friends, a co-founder of the Rocky Mountain Institute (RMI), continues working as one of our climate-change heroes.

Amory Lovins was born on November 13, 1947, in Washington, DC, and is now an environmental consultant, experimental physicist, and 1993 MacArthur Fellow. Working for the past forty years in more than fifty countries, he is widely considered to be one of the world's leading authorities on energy. He is especially adept at devising ways to maximize energy efficiency and developing methods for maintaining sustainable supplies. He is also known as a prolific innovator, specializing in the design of superefficient buildings, factories, and vehicles. In 1982, Lovins and his wife, Hunter, started RMI with the goal of establishing a nonprofit "think-and-do" tank that was both entrepreneurial and independent. Today, the staff of RMI emphasizes using the restorative powers of Earth's natural resources to create a verdant, thriving, and secure world for us all to enjoy.[94]

Over the course of his illustrious career, Lovins has received ten honorary doctorates, briefed nineteen heads of state, provided expert testimony in eight countries, and published twenty-nine books. His books include *Reinventing Fire*,[95] *Winning the Oil Endgame*,[96] *Small is Profitable*,[97] and *Brittle Power*.[98]He also co-authored *Natural Capitalism*[99] with Paul Hawken, another of our climate-change heroes.[100]

Now here's the most encouraging part: When it comes to winning the climate-change battle, Lovins is definitely a "glass-half-full guy." In December 2014, he and RMI joined forces with Sir Richard Branson's Carbon War Room (CWR) to begin double-teaming the search for global climate-change solutions. CWR's ambitious mission statement reads, "We accelerate the adoption of business solutions that reduce carbon emissions at gigaton scale and advance the low-carbon economy."[101]

"Together we can go further, faster," Branson said in a statement announcing his agreement with RMI.[102] According to a report from NBC News, "Executives from both organizations tapped to lead the alliance described it as a marriage between an agile and young entrepreneurial organization full of 'make-it-happen passion' with one that is steeped in analytical rigor, insight and thought leadership."

Announcing the alliance between RMI and the CWR, Lovins and Branson said, "We are closer than ever before to transforming the world to a low-carbon energy system. We are excited to coordinate [our] proven approaches to scaling transformational change worldwide. Together, we will more than double our impact on the scale and speed of the energy transformation to a clean, prosperous, secure, and low-carbon world."[103]

PAUL CRUTZEN: RESEARCHER / SOLVER

The world would be wise to listen to Dr. Paul Crutzen on climate change and global-warming issues. Many environmentalists consider him to be the chief scientific caretaker of life on the planet. In the past, he helped us avoid a global threat—the destruction of our stratospheric ozone layer. If the warnings from him and his fellow winners of the 1995 Nobel Prize in chemistry hadn't come when they did, the Antarctic ozone hole might have proven to be disastrous.

Born on December 3, 1933, Paul J. Crutzen's boyhood and education both took rather circuitous routes to their finishing points. Growing up in Nazi-occupied Holland meant that his early education took place in wartime upheaval. He graduated in 1954 with a degree in civil engineering and initially set out to design bridges and houses.[104]

But Crutzen didn't really want to work in an office; instead, he wanted to be an academic at a university.[105] When he got a chance, he applied for a job as a computer programmer for the Institute of Meteorology at the University of Stockholm. Once there, he found his true callings—studying stratospheric ozone chemistry, developing meteorological models, and earning his doctorate in meteorology.[106]

Crutzen moved to the United States in 1977 to become the director of the National Center for Atmospheric Research (NCAR) in Boulder, Colorado. While there, he studied how burning trees and brush in Brazil affected the atmosphere. The idea at the time was that burning rainforests was releasing carbon monoxide and other carbon compounds into the air and these, in turn, were causing warming of the atmosphere, otherwise known as the Greenhouse Effect.[107]

But Crutzen's research didn't give him the results he expected. According to the *Encyclopedia of World Biography*, "When Crutzen collected samples and did his research on forest burning, however, he found out that the exact opposite was happening. The yearly smoke was actually decreasing the amount of CO_2 in the atmosphere. This discovery intrigued Crutzen, and he went on to study the effects of other kinds of smoke on the atmosphere—especially the smoke that would come from a global disaster such as a nuclear war."[108]

The scientific community was interested when they heard that Crutzen wanted to study how nuclear war would affect the planet. The journal *Ambio* paid Crutzen and John Birks, his University of Colorado colleague, to research the subject.[109] Crutzen and Birks first determined that a global nuclear war would produce a fallout of black carbon soot from the fires burning across the planet. The two scientists then theorized that the carbon soot would absorb as much as 99 percent of the sunlight reaching Earth. This would result in a state of perpetual winter that would destroy all life on Earth. Proposing this nuclear winter theory earned Crutzen the 1984 Scientist of the Year award from *Discover* magazine and the Tyler Award in 1988.[110]

With nuclear capabilities still spread across the globe, the threat of nuclear winter remains with us. According to a 2011 article written by Charles Choi for the National Geographic News, "Even a regional nuclear war could spark 'unprecedented' global cooling and reduce rainfall for years, according to U.S. government computer models. Widespread famine and disease would likely follow, experts speculate."[111]

In 2000, in the International Geosphere-Biosphere Programme (IGBP) *Global Change Newsletter*, Crutzen and biology professor Eugene F. Stoemer, emphasizing the central role of mankind in geology and ecology,[112] proposed using the term *Anthropocene* for the current geological epoch. In regard to this, Crutzen and Stoemer said: "To assign a more specific date to the onset of the Anthropocene seems somewhat arbitrary, but we propose the latter part of the 18th century. . . . We choose this date because, during the past two centuries, the global effects of human activities have become clearly noticeable. This is the period when data retrieved from glacial ice cores show the beginning of a growth in the atmospheric concentrations of several 'greenhouse gases,' in particular CO_2 and CH_4. Such a starting date also coincides with James Watt's invention of the steam engine in 1784."[113]

Steve Connor, science editor of the *Independent*, wrote, "Professor Paul Crutzen, who won a Nobel Prize in 1995 for his work on the hole in the ozone layer, believes that political attempts to limit man-made greenhouse gases are so pitiful that a radical contingency plan is needed. In a polemical scientific essay [that was] published in the August 2006 issue of the journal *Climate Change*, he says that an 'escape route' is needed if global warming begins to run out of control."[114]

Indeed, in 2006, Crutzen first came up with a rather radical solution for helping to stave off the effects of global warming: He suggested using sulfur to alter the chemical composition of Earth's upper atmosphere. The sulfur should then reflect sunlight (and its associated heat) away from Earth and back into space.[115]

As you might imagine, this was a very controversial concept, but the idea received some serious consideration because of Crutzen's known track record of excellence in past atmospheric studies. Crutzen's plan called for the sulfur to either be scattered by balloons designed for high altitude flight or shot into the air by heavy artillery shells. According to the *Independent*, "Such 'geo-engineering' of the climate has been suggested before, but Professor Crutzen goes much further by drawing up a detailed model of how it can be done, the timescales involved, and the costs."[116]

As described in the *Encyclopedia of World Biography*, "[Crutzen's plan] was modeled in part on the eruption of the Mount Pinatubo volcano in 1991. Thousands of tons of sulphur were thrown into the air when the volcano erupted causing temperatures around the globe to decrease. Putting the sulphur into the stratosphere rather than lower down—as in the case of the volcano—would create a year or two of lower temperatures rather than just a few weeks. The project would cost about $25 to $50 billion, but it was Crutzen's belief that this cost is nothing to what global warming is doing to all life on Earth."[117]

In a February 2016 interview with us, Professor Crutzen stated that he agrees that "Geo-engineering is a very risky approach. It has to be further investigated very carefully and should be only the *ultima ratio* [last resort] if the more conservative measures fail. For now, society and politics should focus on cutting GHG emissions. . . . We must make sure that we don't solve one problem and at the same time create new problems that might be even bigger."[118]

Crutzen remains fearless about speaking out; he consistently expresses his concerns about the lack of US leadership in addressing climate change and global warming. He does, however, believe that the December 2015 climate talks in Paris (COP-21) provided a "good basis and promising perspectives that should be pursued" for solving the climate-change crisis.[119]

WALLACE BROECKER: RESEARCHER / PIONEER

Consider this: Solar radiation is trapped in salt water, then gets transported by conveyor belts throughout our planet's oceans before being discharged and sucked up by the Earth's crust. (As described on NOAA's website, "The global ocean conveyor belt is a constantly moving system of deep-ocean circulation driven by temperature and salinity. The great ocean conveyor moves water around the globe."[120]) As the oceans warm, and therefore cannot hold heat, more and more of this solar radiation is released from the water and it slowly cooks the planet and its inhabitants. Then, within a few decades of time, the massive influx of fresh, melting water from the overheated continents cools the ocean waters, plunging the world into a sudden ice age. Does this sound like a science fiction movie? Not if you are Professor Wallace Broecker, perhaps better known as the "Grandfather of Climate Science."[121] Dr. Broecker published a paper in May 1997's issue of *Geological Society of America*, pondering the potential of just such an abrupt shift of climate due to anthropogenic GHG emissions and their effect on the thermohaline current[122]—or the rising and sinking of saltwater that drives the global oceanic conveyor belt.

This paper, while a significant precursor to much of today's current understanding of the relationship between climate change and the global ocean conveyor belt, was not even Wally Broecker's seminal work. Professor Broecker made his biggest mark on history in 1975 when he published a scientific paper entitled "Climate Change: Are We on the Brink of a Pronounced Global Warming?" With this publication, he became the first US scientist to use the term "global warming."[123]

For the past fifty years, Broecker has been studying the mechanisms of climate change—specifically the role our oceans play in the process—and then writing warnings that we should be doing something about the resulting global warming. He has also been spending much of his time researching ways to dispose of the greenhouse gases—most notably CO_2—that are contaminating our atmosphere and causing climate change. Broecker thinks we should be able to do this without significantly changing our lifestyles. Not that he thinks lifestyle changes aren't a good idea; he just believes we are still a long way off from making any lifestyle changes that will really make a difference.

Born in Chicago in 1931, Broecker is one of Earth's greatest living geoscientists. To his colleagues, peers, and admirers, he is a genius and a pioneer. With his ever-smiling countenance and shock of tousled hair, he is simply Wally—the epitome of everyone's best friend.[124]

Broecker earned all of his degrees at Columbia University in New York City, where he also started working in 1959, holding the esteemed title of Newberry Professor of Earth and Environmental Sciences at Columbia's Lamont-Doherty Earth Laboratory since 1977.[125]

The author of more than 450 journal articles and ten books, Broecker is well known for his discovery of the role played by the oceans in triggering the abrupt climate changes that have punctuated glacial time. He is also the first person to recognize the aforementioned "Ocean Conveyor Belt" (which he named)—a critical discovery in the history of oceanography and its relationship to climate systems.[126]

Most recently, Broecker and his colleague Dr. Klaus Lackner, director of Columbia University's Lenfest Center for Sustainable Energy, have been leading the war on anthropogenic climate change by advocating for large-scale carbon sequestration. In his book *Fixing Climate*,[127] Broecker and his coauthor, science writer Robert Kunzig, make a compelling case for using carbon scrubbers to remove carbon from our atmosphere and deposit it back into the Earth's crust where it belongs.[128]

As Broecker explained it in a June 2008 interview with BBC's *Hardtalk* host Stephen Sackur, "We're going to have to learn how to capture the CO_2 and bury it—just like we learned to collect and put away garbage and sewage. . . . We've taken over the stewardship of the planet and with that we have the responsibility to take care of it."[129]

Broecker's approach to solving climate change is heavily tempered by the reality that immediate wholesale changes in how the world functions are not a possibility. As a result, we need to implement measures—call them stopgap, if necessary—to slow down the warming process. Broecker's ideas on how to accomplish this—using a variety of human-created carbon sequestration devices—deserve to be seriously considered. He analyzes one such concept, for example, in a paper he authored called "Does Air Capture Constitute a Viable Backstop against a Bad CO_2 Trip?"[130]

Chapter 24
ADVOCATES / AUTHORS

BILL McKIBBEN: FOUNDER / LEADER

We tracked down Bill McKibben in a windowless makeshift basement conference room in Paris, France. This ad hoc office space was the hub for 350.org's massively successful civil disobedience actions at the COP-21 climate-change conference. At well over six feet tall, signature Red Sox ball cap included, McKibben's presence is commanding yet gentle. We could sense his exhaustion—he and his team had been working around the clock, doing everything (tweeting, texting, emailing, posting on Facebook) they could to raise awareness for strong civil disobedience actions at COP-21 to stop climate change.

When we asked him what brought him to the cause, he said, "I was a journalist and this seemed such a huge story to me—the most interesting possible story. But partway through writing the first book on it, I realized that I was not objective in the strictest sense. I did not want the world to heat up, dry up, and blow away. At some other level, perhaps I was even more objective than in the past. You know, I sort of understood what the basic reality was so . . ." He grew quiet for a moment, then continued, "I mean, there are times when I wished that I hadn't stumbled across all this, because there are other things that I may have wanted to do with my life. But you know, this has been a good place to be engaged in a good fight."[1]

Some of our heroes strategically avoid the words *fight*, *conflict*, *battle* as too incendiary for the politics around climate change. But for McKibben, this is a *fight* and, indeed, it is *personal*. It is becoming apparent that many of the fossil-fuel companies have known about the science for over twenty-five years. In his opinion, they have robbed him of a life that could have been devoted to something different. (What he does now is "rhetorical battle with retrograde congressmen," as he puts it in *Oil and Honey*.[2])

More than that, climate change may rob him of the beauty that satiates and fills his soul. McKibben says, "I think the next big front in the climate fight . . . may be trying to peel back all that we can learn about Exxon and what it and other oil companies knew twenty-five years ago. And I think I take that one fairly personally in a sense because I have been working on this a very long time. I wrote the first book about climate change, so I am getting uncomfortably close to thirty years of steady work on this, and knowing that twenty-five years ago, Exxon could have ended the whole faux/phony debate about it, simply by saying what they knew, makes me aggravated. Because then we could have spent the last quarter century working on solutions. We would not have solved the problem by now, but we would be well on the way. We would have turned the corner. And instead, we wasted what may turn out to be the crucial quarter century of geological history on this."[3] In summary, McKibben believes we have already lost lives, species, and landscapes that could have been saved if we had acted twenty-five years ago.

McKibben is adamant that it is far too late to stop global warming. That is not one of the possibilities at this point. The temperature of the planet is already 1.0 C warmer, and there is momentum that will raise it further than that. For him, the pertinent question now is "Can we still stop it short of damage so severe that it threatens our ability to have civilizations like the ones we are used to?"[4]

There are only so many large physical features on Earth that can absorb carbon. Once you have run through the Arctic, the Antarctic, and the world's oceans it becomes a runaway train.[5] The question now is not "Are we going to make the changes and transitions needed to head toward a renewable future." We clearly are. The question is, "Are we going to do it quickly enough to even begin to catch up to the physics of climate change." This is the question that worries McKibben: "That I am less convinced of, and that is why we have to keep pushing hard!"[6]

And McKibben has been pushing hard—really hard. We told you about 350.org and its inception in chapter 20. The organization has now "organized more rallies than Lenin, Gandhi, and Martin Luther King combined."[7] McKibben calls what he has done with 350.org "the most satisfying work of my life, endlessly difficult and endlessly interesting."[8] He has won countless victories, the biggest of which was Obama's rejection of the Keystone XL pipeline in November 2015, as we discussed in chapter 21.

McKibben takes great hope in the fact that the movement keeps growing and it is working.[9] If one builds big movements, then change starts inexorably to happen. He wants it to continue to grow until it is big enough that everybody's part in it makes sense. His optimism is cautious, however. He explains, "It is anybody's guess whether we can build it in time or not. As I say, we will find out, and we will find out in our lifetimes. It is going to take many generations to win this fight, but we can easily lose it in the next five or ten years if we keep pouring carbon into the atmosphere."[10]

The next phase in the climate fight for McKibben is to get companies to pay for their injustices—which we can all hope will fund the transition toward renewable energy sources. He says, "The most important thing is the mobilization of a big movement. Once that happens it opens up all kinds of space. Now there are lots of politicians who are suddenly moving in the right direction."[11] It is not likely that McKibben and his determined staff at 350.org will relent until the great mobilization has shifted us to one hundred percent renewables and made the polluters pay.

NAOMI KLEIN: VISIONARY / LEADER

In her 2014 award-winning book, *This Changes Everything: Capitalism Versus the Climate*, Naomi Klein writes, "Forget everything you think you know about global warming. The really inconvenient truth is that it's not about carbon—it's about capitalism. The convenient truth is that we can seize this existential crisis to transform our failed economic system and build something radically better."[12] Once Klein realized that engaging deeply with the science of global warming could be a catalyst for forms of social and economic justice which she already believed in, she was hooked.[13]

In the film version of *This Changes Everything*, Klein describes how, at first, she turned a deaf ear to images and discussion about climate change.

"Is it possible to be bored by the end of the world?" she asked of her own original desire to look away from the problem.[14] The moment that made her finally not "click away" was when she attended the 2011 Royal Society Conference at Chicheley Hall, Britain, where scientists were deep in discussion about how to fix climate problems with geoengineering solutions.[15] The real problem became obvious to her—humans have been misleading ourselves with a false story for the past four hundred years. The story we have chosen as a society is that Earth exists for us to pillage and plunder for its resources. That it is a machine and we are its engineers, and we can always fix any problem we create with innovation and human cleverness. Realizing this story needed changing enabled Klein to "stop tuning out the images of a polar bear [in distress]. Unlike human nature, we can change stories."[16] Engaging with climate change began to feel actionable to her.

Prior to fossil fuels, we were bound to nature—ships could sail only when there was wind, and factories only worked if the flow of the water was strong enough to power their equipment.[17] Fossil fuels allowed us to run our machines whenever we wanted—giving us the feeling of dominion over nature. "This story—our dominion over nature—is not quite right. Nature can hit back," explained Klein, referencing Hurricane Sandy and the destruction it wreaked on New York City.[18] If we continue to base our society on nonrenewables, at some point we will fail to renew ourselves.

But Klein thinks that we hesitate to embrace renewables because it requires us to unlearn the myth that we hold total dominion over nature. A shift to renewables forces us to embrace the fact that we are in a relationship with the rest of the natural world.[19] Our society has not functioned that way for several hundreds of years, and to many the concept is daunting, at best, and outright repugnant to others.

Klein is a champion of indigenous rights. She values the way indigenous peoples have built their culture around respecting earth and working with its cycles and rhythms in harmony, rather than trying to control it and make it do their bidding. Native people are slowly teaching non-native people how to live with the land in ways that are not purely extractive.[20] These native ways are under a dual threat—the fossil-fuel extraction processes are destroying the land that indigenous people live off so carefully. At the same time, climate-change effects are drastically altering the resources they depend on, whether via sea-level rise, species extinction, or climate-induced species migration, to name a few examples. For these reasons, native people have been fighting battles against these two forces for a long time.

What is staggering is that some of the poorest and most marginalized people are standing up to the wealthiest and most powerful destructive forces on the planet—and for far too long, they have done it with shockingly little support from the rest of the world.[21] "Their heroic battles are not just their people's best chance of a healthy future . . . they could very well be the best chance for the rest of us to continue enjoying a climate that is hospitable to human life."[22]

While doing everything we can to keep fossil fuels in the ground, we also need to take steps toward a post-carbon economy. Every new public dollar we spend should do more than spur random economic activity—it should transform our energy system and our public sphere so that it meets today's complex needs.[23] Our investments should be thoughtfully designed to bring down emissions and reinvigorate community and local initiatives. To transition away from fossil fuels, Klein claims, "requires breaking every rule in the 'free market' playbook: reigning in corporate power, rebuilding local economies and reclaiming our democracies. Confronting climate change is no longer about changing the light bulbs. It is about changing the world—before the world changes so drastically that no one is safe. Either we leap—or we sink."[24]

ELIZABETH KOLBERT: EDUCATOR / COMMUNICATOR

When it comes to climate change, Elizabeth Kolbert and Al Gore have a great deal in common. In 2006, they both used their considerable literary talents to warn the world in no uncertain terms about the pending ravages of a warming atmosphere. While most people know about Gore's *An Inconvenient Truth*, relatively few have heard of Kolbert's book *Field Notes from a Catastrophe*.[25] But make no mistake, Kolbert's writing opened plenty of eyes and changed lots of minds about the severity of the climate-change/global-warming threat.

In the *Chronicle of Higher Education*, Doug MacDougall wrote about *Field Notes*, "[Elizabeth Kolbert's] research is thorough. She gleaned much of her information from personal interviews and visits to localities around the world.... Kolbert tends not to use alarmist language to argue for a particular viewpoint, choosing instead to let her stories and interviews do the talking.... And by the end of the book, the reader will have no doubt that the problem [global warming] is a serious one."[26]

T. C. Boyle, author of *Drop City*,[27] added, "If you know anyone who still does not understand the reality and the scale of global warming, you will want to give them this book.... The hard, cold, sobering facts about global warming and its effects on the environment that sustains us.... Kolbert's *Field Notes from a Catastrophe* is nothing less than a *Silent Spring* for our time."[28] High praise, indeed!

Born in 1961, Kolbert lived in New York as a child—first in the Bronx and later moving to Larchmont, a suburb located eighteen miles northeast of Midtown Manhattan. She attended Yale University for four years, where she studied literature, then—after winning a prestigious Fulbright scholarship—she moved to Germany to study at the Universität Hamburg.[29]

A career journalist, Kolbert began working in Germany in 1983 as a freelance journalist for the *New York Times*. In 1985, she moved back home to the *NYT*'s Metro desk, where she wrote the *Metro Matters* column from 1988 to 1991. She then served as the paper's bureau chief in their Albany, New York, office from 1992 to 1997.[30]

Since 1999, Kolbert has been a staff writer at the *New Yorker*, where her writing has included book reviews, political profiles—notably of Senator Hillary Clinton and former mayors Rudolph Giuliani and Michael Bloomberg—and copious articles about climate change. Her series on global warming, *The Climate of Man*, which appeared in the *New Yorker*'s spring 2005 issue, won the American Association for the Advancement of Science's magazine journalism award, and also provided the background material for her *Field Notes from a Catastrophe*.[31]

In addition to her work for the *New Yorker*, Kolbert's stories have appeared in the *New York Times Magazine*, *Vogue*, *Mother Jones*, and included in the anthologies *The Best American Science and Nature Writing* and *The Best American Political Writing*.[32] *The Prophet of Love and Other Tales of Power and Deceit*—a collection of her writings—was published as a book in 2004. Most notably, Kolbert's most recent book about Earth's rapidly diminishing biodiversity—*The Sixth Extinction: An Unnatural History*—won the 2015 Pulitzer Prize for General Nonfiction.[33]

In the process of describing the history of the world's mass extinctions, *The Sixth Extinction* combines intellectual and natural history with reporting from field locations all over the world. Throughout this book, Kolbert turns her emphasis away from climate change to focus on another disaster that's currently bludgeoning the natural world—the widespread decline in biodiversity.[34] Bluntly stated, we're currently losing species at a rate of one thousand times higher than unassisted nature was doing before humans came along to contaminate the broth of global life.[35] According to Kolbert, while humans weren't responsible for the first five mass extinctions, our fingerprints are all over the one that's occurring right now.[36]

As Kolbert alludes to in *The Sixth Extinction*, the fifth extinction—the one that wiped out the dinosaurs—was believed to have been caused by a six-mile-wide asteroid colliding with Earth.[37] But this time humans appear to be serving as the asteroid that is threatening to wipe out a significant percentage of the world's current species.

This landmark work—which also first took shape as an article in the *New Yorker*—won many other prestigious awards—including the *New York Times* 2014 Top Ten Best Book of the Year[38]—and it placed number one on the *Guardian*'s list of the 100 Best Nonfiction Books of all time.[39] It was also a finalist for the National Book Critics Circle awards for the best books of 2014.[40] Much high acclaim that we all should pay attention to while we are trying to short circuit Earth's climate-change crisis.

NAOMI ORESKES: HISTORIAN / ORACLE

The instant we walked into Professor Naomi Oreskes's office, her office mates knew exactly whom we sought. Naomi "Engaged Scholar" Oreskes has been lighting up the media. The attention comes from her irrefutable courage, as she throws herself squarely into the climate debate that most career-minded scientists give a wide berth.[41]

Dr. Oreskes is currently a professor of the history of science and an affiliated professor of earth and planetary sciences at Harvard University. She received her doctorate from Stanford University in 1990 in the graduate special program in geological research and history of science.[42] Oreskes's career as a historian led her to an in-depth examination of the role of dissent in the scientific method. To investigate the legitimacy of the climate-science reports, she searched one thousand articles published in peer-reviewed scientific literature over the past ten years—a novel action in regards to global warming.[43] Out of all of the reports that she pulled, not a single paper provided dissent against the Intergovernmental Panel for Climate Change (IPCC) statement: "Most of the observed warming over the past fifty years is likely to have occurred due to greenhouse gas emissions."[44] When Oreskes published these results in *Science* in 2004, titled "The Scientific Consensus on Climate Change," she found herself under immediate political attack. In regards to the article, she said, "It ignited a firestorm. I started getting hate mail. . . . At the same time, Al Gore talked about my paper in *An Inconvenient Truth*. Suddenly, I was a hero to the left because of Al Gore and a demon to the right because I was now part of the conspiracy to bring down capitalism. I thought I'd entered a parallel universe."[45]

The wild backlash to her paper made Oreskes realize that there was something odd going on. She explains how, in the hate mail she received, "I would be accused of all kinds of strange things [e.g., being a Stalinist and a Communist]."[46] That was her first clue that there was something more to the story than just public misunderstanding of the science.[47] "Most scientists thought that this was a problem of scientific illiteracy. . . . People had no idea that the reason there was so much confusion and doubt was because there had been an organized, well financed, well structured, professional campaign to create confusion and doubt," she explained, shaking her head.[48] When she realized that there was enormous public debate about something so important, she knew she needed to tell the full story in a big way.[49]

Dr. Oreskes's critics were scientists, but not climatologists, nor did they study any form of climate science, yet they spoke on the issue of climate change as if they were experts.[50] As we mentioned in chapter 21, these were the same folks who had engineered doubt for the "tobacco wars," as well as ozone depletion. They worked off of a playbook of strategies—"insist that the science is unsettled, attack the researchers whose findings they disliked, demand media coverage for a 'balanced view.'"[51]

Why would they go to such great lengths? These "merchants of doubt," as Oreskes named them, intentionally targeted and undermined science cited to support new government regulations.[52] The battle that this group waged against climate change, a destructive force that threatens the livelihood of all of us, was not about science, but *economics*.[53] These physicists were strong believers in the unfettered free market and felt that without free markets we could not have democracy.[54] What was the link between the topics about which they were doubt-mongering? Oreskes said, "Each was a serious problem that the unregulated free market didn't respond to."[55] To stop any of these problems, tobacco use, acid rain, or climate change, regulation is required—and that is anathema of this group of doubt-mongers.

To compound the problem, the community of climate scientists did very little to speak out against these myths. According to Oreskes, when asked why they didn't do more, the scientists she spoke with said "'We knew it was garbage so we just ignored it.' Well that does not really work you know," Oreskes said. "You have to take out the garbage . . . and that's where the scientific community has been a bit slow."[56] So Oreskes did. Her primary goal was to educate the broader scientific community. The secondary target was journalists, "because journalists were presenting the issue as a big debate and I wanted to say to the journalists, look this is wrong, you are misrepresenting what is actually going on in the scientific community," said Oreskes.[57]

She thinks that the climate challenges we face can be solved, but, as we hear often from our heroes, "the hour is late."[58] To solve the problem, she says, "nearly all economists agree that we need a price on carbon. I would like to see that as a first step, and see how far it takes us."[59] Also, market-based mechanisms have the potential to appeal to a wide variety of people, including those who fear that climate change is an excuse to dismantle capitalism.[60] In the long run though, she thinks we need to change the way we think: "We have deified markets, and made the profit motive not only the dominant one, but the only one that the right wing does not consider suspect. This is a strange state of affairs. Once upon a time altruism was honored, greed was suspect. In the past 30 years, the ideologies of neo-liberalism have turned that on its head."[61]

GUS SPETH: INSIDER / COMMUNICATOR

Hailed by many as the "one of the nation's most influential environmentalists" and now leading the way in "system-changing activism," James Gustave "Gus" Speth went from being the "ultimate insider" to being arrested for activism in front of the White House.[62]

Climate change first caught Speth's attention in 1979, when four revered scientists—Roger Revelle, Charles David Keeling, Gordon MacDonald, and George Woodwell—presented him with a report that highlighted the issue. At the time, he was the chairman of the Council on Environmental Quality (CEQ) under the Carter administration.[63] The report called for government action to limit the buildup of greenhouse gases. But despite this strong warning from scientists nearly forty years ago, Speth thinks climate crises, not the science, will likely be what will force us to take necessary actions to stem climate change. "We are already too late, as you know," says Gus. "We have reached the 1.0 C global average warming and the 400 parts per million mark. We are going to have some climate repercussions that are worse than we have had before. We will have severe repercussions even if we stopped all emissions today. . . . And that is certainly not going to happen."[64]

In 2011, Gus Speth was arrested in front of the White House alongside Bill McKibben and others in a peaceful protest against the Keystone XL pipeline.[65] Speth and McKibben see eye to eye on many issues—a relationship that a few days in jail helped galvanize. Speth released the best statement of the arrest in Washington, DC: "I have held a lot of important positions in this town, but none seem as important as this one."[66]

For Speth, the key solutions involve transitioning toward a new economy, working with local governments, community building, and divestment. He says, "As the saying goes, 'You have got to walk on both legs,' and the leg we can put forward first is to try to do things like our president, at long last, is trying to do—and hopefully more. . . . But in the somewhat longer run, we need to walk on that other leg, and that is to transform the system so that it is not creating such enormous pressure to increase or not decrease emissions."[67]

Having tried to work within the system, and having become somewhat frustrated with the pace of progress, Speth has dedicated himself to trying to change the system: "I am working now mostly on what we call the 'Next System Project' [www.thenextsystem.org], which is an effort to promote really deep systemic change."[68] To transform the system, we need a drastic paradigm shift, beginning with our relationship to the environment. He defines an environmental issue not as "air pollution, water pollution, climate change" and so on, but as "something that has a big effect on environmental outcomes, on our prospects of leaving a good environment to our children and grandchildren."[69] Therefore, an environmental issue "includes things like the health of our political system and sustaining our failing democracy and not just sustaining our natural areas."[70]

Speth takes issue with today's capitalism. It dominates as a ruthless system that is driven by profit at all costs and allied with a political system that seeks to largely project national power around the world. We need to change this to a system that gives true and honest priority to people, place, and planet—not to profit, GDP and national power. "One thing that is critical to this [Next System Project] is that we have a functioning political system that is truly democratic and that that democratic political system has a huge say in the direction of investments in the country," Speth says.[71]

In addition to a belief that the system needs to be changed, Speth thinks that the people who corrupted it need to be held accountable. Bill McKibben and 350.org have launched the divestment effort, which Gus fully supports. He thinks the divestment movement highlights the "huge corporations which have systematically distorted public viewpoints, knew what they were doing, and that they really were destroying the climate, and determined to ignore that and makes as much profit as they could in the meanwhile."[72] Additionally, these corporations were determined to stop the public from doing anything about their destruction through their political power—and they succeeded. Another hopeful way to pinpoint the transgressions of the fossil-fuel companies is that people have begun to quantify the cumulative emissions per corporation, "so we know how much each corporation has contributed to the problem, and that is going to lead to legal actions against the corporations, to hold them accountable."[73]

Speth says, "If we want to save the planet's climate for children and grandchildren, we've got to act in a dramatic and drastic way, starting now, and wring the fossil fuels out of our energy system over the next thirty-five years."[74] With Congress holding up significant action at the federal level, we need to act at the state, town, city, and individual level. "Communities are coming together and focusing on resilience and sustainability, focusing on the level of care that they give each other and focusing on trying to build stronger local economies," he explains.[75] Equally important, we have to build up a "mighty political force." He was inspired to see the beginnings of that with the 400,000-person climate and climate justice march in September 2014 in New York.[76]

When you add government initiatives, system change in theory and practice, and divestment and litigious initiatives to all of the positive things that you see going on in communities across the country, Gus Speth has some hope: "People [are] taking responsibility, changing their lifestyles and other things. It is a slow process, but it is changing and it is moving in the right direction. That is the biggest ground for hope that we have."[77]

JOE ROMM: COMMUNICATOR / SCIENTIST

Joseph Romm's unique background as a scientist and supercharged communicator has earned him the moniker "America's Fiercest Climate Change Blogger." He makes the bold but apt claim that climate change will have a bigger impact on all of humanity than the Internet has had:

> Imagine if you knew a quarter-century ago how information technology and the Internet were going to revolutionize so many aspects of

> your life.... It turns out that we have such advanced knowledge of how climate change will play out over the next quarter-century and beyond. ... Climate change is now an existential issue for humanity. Serious climate impacts have already been observed on every continent. Far more dangerous climate impacts are inevitable without much stronger action than the world is currently pursuing, as several major 2014 scientific reports concluded.[78]

Joe Romm has been deeply involved in climate science, policy, and solutions for over twenty-five years, ten of which he has devoted to communicating the science to a general audience.[79] His interest in climate change was piqued in the mid-1980s while studying for his physics doctorate at the Massachusetts Institute of Technology. His thesis focused on the physical oceanography of the Greenland Sea, which drew him to Scripps Institute of Oceanography for a portion of his graduate work. "The impact of climate change on the oceans was already a concern at Scripps in the 1980s," Romm writes in his book *Climate Change: What Everyone Needs to Know*.[80]

For five years in the mid-1990s, Romm was acting assistant secretary of energy for Energy Efficiency and Renewable Energy in the US Department of Energy. Currently, Romm is the founding editor of *ClimateProgress* and chief science advisor for the Showtime TV series *Years of Living Dangerously*.[81] He has written nearly a dozen books on communication, clean energy, pollution reduction, and hydrogen fuel—our favorite title is *Language Intelligence: Lessons on Persuasion from Jesus, Shakespeare, Lincoln, and Lady Gaga*.[82]

Dr. Romm is the first to admit that we have already lost an important decade (or two), and that now it is going to be a major race to avert catastrophe. He takes solace in the agreement in Paris, as a start. It gives Romm hope "that 185 countries have made substantial commitments to slow, or actually reverse, their carbon pollution trends and to increase funding for development and deployment of clean energy."[83] He makes it clear that if we want to avoid high risks of catastrophic impacts, we need to deploy renewables and related energy efficient technologies at a much higher rate than we are now. "It is going to be the greatest race in the history of humanity between the impacts of the warming we get and the rate of the warming vs. the technology and the political will to go to zero. We do need to take the entire planet to zero net emissions before 2100. And obviously that is an epic challenge, but there is nothing in principle that stops it," he says emphatically.[84] Romm is a big fan of Mark Jacobson's work. Jacobson has certainly shown that the United States and other countries can absorb vastly more renewable electricity than we currently are, and that one can have a straight pathway to a carbon free electric grid.[85]

Romm has a unique perspective as a scientist and a journalist. He points out the incongruities of the scientific method and the need to speak in emotional and moral terms about the issues that arise from climate change. Excluding your personal beliefs from your scientific research is important to stay objective, but, he explains, "if someone gives you a scientific lecture whose obvious implication is that a lot of people are going to suffer needlessly unless we take a bunch of action . . . but then does not say that we have a moral obligation to do something about this data . . . well, that creates a big disconnect between the science and how it should be applied."[86]

Improved communication by scientists such as Joseph Romm has helped the public awareness and understanding of climate change grow in recent years. As Dr. Romm suggests, scientists are not inherently great at communicating—they are more comfortable analytically, dealing with numbers and facts: "There used to be an entire profession whose job was to understand what scientists said and then explain it to the public but that profession, which my father was a part of, the journalism profession—particularly science journalism—kind of crashed and burned."[87]

Romm's *ClimateProgress* blog is backed by the Center for American Progress Action Fund—in other words, he does not get paid based on the number of "clicks" he can collect. He has no pressure to use sensational journalism to generate buzz, and his posts are always bold, honest, and factual. Of course, his background as an MIT physicist gives him clout and credibility when it comes to interpreting the data. In turn, the public, most specifically the cross-section of the public that wants in-depth knowledge on climate change, has gained a trust in his reporting.[88]

When we read Dr. Romm's posts, we know that they are informed by peer-reviewed science and documented sources. Gaining trust through honest, non-sensationalist online journalism is becoming a lost art, but the success of Romm's blog shows that it can be done. And his immense success as a blogger promises to continue. As he says, "Climate change and energy touch so many aspects of our lives that I certainly never run out of things to write about."[89]

PAUL HAWKEN: INNOVATOR / VISIONARY

When it comes down to personal ingenuity and plans for solving climate change, Paul Hawken stands out. Entrepreneur Hawken has transformed a financial empire built on selling garden supplies and materials—through the once-world-famous Smith & Hawken Company that he cofounded—into Project Drawdown, the world's most ambitious undertaking for finding and testing solutions to our climate dilemma.

As Hawken envisions it, Drawdown will culminate in a book that details the costs and benefits of scores of climate solutions—from light bulb technology to tropical forest carbon sinks and alternative strategies for livestock grazing and crop production.[90] For each potential solution, Hawken and his team of scientists will "do the numbers." This will provide detailed, science-based data and econometric models showing how each potential solution plays out—based on current technology. It will also demonstrate how each solution will likely evolve over the project's thirty-year frame of reference.[91]

In 1966, Hawken took over a small retail store in the city of Boston called Erewhon (after Samuel Butler's 1872 utopian novel) and turned it into the Erewhon Trading Company, a natural-foods wholesaler. Next, with Dave Smith, he cofounded the Smith & Hawken Garden Supply Company in 1979—a retail and catalog business.

In 1999, Hawken coauthored a book with Amory and Hunter Lovins entitled *Natural Capitalism: Creating the Next Industrial Revolution.*[92] *Natural Capitalism*—which has been translated into twenty-six languages—popularized the idea that Earth's natural resources should be considered "natural capital," since they provide "ecosystem services" from which humans derive such benefits as clean water and waste decomposition. Then, in 2008, he cofounded Biomimicry Technologies with biologist Janine Benyus, the author of *Biomimicry: Innovation Inspired by Nature*.[93]

In 2007, Viking Press published Hawken's *New York Times* bestseller *Blessed Unrest: How the Largest Movement in the World Came into Being and Why No One Saw It Coming*.[94] The book is about the many nonprofit groups and community organizations dedicated to many different causes, which Hawken calls the "environmental and social justice movement."[95]

In an interview with us, Hawken elaborated: "*Blessed Unrest* describes what I call humanity's immune response to ecological degradation, economic disease, and political corruption. All three are intimately intertwined with global warming. When I was doing the initial research [for this book], our institute was cataloging the more than two thousand different types of nonprofit organizations in the world according to their purpose, and month after month we saw the climate movement emerge, grow, and differentiate."[96]

Project Drawdown, however, is Hawken's piece de resistance. The project is aimed at reducing—not just stabilizing—greenhouse gas (GHG) concentrations in the atmosphere in order to reverse rising global temperatures. Drawdown grew out of Hawken's frustration with actionable, scalable solutions that would make a meaningful dent in the atmosphere's growing accumulation of GHG. As he saw it, the solutions that had been put forward over the years were all seemingly out of reach—involving either ungodly amounts of solar and wind energy or the mass adoption of futuristic, unproven technologies.[97]

In a conversation with GreenBiz's Joel Makower, Hawken recalled, "It made me feel like this is intractable, that it requires such Promethean work by such mammoth institutions—with policy changes that are more than structural. It made me feel like it wasn't possible to address climate change, rather than giving me hope."[98] In climate-change activist Bill McKibben's seminal 2012 *Rolling Stone* article, "Global Warming's Terrifying New Math," Hawken asked, "Why aren't we doing the math on the solutions? Somebody should come up with a list and see what it requires so you get to drawdown."[99]

In 2013, Hawken began teaching at San Francisco's Presidio Graduate School, alongside climate activist and entrepreneur Amanda Joy Ravenhill. "One day we were just riffing, and we started talking about drawdown and we said, 'Let's do it. No one else is doing it,'" Hawken recounted. Today, Ravenhill is Project Drawdown's executive director and, with Hawken, the project book's coeditor. Together, the two have recruited more than eighty advisors, partners, scientists, government agencies, and participating universities—plus another two hundred graduate students—to work on the project.[100]

Hawken further described his Project Drawdown process for us in a February 2016 email interview: "[In Project Drawdown] we are filling this void by doing the math on the atmospheric and financial impacts of state-of-the-shelf solutions if deployed globally and at scale over the next 30 years. State-of-the-shelf refers to techniques that are widely practiced, commonly available, economically viable and scientifically valid."[101]

He continued: "In Drawdown we identify solutions that are already in place. But we also describe what we call 'coming attractions,' solutions so new and incipient that we cannot as yet fully measure and map their impact. Here we see genius and brilliance and humanity at its best."[102]

True to Hawken's nature, he's not likely to be satisfied with simply creating a book, however ambitious and meticulously detailed. Instead, Project Drawdown's plans extend in several directions: The solutions and calculations will be contained in a publicly available database—along with the means for individuals and groups to create customized applications. There are also plans for accompanying educational curricula developed by the National Science Foundation. And there will possibly be some media projects based on the work.[103]

In our interview, Hawken concluded,

> There are many reasons to believe [that climate change can be solved]. In "Drawdown," we identify over one hundred of the most substantive solutions that are in place and expanding globally. We see in our models [that] the moment in time when greenhouse gases decline on a year-to-year basis in the upper atmosphere is possible within three decades. "Drawdown" is the only goal that makes sense for humanity.

> And it is eminently doable. By collectively drawing carbon down, we lift up all of life.[104]

As author Joel Makower concludes in his GreenBiz article, "It's easy, in today's divisive and toxic political environment, to view 'Project Drawdown' as too good to be true—a quixotic quest for an unattainable goal. But there's something simple and sane about the project's collective ingredients: unabashed optimism tempered by sharp-pencil calculations, a bold goal undergirded by scientific pragmatism, immediacy coupled with a 30-year horizon, all leveraging the wisdom of a very smart crowd."[105]

Chapter 25

POLITICIANS / ADVOCATES

AL GORE: COMMUNICATOR / LEADER

If we were asked to select a poster boy for the climate-change movement in the United States, that person would be former senator and vice president Al Gore, Jr. In a highly unusual move among typically image-conscious politicians, Mr. Gore stood up and—with his PowerPoint presentation that became the 2006 Oscar-winning documentary *An Inconvenient Truth*—told the world that climate change was real, that humans were responsible, and that we needed to take action immediately to preserve our quality of life on the planet. In the process, he endured a barrage of hate mongering and insults, completely unbefitting to his mission—to maintain a habitable home for us all.

Because of his heavy involvement in many facets of the US environmental movement, we have already talked a lot about Mr. Gore in part 3. So we are going to focus on giving you some of his background here, with an emphasis on how he became so ingrained in the climate-change movement in the first place.

As a powerful figure in the Washington, DC, establishment, Al Gore Jr. has been involved in politics his whole life. He was born on March 31, 1948, and raised in Washington, DC, where his father, Al Gore Sr., served on Capitol Hill from 1939 to 1971—first as a US representative and then as a senator from the state of Tennessee. As a result, the junior Gore spent his boyhood attending prestigious private schools in the District of Columbia, with summers working on the family farm in Carthage, Tennessee.[1]

As a student at Harvard University during the 1960s (where he roomed with future actor Tommy Lee Jones), Gore stumbled into the issue of climate change and associated global warming. In his senior year at Harvard, he took a class with oceanographer Roger Revelle, who sparked his interest in global warming and other environmental issues.[2] Gore grasped the climate-change science quickly and, as his political star rose, never relented in his determination to alert people that we're baking our planet and ourselves with our lust for fossil fuels.[3]

Beginning his career in public service in 1977, Gore was elected to represent the state of Tennessee in the House and then in the Senate, serving from 1977 to 1993. Early on, he became one of the first politicians on Capitol Hill to grasp the seriousness of climate change. In fact, he held the first congressional hearings on the subject in the late 1970s.[4]

Gore also cosponsored hearings on toxic waste in 1978–79. Then he organized more hearings on climate change and global warming in the 1980s. As a US senator in 1989, Gore published an editorial in the *Washington Post*, in which he lamented, "Humankind has suddenly entered into a brand new relationship with the planet Earth. The world's forests are being destroyed; an enormous hole is opening in the ozone layer. Living species are dying at an unprecedented rate."[5]

Then, in 1990, Senator Gore presided over a three-day conference involving legislators from more than forty-two countries. The overall objective was to create a "Global Marshall Plan," under which developed nations would assist less developed countries to grow economically while still providing environmental protection.[6] In 1992, he wrote his first book, *Earth in the Balance*,[7] which became the first book written by a sitting senator to make the *New York Times*' bestseller list since John F. Kennedy's *Profiles in Courage*.[8]

While serving as vice president in the Clinton administration, Gore launched the Global Learning and Observations to Benefit the Environment (GLOBE) program on Earth Day 1994.[9] This was an education and science initiative that, according to *Forbes* magazine, "made extensive use of the Internet to increase student awareness of their environment."[10] Then, in the late 1990s, Gore strongly pushed for the passage of the Kyoto Protocol, thereby supporting reduction in greenhouse gas emissions.[11] As we have previously discussed, despite Gore's staunch efforts, the United States totally botched the Kyoto Protocol.

In 2000, Gore famously lost one of the closest presidential elections in US history to George W. Bush. Having the election to become the world's most powerful leader in the palm of his hand, only to see it ripped from his grasp, initially devastated Gore.

While fate may have kept Al Gore from the US presidency, he was able to channel his energy into a genuine cause he felt passionate about—climate change. In 2005, he founded the Climate Reality Project (originally the Alliance for Climate Protection), a nonprofit organization devoted to solving the climate-change crisis. Climate Reality works to spread the truth and raise awareness about climate change using a combination of grassroots leadership training, digital communications, global media events, and issue campaigns.[12] Today, Gore serves as the group's chairperson and works with the organization to promote awareness of the ongoing dangers posed by global-warming pollution and to develop solutions for climate change.[13]

Gore also created a global media brand around his *An Inconvenient Truth*, which became—at the time—the fourth highest-grossing documentary film in US history. A 2006 book with the same title became a bestseller.[14]

In October of 2007, Gore was named as joint winner of the Nobel Peace Prize, together with the UN's Intergovernmental Panel on Climate Change (IPCC). At the ceremony, he was recognized for being "probably the single individual who has done most to create greater worldwide understanding of the measures that need to be adopted to combat global warming."[15]

Gore has remained busy traveling the world in recent years. He speaks and participates at events aimed toward climate-change awareness. As we previously mentioned, he says he has presented his keynote presentation on climate change, *An Inconvenient Truth*, at least a thousand times. And almost always he receives a standing ovation.[16]

BOB INGLIS: REPUBLICAN LEADER

Bob Inglis is executive director of the Energy and Enterprise Initiative (E&EI), a conservative Republican, and head of RepublicEn, a commu-

Our hands are needed to ensure the future of Earth's inhabitants. (p. 431) © *CHOATphotographer/Shutterstock.*

2012 postage stamp featuring Lady Bird Johnson, First Lady and founder of the Highway Beautification Act. (p. 192) © *neftali/Shutterstock.*

Sign depicting the debilitating effects of the OPEC Oil Embargo in 1973. (p. 224). © *Everett Historical/Shutterstock.*

Greenpeace demanding the use of renewable energy sources, on their vessel *Rainbow Warrior*, June 10, 2014, Valencia, Spain. (p. 221) © *Rob Wilson/Shutterstock.*

Climate scientist Dr. James Hansen, a global-warming messenger for over four decades. (p. 339) © *Joyce Vincent/Shutterstock.*

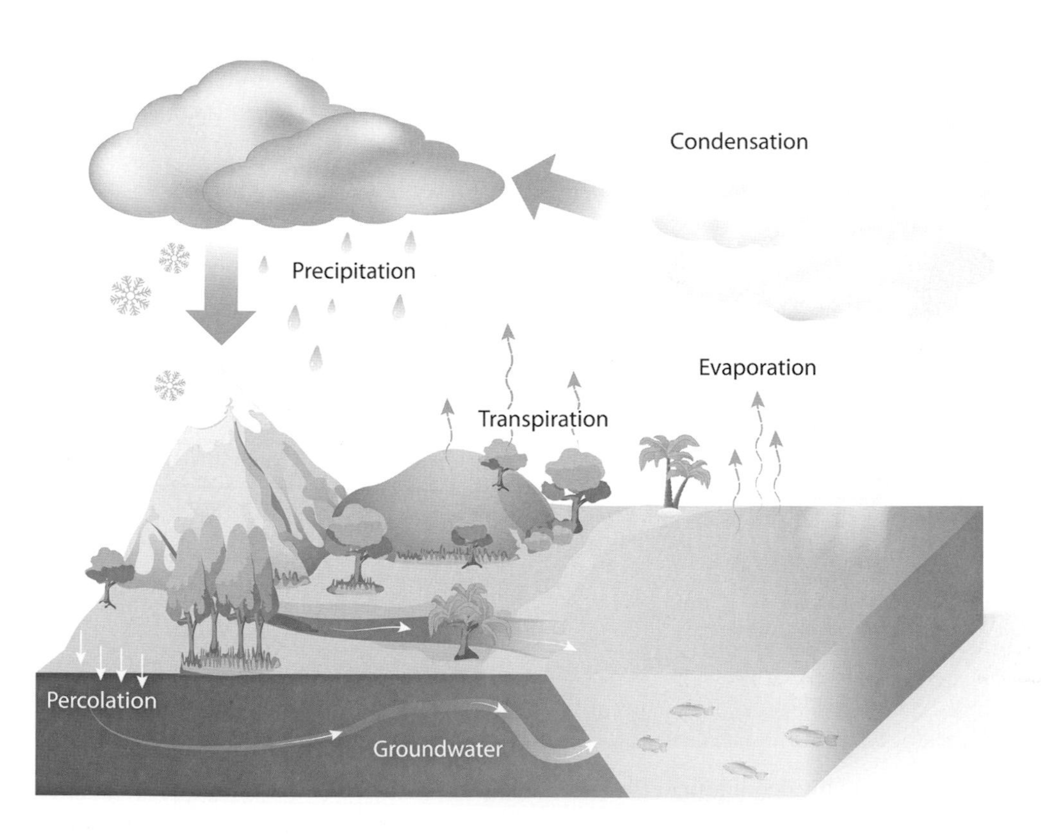

Diagram of the basic water cycle. (p. 46) © *Designua/Shutterstock.*

Former president Bill Clinton, defender of the Endangered Species Act and protector of our waterways. (p. 258) © *R. Gino Santa Maria/ Shutterstock.*

Former vice president Al Gore, one of our most dedicated and determined climate-change heroes. (p. 382) © *Frederic Legrand–COMEO/ Shutterstock.*

An icon of the frozen north, the polar bear—now established by the federal government as a threatened species—jumps between ice floes. (p. 295) © *FloridaStock/Shutterstock.*

MOSE vessel used to set up high-tide gates in the flood-prone city of Venice, Italy. (p. 52) © *chris kolaczan/ Shutterstock.*

Oil sands development taking place in Northern Alberta, Canada. (p. 78) © *meunierd/ Shutterstock.*

Part of the devastation caused by the July 2013 rail disaster in Lac Megantic, Quebec, Canada. (p. 77) © *elvistudio/ Shutterstock.*

In 2014, over 400,000 participated in the People's Climate March in New York City. (p. 321)
© *andyparker72/Shutterstock.*

Professor Thomas Lovejoy, defender of forests and protector of biodiversity, at his Paris hotel during COP-21. (p. 341)
© *Budd Titlow/NATURE-GRAPHS.*

Keystone XL Pipeline demonstrators at the White House in March 2014. (p. 373)
© *Rena Schild/Shutterstock.*

Divest Fossil Fuels demonstrators march in Toronto, Ontario, Canada, in March 2015. (p. 436) © *arindambanerjee/Shutterstock.*

COP-21 sign in Paris, France, in December 2015. (p. 331) © *ricochet64/ Shutterstock.*

Authors Budd Titlow and Mariah Tinger attending COP-21 in Paris, France. (p. 331) © *Budd Titlow/ NATUREGRAPHS.*

President Barack Obama, a staunch climate-change advocate, speaking at COP-21 in Paris, France. (p. 331) © *Frederic Legrand COMEO/ Shutterstock.*

The authors enjoyed the upbeat, festive atmosphere at COP-21—peppered with costumed emissaries such as this man with an S.O.S. Earth. (p. 331) © *Budd Titlow/NATUREGRAPHS.*

Power plant with both an array of solar panels and a field of wind turbines. (p. 443) © *taraki/Shutterstock.*

A wind farm: an iconic symbol of our renewable energy future. (p. 443) © *WDG Photo/Shutterstock.*

Tom Steyer, environmental activist and president of NextGen Climate, at Lincoln Center in New York City. (p. 401) © *Debby Wong/ Shutterstock.*

Actor Mark Ruffalo, climate activist and one of the founders of the Solutions Project, holding a poster for clean energy. (p. 393) © *Massimiliano Marino/Shutterstock*

nity of conservatives committed to action on climate change. He received the 2015 JFK Profile in Courage award for reversing his position on climate change even though he knew it would jeopardize his political career.[17] Indeed, his stance as a conservative who supports a revenue-neutral carbon tax for emissions to stop climate change is largely credited with sinking his election. "I had committed various heresies, but the most enduring heresy was saying 'climate change exists, let's do something about it,'" Inglis explains.[18]

Inglis had served six terms as a very popular state representative in South Carolina, one of the most conservative states in the country. Not only did he lose his subsequent election, his stance on climate change deeply affected his position within his party. As Bob related to us in a personal interview, conservatives call him the "'Al Gore of the Republican Party'—and they don't mean it as a compliment."[19] They use his name as cautionary shorthand—as in, "Don't be the next Bob Inglis"—for supporting climate-change efforts.[20]

As we mentioned in chapter 21, Inglis was unfazed. He had a vision for a new need in his party. He founded E&EI to jump start an "environmental right" movement, geared at creating a metaphorical "song to play on the radio,"[21] by which Inglis meant a message that resonated with right-wing conservatives.

E&EI's environmental message is catching on, especially with young Republicans:

> Folks that readily agree with us are young conservatives, college republicans, federalist societies of law schools and energy clubs of business schools. . . . I think it's because those millennials see themselves as owning the future and perhaps their grandparents are more of the thought that there is too much change. So we find that young conservatives are pretty quick to say "there is a way that the free market can fix this problem" and they feel a responsibility to take on the task.[22]

Generally, these younger people support putting a price on carbon dioxide, as long as it is revenue neutral. We mentioned Inglis's strategy in chapter 21, but it bears repeating: "Tax something you want less of—which is emissions—and do not tax something you want more of—

which is income. It's a no-brainer."[23] The other imperative for conservatives is that the tax would be border adjustable, meaning that it would be removed on exports but imposed on imports, in a way that is compatible with the World Trade Organization. As a result, the free enterprise system will kick in to solve climate change.

Inglis suggests that by revealing the costs of burning fossil fuels and making all emitters pay and be accountable, that both consumers and producers of carbon will look for noncarbon alternatives. The pressure for better renewable alternatives will drive the technology and demand, which the market will rise to meet. Bob provides an apt, albeit colorful, analogy: "As long as New York City was scooping poop behind horses, people weren't as interested in Henry Ford's new cars. . . . If you had to put a bag on the back end of your horse and pay to dispose of it, well, Henry Ford's cars would have sold a lot faster. And so, that is how we are in climate change . . . similar to what Ronald Reagan said of government—it's an alimentary canal with a voracious appetite at one end and no sense of responsibility at the other."[24] Choose any analogy you prefer, but the crux of Inglis's sentiment is legitimate: polluting our global air, water, and land has never been acceptable, but it is even less so now, with so much at stake.

SHELDON WHITEHOUSE: CONGRESSIONAL LEADER

There is a lot to admire about Senator Sheldon Whitehouse. The two-term Democratic senator from Rhode Island has assumed the role of our climate-change champion in Congress. As he explains it, he has to remain vigilant to protect his constituents, stating that "there's a very good reason Rhode Island is called the 'Ocean State.'"[25]

Just about everything that happens in Rhode Island is strongly influenced by either the Atlantic Ocean or Narragansett Bay. This means that climate change is much more than a looming future catastrophe. Senator Whitehouse knows that sea-level rise is happening right now throughout Rhode Island and low-lying portions of other coastal states. Beaches, tidal flats, and coastal wetlands are being inundated. Coastal birds and marine mammals are losing their nest sites and home territories and are being forced to relocate. Fisherman are losing their breeding grounds and harvesting territories. Homes and businesses are being undercut and tumbling into the water.

For the past three years, Senator Whitehouse has been introducing legislation aimed at slowing the planet's warming, and he has routinely been presenting climate-change speeches on the Senate floor. Unfortunately, very few of our faithful elected officials would be able to vouch for this fact. According to *Agence France-Presse*, an international news agency based in Paris, Whitehouse usually gives these speeches to an empty or near-empty chamber. But whenever he speaks to the Senate—whether or not anyone is in attendance—he always displays a green sign, telling his colleagues that it's "Time to Wake Up."[26]

On January 27, 2016, Senator Whitehouse marked his 125th climate-change presentation on the Senate floor. Only this time there was a twist. This time, instead of simply warning his colleagues about the dangers of climate change, his message had an upbeat tone, inspired by the ongoing debate over the Energy Policy Modernization Act (EPMA).[27]

Noting that the EPMA may become our first comprehensive energy-efficiency legislation since 2007, Whitehouse told the Senate, "The World Economic Forum released its 'Global Risks Report 2016,' which for the first time ranked an environmental risk—climate change—as the most severe economic risk facing the world. The report found that a failure to deal with and prepare for climate change is potentially [our] most costly risk over the next decade."[28]

To date, few of Whitehouse's fellow senators seem to have been listening to him. More than 56 percent of Republicans in the 114th Congress don't believe in human-caused climate change. Some members of Congress—such as Senator James Inhofe (R-OK)—make fun of the idea that climate change might be a problem at all.[29]

Whitehouse isn't discouraged in the slightest. He told *Morning Consult*, "If I look back 20 years from now and I can't say I did everything possible, I'll never be able to live with myself."[30]

All we can say about this is, "Good on you, Senator Whitehouse!" (What a great name if he decides to run for President some day!) It is a major relief that someone up there on the Hill has the courage to take a stand, even though it may not be politically expedient. Now if we can just find a few hundred more like him, we'll get somewhere in solving this crisis!

Chapter 26
ARTISTS / CELEBRITIES

JAMES BALOG: EXPLORER / COMMUNICATOR

James Balog has given a visual both beautiful and devastating to climate change. He squeezed in an interview with us just before his trip to touch the glaciers of Mount Kilimanjaro—a trip he had been anticipating and planning for fifteen years.[1] He was in a race against time to visit this glacier, a recurring theme for him: the frozen subject matter of his Extreme Ice Survey (EIS) keeps disappearing. He has devoted his recent work to capturing these glaciers before they melt away permanently.

Balog is the founder and director of Earth Vision Institute, a *National Geographic* photographer and geomorphologist, and also the founder of the EIS, which is the most wide-ranging, ground-based photographic study of glaciers ever conducted. He has been photographing the Anthropocene "since twenty years before it was given a name."[2]

Like several of our new heroes, James Balog was once a skeptic about climate change.[3] The catalyst that initiated his transition to a climate-change believer was the realization that there were concrete measurements of ancient climates trapped in the ice cores of Greenland and Antarctica. These cores held an actual empirical record of how the atmosphere had changed. "The climate change story was not about computer models," Balog told us. "When I understood it to be an empirical science, an actual tangible collection of evidence, that is what really got me fired up."[4] What sent him on his quest to photograph the melting glaciers, however, was an article he read fifteen years ago about the vanishing snows of Kilimanjaro. This kindled his interest and got him excited to see the glaciers—which have receded significantly since he read that article.

His EIS captured shots of twenty-three glaciers in Antarctica, Greenland,

Iceland, Canada, Austria, Alaska, and the Rocky Mountains of the United States. Since the project began in 2007, through the time-lapse photography we described in chapter 20, Balog has captured images of the world's glaciers retreating or shrinking—which 95 percent of them are doing.[5] The documentary *Chasing Ice* shows this happening in an incredibly dramatic fashion.

Balog's cameras make the invisible visible, and, if seeing is believing, the images Balog has collected prove that we are losing glaciers permanently and rapidly. The loss of this frozen ice is turning into sea-level rise, directly attributing to changing precipitation and temperature patterns.[6] Balog says there is no significant scientific dispute about this: "It's been observed, it's measured, it's bomb-proof information."[7] He refers to these glacial retreats as "the canaries in the coalmine," saying their rapid melting should be setting off warning bells for the world.[8]

In *Chasing Ice*, Balog's resiliency and his determination to capture glacial retreat and communicate its importance is highlighted by his persistence in the research—despite dealing with a serious knee injury and several surgeries. When asked why he kept going, he said, "We are fundamentally a species that works in favor of its survival—we self-propagate. The more emotional and intellectual understanding we have of how rapidly the world is shifting around us, the more likely we are to take the actions necessary to alter course."[9]

We need a paradigm shift—a demand for the technological and political will to help us incrementally peel away from fossil-fuel use where we can. According to Balog, for change to begin happening, we need to take all of the different things that people know how to do and apply them to climate change. "If everybody does a piece of that activity—whether it is to engineer wind turbines, put photovoltaics on the roof of your house, caulk your windows, put a smart thermometer in your house, change to a different car, or go to Washington and try to influence that crazy policy machine—it all keeps rippling out. Eventually it makes a new story that society absorbs and understands," he says.[10] This story eventually becomes the new paradigm and creates a new future.

Chasing Ice has enraptured audiences since November 2012. It won the 2012 award for Excellence in Cinematography at the Sundance Film Festival and was shortlisted for the 2013 Academy Awards. It has also been featured on the ABC, NBC, CBS, and PBS television networks. Along with making new films that highlight climate change, Balog will continue to use glacier imagery from the EIS to inspire us and show how much we are losing by waiting to act.

JAMES CAMERON: PRODUCER / ENVIRONMENTALIST

Many of us know Director James Cameron through his litany of Hollywood hit movies. From *Terminator* to *Titanic* to *Avatar*, Mr. Cameron is responsible for producing and directing some of the biggest blockbusters of all time.

But how many of us realize that Cameron is arguably Hollywood's most knowledgeable and passionate activist for environmental and conservation causes? He intentionally does not play up the environmental sentimentality of his "entertainment movies"—they are, after all, carefully crafted to appeal to the general public and generate revenue for studios. But his love and concern for the natural world and resource conservation is always there, no matter how subtly, in everything he does.

As the father of five children, though, Cameron really feels strongly about climate change, and he is capable of putting the gloves on and becoming a heavy hitter. Such is the case with his 2014 Showtime documentary series, *The Years of Living Dangerously* (*YLD*), for which Cameron served as executive producer and won an Emmy Award.

Born in 1954 in a small town in Northern Ontario, Canada, Cameron always knew exactly what he wanted to do with his life. And at age seventeen, when his family moved to Southern California, he dove right in creating sci-fi movies and trying to convince Hollywood studio executives to support his products. Of course, we all now know that his diligence paid off in a big way, making him one of Tinseltown's all-time greatest producers and directors. Around the same time as his move to Southern California, young James learned to scuba dive. "[I] spent hundreds of hours marveling at the biodiversity of the coral reef ecosystem," he says. "To think that all of that could be gone—literally gone—in my kids' lifetime, that is shocking."[11]

Each of the nine episodes of *YLD* focuses on a human—rather than natural—event occurring somewhere in the world. Cameron believes that the key to producing climate-change converts lies in telling intriguing human stories that resonate with the public—by involving real effects on real people's lives. With this in mind, he solicited the assistance of celebrity actors in addition to scientists and journalists, having them serve as the correspondents for the episodes of *YLD*.[12]

Cameron's logic here made perfect sense. He did not want to use his cadre of A-List celebrities simply as "talking heads." Even though the celebrities could not provide any technical expertise (since they were not climate scientists), Cameron wanted them to "roll up their sleeves and get in there as investigative journalists"—passionately displaying their concern for the causes they were covering.[13] And that's exactly what happened throughout *YLD*. Matt Damon, Harrison Ford, Arnold Schwarzenegger, America Ferrera, Olivia Munn, and the rest provided their services pro bono and worked hard, committed to the causes they were covering.

For the second season of *YLD*—which is set for release on October 30, 2016—Cameron is moving to the National Geographic Channel. Here, he will have access to a much bigger audience—which he hopes will reach a lot more people who are not already in the choir. He urgently wants to reach mainstream television audiences with his message that it's now time to stand up, be counted, and do something to help solve climate change. And once again, Cameron is using his Hollywood clout and financial resources to get the job done right.

According to Cameron,

> Government is not the answer—I don't have much faith in the political process. I believe I can be most effective at the grassroots level, inspiring and informing people, and using my cinematic skills to make a point. . . . Our leaders don't lead; they follow the polls. Until they get a mandate from the grassroots up, they are not going to do anything, because the entrenched interests have too much of a lock on things through lobbying dollars and the campaign funding process. There are too many people making too much money digging up hydrocarbons and burning them.[14]

Cameron also firmly believes that switching the world to plant-based diets is another primary solution to the long-term climate-change dilemma.[15] Animal agriculture is one of largest sources of GHG emissions in the world. If we all start eating only plants—which, as a vegan, he himself does—the world will start solving our climate-change crisis immediately.

MARK RUFFALO: ACTOR / ORGANIZER

A meme circulated on the Internet depicts the Incredible Hulk gritting his teeth, next to the quote, "The Credible Hulk: You wouldn't like me when I am angry because I always back up my rage with facts and documented sources."[16] Mark Ruffalo, the actor who played Bruce Banner (the Hulk) in Marvel's *The Avengers*, witnessed injustices that moved him, maybe not quite to the level of a raging hulk, but certainly in a way that he could not turn from.

To learn more about Mark Ruffalo and his work for climate-change solutions, we spoke with Sarah Shanley Hope, the eloquent and passionate executive director of the Solutions Project (TSP), which we talked about in chapter 21. As we mentioned then, TSP is the nonprofit group that Ruffalo co-founded, dedicated to the transition to one hundred percent renewable energy (wind, water, and solar) solutions.

Ms. Shanley Hope described how Ruffalo, at a town hall meeting in Pennsylvania, first felt clearly that he must take a stand against the reliance on fossil fuels, when community members spoke of how terribly fracking was polluting their water. She says that, for Mark, "Being face to face with the dignity and resilience, but also the powerlessness from a political perspective, of these people who were being so immediately affected in life and death consequences . . . he could not turn away from the inhumanity of the situation. And so he stepped into his humanity. That is the kind of leader he is. He shows up for people. He understands the power of the platform that he is fortunate to have through a successful acting career, and so he is constantly bringing that asset."[17]

Mark Ruffalo has placed himself deeply into the fight against fossil fuels in a way unmatched by many celebrities other than Leonardo

DiCaprio. He and Leo marched with the indigenous, First Nations leadership in the New York City climate march two years ago, knowing that their presence would draw much-needed media attention to their cause.[18] Ruffalo used his presence to call attention to Detroit a year and a half ago, when the water was shut off in low-income African American communities—one of the only people from outside of the state to attend.[19]

The Solutions Project has countless pictures of Mark Ruffalo alongside workers, families, and community members, fighting for their rights to clean, affordable, healthy communities and resources such as air, water, and energy—along with tweets and blog posts that show his concern for the issues faced by the people he stands with. Sarah Shanley Hope extols Ruffalo's commitment: "I mean you talk about a leader who is giving 100 percent for 100 percent—giving everything he can for this beautiful diversity of people we have in our country—Mark Ruffalo is one of those people doing that. . . . He is a rare celebrity for sure, but also a rare human being in general."[20]

Using their large social media influence, Sarah asserts that TSP entreats everyone to give one hundred percent for climate-change solutions: "How we achieve the transition to 100 percent [renewable energy] is really about it being led by 100 percent of people, and is for 100 percent of people . . . tapping into engineers, artists, teachers, cooks . . . everyone has a role to play in this transition."[21]

TSP employs an ecosystem approach, working across sectors to find solutions. "It is not the people by themselves," explains Shanley Hope. "It is the people leading the call and bringing that moral clarity, that energy that only parents fighting for their kids, or neighbors fighting for their community, bring to any problem, and then it is matching that people-power with the innovation and energy of business and the policy levers and the infrastructure to scale that only government has."[22]

According to Ms. Shanley Hope, TSP focuses on "positive, attractive, solutions-oriented stories and calls to action that keep us all lifted up, hopeful, and squarely focused on the choice that we actually have right now."[23] At the individual level, TSP strives to tap into our human energy, spirit of perseverance, resilience, creative thinking, and commitment to the safe, healthy, thriving future of our children and children's children.

And it is working. "If you start to look at those places that are adopting policies, removing those political barriers to transition—people in those states [such as California, New York, Arizona, Nevada, and Iowa] are starting to have greater access to affordable, clean energy . . . and so those solutions are being adopted at ever-increasing scale and speed."[24] TSP clearly illustrates that the economic benefits from switching to one hundred percent renewables can be achieved without subsidies or incorporating the externalities (hidden costs, such as pollution clean-up and human health damages) from fossil fuels. (We discussed this to a greater extent when we talked about Mark Jacobson in chapter 23.)

LEONARDO DiCAPRIO: ACTIVIST / ACTOR

We all know Leonardo DiCaprio as arguably the finest actor of his generation. Because of his insistence on taking on something more than run-of-the-mill, "pretty-boy" rom-com roles—which he could do in his sleep—DiCaprio is best-known for his edgy, unconventional acting. No doubt, most of us have seen at least a few of his neoclassic movies, including *Titanic*, *The Great Gatsby*, *The Aviator*, *Django Unchained*, *Catch Me If You Can*, *J. Edgar*, *The Departed*, and *The Wolf of Wall Street*.

But, as Suzanne Goldenberg writes, "Leonardo DiCaprio was a climate champion long before he wrapped himself in a horse carcass, vomited up raw bison liver, and risked hypothermia for his 2016 Oscar-winning role in *The Revenant*."[25] In fact, he once considered quitting Hollywood to take up protesting the fossil-fuel industry, becoming a full-time environmental activist.

DiCaprio's comments in a recent interview in the *Sunday Times* tells us everything we need to know about his sincerity on this issue: "I had a friend say, 'Well, if you're really this passionate about environmentalism, quit acting.' . . . But you soon realize that one hand shakes the other, and being an artist gives you a platform. Not that necessarily people will take anything that I say seriously, but it gives you a voice."[26]

An only child, born on November 11, 1974, in Los Angeles, the young Leonardo would regularly wander over to the nearby Natural History

Museum for his weekend entertainment. There, he became mesmerized with the natural history and wildlife epics often presented in the IMAX theater. This background instilled in him his life-long love and support for environmental causes throughout the world.

A visit to his website tells us even more about how serious DiCaprio is about environmental activism. In addition to praise for his films, the thing that most stands out is the Leonardo DiCaprio Foundation (LDF). Operating through collaborative partnerships, the mission of LDF is "the long-term health and wellbeing of all Earth's inhabitants."[27]

The 2016 Oscar stage was probably the biggest audience to date for DiCaprio's climate message. But those who have worked with him behind the scenes for years claim that he has been passionate about the issue for a very long time. Recently, he has poured both his celebrity and his money—almost $60 million through the LDF[28]—into supporting UN climate negotiations and spreading public awareness about the dangers of climate change.[29]

In September 2014, DiCaprio joined with an estimated 400,000 peaceful protestors from more than 150 countries in the streets of Manhattan and then gave an impassioned speech to the UN about the dangers of climate change. In December 2015, DiCaprio held a private discussion with Ban Ki-Moon, the UN Secretary General—on the sidelines of the Paris climate negotiations (COP-21)—before dropping in at the World Economic Forum in Davos-Klosters, Switzerland, to pick up another climate-change award in January 2016.[30]

Other actors have also dedicated much of their time to climate change and other environmental causes. But DiCaprio is a special case because of the level of fame he has achieved in Hollywood. Enric Sala, explorer-in-residence for *National Geographic*, talking about DiCaprio, says, "There are many foundations and non-governmental organizations interested in oceans and many do great work. He has a megaphone that nobody else on the planet has. He is so respected and admired and influential all around the world—from the general public to heads of state—so when he says something people listen."[31]

"I am consumed by this," DiCaprio told *Rolling Stone*. "There isn't a couple of hours a day where I'm not thinking about it. It's this slow burn. It's not 'aliens invading our planet next week and we have to get up and fight to defend our country,' but it's this inevitable thing, and it's so terrifying."[32]

After his February 29, 2016 Oscar win and heartfelt speech, DiCaprio announced that his next big project will be a new documentary about climate change. So far, production for this project has taken him to several frontlines of the pending crisis—including Arctic ice fields, Indian floodplains, and the playgrounds of Miami's South Beach.[33] Leonardo DiCaprio has no intention of giving up.

NEIL DEGRASSE TYSON: SCIENTIST / COMMUNICATOR

What do you say about a man who—during his educational years—was captain of his college wrestling team; rower on his collegiate crew team; ballet, jazz, and ballroom dancer; natural history essayist; and doctor of astrophysics and cosmology. And—since obtaining his degrees—has become a columnist; multi-published author; public television show host; planetarium director; and a NASA Distinguished Scientist? When we first read Dr. Neil deGrasse Tyson's bio, we knew that this was someone who could help us solve the climate-change crisis.

Born on October 5, 1958, in New York City, Neil deGrasse Tyson became hooked on astronomy when he made his first visit to the Hayden Planetarium—a facility he now directs—at the age of nine. Truly a contemporary renaissance man, Tyson not only has had an asteroid named after him but was also voted the Sexiest Astrophysicist Alive by *People* magazine in 2000.[34] Educated as an astrophysicist and cosmologist, he epitomizes the oft-used cliché "rocket scientist." But his skills as a science communicator and host of such PBS shows as *Nova ScienceNow*, *Origins*, and *Cosmos: A Spacetime Odyssey* are exceptional.

Tyson has also written numerous books on the universe and our purpose within it, such as *Death by Black Hole and Other Cosmic Quandaries*,[35] *Origins: Fourteen Billion Years of Cosmic Evolution*,[36] *Space Chronicles: Facing the Ultimate Frontier*,[37] and his memoir, *The Sky Is Not the Limit: Adventures of an Urban Astrophysicist*.[38]

Tyson's knowledge of a variety of scientific fields is vast. But after watching many of his shows on PBS, we have to say that his greatest talent lies in getting his messages across with a deft combination of humor, rhetoric, and candor. The popularity of Tyson's YouTube videos is a proof positive of his mass appeal. Star formation, exploding stars, dwarf galaxies, and the structure of our Milky Way are among his many professional research interests, and he regularly works with data from the Hubble Space Telescope, as well as from telescopes in California, New Mexico, Arizona, and the Andes Mountains of Chile.[39]

Using his quick brain, acerbic wit, and media connections, Tyson has never backed away from a good fight. Just ask any of the creationists he's verbally sparred with on the air. And he is especially adroit when it comes to expressing his views on climate change.

In "The Lost Worlds of Earth," episode nine of *Cosmos: A Spacetime Odyssey*, Tyson said, "We're dumping carbon dioxide into the atmosphere at a rate the Earth hasn't seen since the great climate catastrophes of the past, the ones that led to mass extinctions. We just can't seem to break our addiction to the kinds of fuel that will bring back a climate last seen by the dinosaurs, a climate that will drown our coastal cities and wreak havoc on the environment, and our ability to feed ourselves."[40]

From an educational perspective, Tyson provides the absolute best demonstration of the difference between *weather* and *climate change* we have ever seen, in an astonishingly simple video showing him walking a typically curious dog along a beach. As the dog veers randomly from side to side on its leash, Tyson explains that the dog's erratic movements are like *weather*—which may vary wildly from day-to-day. Meanwhile, he says, his own straight-line movement mimics climate change, which captures the center-line of the periodic weather patterns. In the final analysis, Tyson says, "you have to keep your eye on the man, not the dog."[41]

Finally we offer this pithy—and almost fatherly—Tyson admonition to nonbelievers: "The problem that I see is that if you remain in denial, then you are not at the table discussing reactions to anthropogenic climate change. So, we're losing time here, which is to say we're causing climate change. Now, let's go back in the room and debate what to do about it."[42]

While Dr. Tyson's scientifically based put-downs of unfounded rhetoric may come across to some as arrogant, they may be just the words we need to educate the general public about the potential pending ravages of climate change.

NORMAN LEAR: PRODUCER / PROMOTER

Anyone over sixty probably knows Norman Lear—or at least his creative products—very well. Throughout the 1970s, Lear's television productions helped us define and understand the culture and world in which we lived. Lear's shows—many instant hits and now considered classics—included *All in the Family*, which ran for nine seasons, *Maude*, *Sanford and Son*, *Good Times*, *The Jeffersons*, *One Day at a Time*, and *Mary Hartman, Mary Hartman*. While definitely making us all laugh, Lear's socially poignant narratives also made us reflect on our place in society and how we could all function better as bits of cloth in an interwoven fabric.

As you might expect from his shows, Lear spreads his progressive thinking and liberal activism in everything he does. We are including Lear here not because of the strength of his association with environmental causes—although he has espoused many—but because of something he and his wife, Lyn, started in 1989, together with Cindy and Alan Horn.

The Environmental Media Association (EMA) is a nonprofit organization that "mobilizes the entertainment industry in a global effort to educate people about environmental issues and inspire them into action. EMA relies on the simple but powerful concept that, through television and film, the entire entertainment industry can influence the environmental awareness of millions of people."[43]

Today, EMA continues to positively affect how the general public receives and understands environmental information. They do this by weaving key environmental messages into programming and working with celebrities to emphasize positive environmental role models.[44] EMA also regularly joins forces with environmental groups, creates public service announcements, and educates the entertainment industry about environmental issues.

Every year, the EMA Awards honor those in the entertainment industry who have successfully incorporated environmental messages in their work. In addition, EMA's Green Seal Award recognizes "environmentally responsible production efforts behind-the-scenes, such as set construction, energy usage, resource conservation, recycling and purchasing policies."[45]

Since 2014, the Lears have been opening their conference rooms and their Hollywood home so that climate-change heroes—such as James Hansen and Bill McKibben—can speak to Hollywood writers, directors, and producers. The idea is to get the mainstream media—namely primetime television and studio-produced movies—to pay more attention to climate change in their big budget productions. The Lears understand that putting climate change, tactfully and tastefully (i.e., no doomsday scenarios), at the forefront of the entertainment industry is one of the best ways to get the American public to pay attention and start clamoring for something to be done.

Chapter 27
BUSINESSPEOPLE

TOM STEYER: STRATEGIST / ORGANIZER

If you need a person to go toe to toe in the climate-change arena with the Koch brothers and their network of lobbyists and naysayer politicians, Tom Steyer's your man. In fact, every time we read anything about what Steyer is doing, we feel proud that he is on our side. It is not every day that you can find a self-made billionaire who has the courage to take a stand for the natural environment. And that's what Steyer seems to be doing now, all the time.

Born in 1957 in New York City, Tom Steyer attended the Buckley School in Manhattan, Phillips Exeter Academy in New Hampshire, and then graduated from Yale University summa cum laude in economics and political science. He was also elected to Phi Beta Kappa and was captain of the Yale soccer team. He then matriculated to California, where he obtained his MBA from Stanford Business School and was an Arjay Miller Scholar, graduating in the upper 10 percent of his class.[1]

Steyer made his considerable fortune—estimated at $1.4 billion—through shrewd investments at Farallon Capital Management, the hedge fund he founded in 1986. In 2012, he stepped down as co-managing partner of Farallon, however, to focus on environmental activism. He said that he "no longer felt comfortable being at a firm that was invested in every single sector of the global economy, including tar sands and oil."[2]

Steyer didn't take a break after leaving Farallon. Instead, he and his wife—Kat Taylor—used some of their money to found two renewable energy research institutions at Stanford University: the TomKat Center for Sustainable Energy and the Steyer–Taylor Center for Energy Policy and Finance.[3]

Steyer went on to found, with several other businesspeople, Advanced Energy Economy, a group working toward policy that supports the clean energy sector; to join with Next Generation, which addresses family policy and climate; and—most importantly for our purposes—to found NextGen Climate, which takes political action on climate issues. Tom and Kat have also joined people like Warren Buffett and Bill and Melinda Gates in the Giving Pledge—a promise that they will, during their lifetimes or at their death, donate the majority of their wealth to charitable causes and nonprofit organizations.[4]

In the 2014 election cycle, Steyer spent $74 million of his own money, backing candidates he believed in, and NextGen spent another $11 million. Unfortunately, most of the candidates he backed lost their races. In 2016, Steyer hopes for better success. He wants to help the Democrats stay in the White House, while also taking back the Senate.[5]

NextGen Climate has asked that all presidential candidates, whatever their party, to create a detailed plan explaining how they would get the country to a 50 percent mix of renewable energy sources by 2030. "Not agreement with the science, not with vague good intentions, but detailed plans," Steyer said.[6] "We think energy from the people who were pushing on Keystone [XL pipeline] should [now] go to demanding people that are running for president or statewide office [to offer] solutions,"[7] said Steyer. "Offering concrete plans . . . is one of the most powerful ways leaders can move conversation on climate change forward," he added.[8]

Another part of Steyer's 2016 plan is to show how many of the Republicans seeking office are connected with his adversaries Charles and David Koch. The Kochs are, of course, the billionaire brothers who spent more than $100 million in the 2014 midterms supporting GOP candidates who opposed climate change and other environmental regulations. "With each dollar that they're spending," Tom Steyer's top strategist Chris Lehane says, "they're building a political system that is responsive to their economic bottom line—not the majority of the people, not the will of the people. . . . And, in effect, the Kochs and their allies are creating what we see as a new political party. Instead of the Grand Old Party [GOP], the Republicans now have a new Koch Republican Party, the party of Big Oil."[9]

According to NextGen, future elections are a choice between Big Oil—which puts its own gain over the wellbeing of the planet—and a sustainable world that can be found through reducing carbon emissions and converting to renewable energy sources. "[Upcoming elections represent] an electoral crossroads," Lehane says. "With the window closing on the time we have to address climate change for our kids, 2016 represents the last best chance to move the politics on this issue."[10]

JEREMY GRANTHAM: STRATEGIST / VISIONARY

Jeremy Grantham is a bubble guy. As cofounder and chief strategist of Grantham, Mayo & van Otterloo (GMO) LLC, an investment management firm, he is a highly knowledgeable investor in a variety of stock, bond, and commodity markets. But his work on predicting investment bubbles makes him particularly interesting to listen to on environmental issues. As he has followed commodities, he has watched many of them crash due to environmental challenges. He does not call this a "bubble" in the strict definition of the word, but he knows that it spells disaster—not just for the environment, but for the economy and, in fact, for our survival as a species. Grantham writes, "I am a specialist in investment bubbles, not climate science. But the effects of climate change can only exacerbate the ecological trouble I see reflected in the financial markets—soaring commodity prices and impending shortages."[11]

Overuse has created shortages in elements that we have become accustomed to having on hand, Grantham writes. For example, several of the minerals used in fertilizers will soon be in short supply, and no one knows how that will affect the agricultural industry. Put this together with the damage climate change has caused to agriculture—for example, through reduced grain harvests caused by extreme weather—and we may soon have a very serious problem.[12] "Recognition of the facts is delayed by the frankly brilliant propaganda and obfuscation delivered by energy interests that virtually own the US Congress. (It is not unlike the part played by the financial industry when investment bubbles start to form . . . but that, at least, is only money). We need oil producers to leave eighty percent of proven reserves untapped to achieve a stable climate. As a former oil analyst, I can easily calculate oil companies' enthusiasm to leave eighty percent of their value in the ground—absolutely nil," writes Grantham in an article for *Nature*.[13]

Jeremy Grantham admits that he often despairs about the direction we are headed. "Climate change is, second only to population, the single biggest problem in food. Without population, of course, we could have cruised through this. If we had a billion people on the planet, we probably would not be having this conversation—we would not need any hero. But with 7.25 billion people behaving badly often and the rich people behaving worse. . . ." He trailed off when he spoke with us, his voice tinged with frustration.[14]

Grantham is known as the behind-the-scenes funder who underwrites a large portion of the US and British green movements.[15] In addition to his day job at GMO LLC—and the founding of the Grantham Research Institute on Climate Change and the Environment, as we talked about in chapter 21—Grantham has also founded the Grantham Foundation for the Protection of the Environment with his wife, Hannelore. The Grantham Foundation "is to climate change what the Gates Foundation is to malaria."[16] Grantham puts nearly everything he earns back into the Grantham Foundation, which is managed by executive director Ramsay Ravenel. Organizations the Foundation supports include 350.org, the Sierra Club, the *Years of Living Dangerously* television series, the World Wildlife Fund, and scores of major programs at universities around the world.[17]

Grantham's first act of civil disobedience in all his seventy-plus years was a Sierra Club protest against the completion of Keystone XL pipeline at the White House in February 2013—and he was nearly arrested.[18] Grantham was not the only hedge fund manager to oppose the pipeline—Tom Steyer, whom we have discussed, is the other. Grantham opposed the pipeline both for economic and for environmental reasons. He called it poisonous and disastrous for the environment.[19] "I have told scientists to be persuasive, be brave, and be arrested, if necessary, so it only seems proper to do this," he said of his decision to join the protest.[20]

Indeed, he wants scientists to speak out more loudly about climate change:

> I have yet to meet a climate scientist who does not believe that global warming is a worse problem than they thought a few years ago. The seriousness of this change is not appreciated by politicians and the public. It is crucial that scientists take more career risks and sound a more realistic, more desperate, note on the global-warming problem. Be arrested if necessary. This is not only the crisis of your lives—it is also the crisis of our species' existence. I implore you to be brave.[21]

For Jeremy Grantham, the biggest solution would be to fix our broken political and economic systems. He argues that we desperately need enlightened government and leadership by the United States, beginning with new regulations to limit US campaign contributions and political spending. Our leaders need to think long-term, he says, incorporating environmental impacts, health impacts, and social justice issues into their decisions, rather than their current focus on short-term profit. We need a carbon tax on polluters and that money needs to be returned to individuals—a similar viewpoint to the one we discussed in James Hansen's chapter—and an energy policy that forces fossil-fuel companies to leave over two-thirds of their proven reserves in the ground.[22] We know this, and yet, "we collectively spend 650 billion dollars a year to find new oil that we absolutely cannot burn without going dangerously beyond the limit. . . . With capitalism in its current form, we are reaching levels of inequality that may lead to extreme social instability. Much worse, on our current path, we will erode our soils, poison our planet's air and water and cook our collective goose."[23]

Grantham says that he has a social contract with his colleagues to talk about finance, and they are perplexed when he talks about issues of social justice, the environment, climate change, or resource limitations, and tend to write him out of their reports. "And that is a price I pay," he says.[24] The truth is, his colleagues need to start paying attention soon, or they will be the ones paying the price. They are likely to lose money if they maintain investments in the fossil-fuel industry—an industry that has to switch to renewables or fail, both to maintain our existence on this planet and by decree of the Paris Agreement, which limits fossil-fuel emissions. Not only might they lose money in stranded (i.e., worthless) fossil-fuel assets, they may also lose clients from a growing population that values social justice and environmental issues, and therefore pulls their business from investors who do not share the same values. A wise investor should optimize seven to ten year gains, instead of focusing on short-term gain—it rewards the investor with larger growth, but more importantly, favors investments that are better for the environment and people.[25] Long-term investments are more likely to take into account factors such as limited resources and business sustainability.

BILL GATES: INVESTOR / ENTREPRENEUR

Billionaire, investor, and founder of Microsoft, Bill Gates strongly believes that human beings have the intellectual capacity to invent our way out of the climate-change dilemma. "His voice carries enormous credibility about how technology can be used to solve global warming," says Fred Krupp, head of the Environmental Defense Fund.[26] In fact, Gates's approach goes way beyond converting to existing renewable energy supplies of wind, water, and solar (WWS) as a solution. He's talking about stimulating ingenuity to come up with an "energy miracle," based on research and development projects that will seek completely new technologies for revolutionizing the power sources that run the world.[27]

Gates's arguments for promoting intensive R&D to find new technology solutions were so persuasive that in June 2015 he travelled to France and convinced President François Hollande—as president of the host country—that energy innovation had to be a top priority during the December 2015 climate change summit (COP-21).[28] This conversation, and later conversations with White House staff members, led to the creation of the Breakthrough Energy Coalition (BEC), one of the biggest public-private partnerships that has ever tackled climate change.[29]

On the opening day of COP-21, Gates, together with President Hollande and President Obama, introduced the BEC. The group unites twenty-eight of the world's wealthiest businesspeople—with collective holdings totaling more than $350 billion—who will work with governments to fund innovations in clean energy.[30] The United States, China, and seventeen other countries pledged at that same time to double their spending on energy research over the next five years.[31]

One of the ideas Gates is interested in is solar-chemical technology—or "artificial photosynthesis"—which would convert the sun's energy into hydrocarbons that can be stored and used as fuel. The *Washington Post* reported that Gates spoke at a Paris news conference during the COP climate talks, saying, "There are about twenty different things like that. They could give us solutions in which the premium you pay for clean, reliable energy is actually gone, so rich countries can go to zero [emissions], and countries such as India that want to provide electricity so their citizens can go at full speed, without having to choose between development and being green."[32]

Gates knows that these new technologies will be more important for the long term rather than the immediate future, but he also knows time is short. Because of this he says, "We need to move faster than the energy sector ever has—at an unnaturally high pace! Historically, it takes more than fifty years before you have a substantial shift in energy generation, but we need to do it [much] more quickly!"[33] Which is exactly why Gates and the BEC have committed to focusing their resources on making a difference for the future of Earth.

ELON MUSK: ENTREPRENEUR / VISIONARY

If you set out to design the perfect combination of climate-change activist, alternative-energy researcher, and wealthy entrepreneur, the result might look a lot like Elon Musk. Musk, one of the world's top 100 wealthiest men, puts his money where his mouth is. After cofounding the online payment company PayPal, Musk turned his attention to either founding, investing in, or managing a number of high-profile businesses focused on using sustainable energy ideas to solve human problems.

Born on June 28, 1971, and raised in Pretoria, South Africa, Musk moved to Canada just before his eighteenth birthday. After obtaining his Canadian citizenship, he moved to the United States and earned his Bachelor of Science degree in physics from the University of Pennsylvania. This was followed by another BS in economics, from the prestigious Wharton School of Business.[34]

At age twenty-four, Musk moved to California, intending to pursue a doctorate in applied physics at Stanford, but he dropped out after just two days. He decided the doctoral program wasn't what he wanted to be doing with his life. He was ready to pursue the entrepreneurial passions and ideas he had for the Internet, renewable energy, and outer space.[35]

Musk is probably most famous for his involvement as a primary investor and then CEO of Tesla Motors, which specializes in the design, manufacture, and sale of luxury, and now—with the Model 3 due for release in 2017 and priced around $35,000—mainstream electric cars.[36] The Tesla Roadster—the world's first fully electric sports car—was the company's first vehicle to garner widespread attention.[37] Next came the Model S, which was a fully electric luxury sedan, and the Model X, a crossover SUV. In 2015, the Model S was the world's bestselling plug-in electric vehicle.[38]

In 2015, Tesla announced the formation of Tesla Energy—a collection of batteries for homes (Powerwall) and businesses or utilities (Powerpack).[39] The Powerwall uses electricity generated by solar panels during the day to charge its battery, which will then have sufficient capacity to power most homes at night. If families need even more energy at night, they can use multiple Powerwalls.[40]

The Powerpack is a much bigger unit, designed for commercial use. Musk describes the Powerpack as being "infinitely scalable," with as many batteries as necessary linking together to give it the ability to meet a range of industrial energy needs.[41] In a Powerwall promotional video, Musk points to a modified Keeling Curve graph, which shows Earth's CO_2 levels going back down, saying, "This is the future we could have: it goes to zero, no incremental CO_2. That's the future we need to have. And the path that I have talked about [with solar panels, Powerwalls, and their factories], that's the only path that I know that can do this. And I think that it's something that we must do, and that we can do, and that we will do."[42]

Some of Musk's other ventures involve quite a bit of speculative technology. For example, in 2002, he founded the Space Exploration Technologies Corporation (SpaceX), an aerospace manufacturer and space transport services company. The goal of SpaceX is to create the technologies to reduce the costs of space transportation and enable the colonization of Mars.[43] He is also the originator of the concept for Hyperloop, a theoretical system of travel that would carry passengers—at speeds of over 700 mph—through two massive tubes stretching between Los Angeles and San Francisco[44]

According to his biographer, Ashlee Vance, Musk wants to "save the world from self-imposed or accidental annihilation."[45] Many of Musk's concepts are still just in the idea stage, and, as you might imagine, plenty of people think his ideas are crackpot. But most good ideas started this way. How many people do you think told Ben Franklin, Thomas Edison, or the Wright Brothers how crazy their ideas were—at first, anyway? Many of the great people in history were called "lunatics" before they became "visionaries"—and Elon Musk has a proven track record of showing just how visionary his ideas can be.

Chapter 28

RELIGIOUS / GRASSROOTS ORGANIZERS

MICHAEL BLOOMBERG: LOCAL / GLOBAL

Michael Bloomberg, three-term New York City mayor, is a feisty politician who does not back down from making politically unpopular decisions when he believes they're the right things to do. And a big part of Mayor Bloomberg's initiative package during the twelve years he held that position (2002 to 2013) involved putting the Big Apple right in the center of the climate-change arena.

Born on February 14, 1942, in Brighton, Massachusetts, and then raised in Medford, Massachusetts—both Boston suburbs—Bloomberg attended Johns Hopkins University and Harvard Business School before becoming a partner at Salomon Brothers, a Wall Street investment bank, in 1966.[1]

After Salomon Brothers was bought in 1981, Bloomberg left to build his own company, Bloomberg LP, which revolutionized the way securities data was stored and used. Enormously successful, the company soon branched out into the media business, and today has more than one hundred locations worldwide. Now a wildly successful billionaire, Bloomberg has been able to focus his talents and fortune on philanthropic matters that he feels passionate about, including education, medical research, the arts, and the environment.[2]

With a net worth estimated at around $40 billion, Bloomberg is now ranked as the world's eighth wealthiest person, which no doubt played a role in his willingness to take politically inexpedient stands on a variety of issues.[3] Although an elected Republican, Bloomberg primarily supported liberal and progressive social issues throughout his tenure as mayor. He was and is a staunch supporter of gay rights—including marriage equality—gun control, a woman's right to choose, and immigration reform.[4] But oddly, Bloomberg may be most famous nationally for a city-wide ban on super-sized sweetened drinks—sometimes called the "Big Gulp Initiative"—that he attempted to pass in 2012.[5]

Bloomberg has always been a dedicated environmentalist and climate-change advocate. In fact, he has always practiced his proclivity for maintaining environmental quality—consistently doing so at the local, national, and international levels. At the local level, during his second mayoral term, Bloomberg unveiled his PlaNYC—"For a Greener and Greater New York"—on April 22, 2007. The goals of PlaNYC were fighting global warming, protecting the environment, and preparing for the projected one million additional people expected to be living in the city by the year 2030.[6]

In only six years—under Bloomberg's guidance—the city reduced greenhouse gas emissions by 19 percent. Plus "The Big Apple" was projected to achieve a 30 percent reduction in GHG's—well ahead of the PlaNYC 2030 goal.[7] Also as part of PlaNYC, Bloomberg launched the Million Trees NYC Initiative in October 2007. In November 2015, the city planted its one-millionth tree—two years ahead of the initiative's original ten-year schedule.[8]

Initiating another of his local environmental programs, Bloomberg convened the New York City Panel on Climate Change (NPCC) in 2008. His emphasis this time was on educating and preparing the city's residents about the harsh realities of climate change and the need to take immediate action—both individually and collectively.[9]

After the city was severely damaged by Superstorm Sandy in October 2012, Mayor Bloomberg unveiled his Special Initiative for Rebuilding and Resiliency (SIRR). The $20 billion initiative laid out extensive plans in June 2013 to protect New York City against future impacts of climate change.[10]

Then, in September 2013, Bloomberg announced that his administration's air pollution reduction efforts had resulted in New York City's best air quality in more than fifty years.[11] By phasing out heavily polluting heating oil, the city's Clean Heat Program was credited with the majority of the air quality improvement[12]—which, in turn, has been credited with saving the lives of an estimated eight hundred residents a year.[13]

At the national level, Bloomberg has consistently pushed for transitioning the United States' energy usage from fossil fuels to clean energy. Beginning in July 2011, he donated a total of $80 million through Bloomberg Philanthropies to the Sierra Club's Beyond Fuel campaign, which is focused on retiring half of America's fleet of coal-fired plants by 2017.[14]

In October 2013, Bloomberg provided a $6 million grant through Bloomberg Philanthropies to the Environmental Defense Fund in support of strict regulations on fracking in the fourteen states with the heaviest natural gas production.[15] Next—aligning with former Treasury secretary Hank Paulson and hedge fund billionaire Tom Steyer—Bloomberg Philanthropies launched the Risky Business Initiative. This joint venture was focused on quantifying and publicizing the economic risks the United States faces from the impacts of climate change in order to convince the business community that more sustainable energy and development policies were needed.[16]

In January 2015, Bloomberg Philanthropies partnered with the Heising-Simons Foundation to launch the $48 million Clean Energy Initiative. This joint venture supports sustainable, state-based solutions aimed at ensuring America has a clean, reliable, and affordable energy system.[17]

Internationally, from 2010 to 2013, Bloomberg served as the chairman of the C40 Cities Climate Leadership Group, which is a network of the world's biggest cities working together to reduce carbon emissions.[18] In this capacity, Bloomberg worked closely with former president Bill Clinton to merge C40 Cities with the Clinton Climate Initiative. By combining forces, this joint venture focused their efforts on amplifying the global fight against climate change.[19] Bloomberg now serves as the president of the board of C40 Cities.[20] (We talked more about the C40 Cities in chapter 20.)

Finally, on June 30, 2015, Bloomberg and Anne Hidalgo—the mayor of Paris—joined forces to announce the creation of the Climate Summit for Local Leaders. This collaboration was officially launched on December 4, 2015 as part of COP-21 in Paris.[21] Also during COP-21, the governor of the Bank of England and chair of the Financial Stability Board—Mark Carney——announced that Bloomberg will lead a new global task force, charged with the objective of helping industry and financial markets understand the growing risks of climate change.[22]

We need more people like Michael Bloomberg in the climate-change fight. He has the financial and political clout to build hard-hitting grassroots organizations throughout the United States and across the world. As the majority of our current heroes have emphasized, the climate-change movement has to begin at the local, state, and regional levels, building until it reaches crescendos that cannot be ignored at the national and international levels.

VAN JONES: STRATEGIST / VISIONARY

Van Jones is a spirited and passionate speaker. With a law degree from Yale, a zeal for public speaking, and eloquent words to get his point across, his communication skills are widely recognized. But it is his decades of commitment to the underserved, low-income communities that gives him the understanding to drive home his powerful message.

For Jones, the environment is central to the kind of social justice he cares about.[23] When I (Mariah) saw Van Jones speak in Boston nearly a decade ago, he asked the audience to picture a wealthy social elite eating organic food in lotus position on a yoga mat, Prius parked in front, telling us that we need to save the polar bears. This is *not* the kind of environmentalism that Van Jones promotes, although he does not deny that this kind of activist is an important piece of the puzzle.

What Jones cares about most is how the environment affects people in the poorest part of town, where environmental issues mean a cancer cluster due to the mishandling of toxic waste, or increased rates of asthma and other respiratory issues due to increased air pollution from the power plant down the road. It is unequivocally clear to him that decades of shortsighted economic and environmental policies—from both Democrats and Republicans—present a double danger to humanity on a massive scale.[24]

The future looks grim for all of us, but the poorest are the most at risk. Jones has devoted his career to justice for all people, and he has a long list of honors and awards that highlight his accomplishments, including being part of *Rolling Stone*'s 2012 list of Twelve Leaders Who Get Things Done and *Time*'s 2009 list of 100 Most Influential People in the World.[25] His initial focus was reducing police brutality—starting a hotline for victims of such injustices—which eventually became the Ella Baker Center for Human Rights. In 2005, the Ella Baker Center expanded its vision from the immediate concerns of policing to a focus on job, wealth, and health creation.[26]

The Ella Baker Center's Green-Collar Jobs Campaign was Van Jones's first concerted effort to bring together his desire to improve racial and economic equality while addressing environmental concerns.[27] This effort was expanded nationally in 2008, when he launched Green for All, a nongovernmental organization focused on creating green pathways out of poverty in America.

Jones's book *The Green Collar Economy* became the first environmental book authored by an African-American to make the *New York Times* bestseller list.[28] Following this, Jones was appointed to President Obama's administration as a special adviser for green jobs—the first person to ever serve in the White House with "green jobs" in their title.[29] Columnist Chadwick Matlin described Jones's new position as "switchboard operator for Obama's grand vision of the American economy; connecting the phone lines between all the federal agencies invested in a green economy."[30]

In *Green Collar Economy*, Jones walks his reader through the environmental timeline that we present in this book, but he offers a different perspective, one that we honor and support. In Jones's book, he looks at how the mainstream environmental movement has a "sad history of racial blind spots and class exclusion."[31] In the first wave of environmentalism, "Conservation," the activists of the day "set the enduring pattern for most conservationists' racial politics: 'Let's preserve the land we stole [from the Native Americans].'"[32] Jones asserts that, though there is no question that the environmentalists of this time stood up for the most vulnerable places, they did not stand up for the continent's original environmental stewards. This failure to honor the contributions and humanity of the land's most vulnerable people marred the impact of the original movement.[33]

About the second wave, "Regulation," Jones says, "This wave, too, was affluent and lily white. As a result, it developed huge blind spots to toxic pollution concentrating in communities of poor and brown-skinned people."[34]As we have discussed in this book, there were many excellent legislations passed during the 1970s to ban the rampant dumping of pollutants into our air, water, and land—but the fact that much of this legislation proceeded with "minimal involvement from people of color practically ensured that racially imbalanced outcomes would result," writes Jones.[35]

Where we currently stand with the environmental movement is that we have the environmental justice movement, which defends the poor and vulnerable, and mainstream environmentalism, which fails to embrace the causes of environmentalists of color.[36] The current wave—"Investment," meaning investing in solutions for the future—will succeed if it includes all people, regardless of race, color, or class. "If the mostly white global warming activists join forces with people of color, the United States can avoid both eco-apocalypse and eco-apartheid—and achieve eco-equity," writes Jones.[37]

To achieve "eco-equity," Jones recommends three major actions to move us out of an era of American capitalism and unregulated growth that is trashing our planet and its people. The first is to return to producing things here, placing our faith in the skilled American holding a tool (vs. shopping with a mouse).[38] The second is to go back to relying on smart savings and thrift, including the conservation of our natural resources (such as nonrenewable energy). The third is to start honoring Earth instead of plundering it. At the center of all three of these efforts is "a crash program for energy independence based on clean energy and energy efficiency. By producing renewable energy and other green products here, while better conserving our monetary and natural resources, we can create secure pathways to more work, more wealth, and better health for millions of Americans," Jones explains.[39]

A caulking gun and a clipboard are key tools for the green collar jobs that Jones envisions will move us all to a better place. He explains, "When you think about the emerging green economy, don't think of George Jetson with a jet pack. Think of Joe Sixpack with a hard hat and a lunch bucket, sleeves rolled up, going off to fix America. Think of Rosie the Riveter, manufacturing parts for hybrid buses or wind turbines. Those images will represent the true face of a green-collar America."[40] To combat climate change, we need to weatherize buildings, install solar panels, manufacture wind turbine parts, plant and care for trees, and construct solar farms, wind farms, and wave farms. These types of green-collar jobs should match Jones's definition "a family-supporting, career-track job that directly contributes to preserving or enhancing environmental quality."[41]

Stopping climate change will require a World War II–level of mobilization and will require massive support at the social, cultural, and political levels.[42] In an increasingly nonwhite nation, racial inclusion is paramount to the success of these efforts. "To give the Earth and its people a fighting chance, we need a broad, populist alliance—one that includes every class under the sun and every color in the rainbow," says Jones.[43] Van Jones believes that once the green economy is no longer just a place for affluent people to spend money and becomes a place for ordinary people to earn and save money, nothing will stop it. And then we will rise to the dual challenge of growing the economy without hurting the Earth.

SALLY BINGHAM: ORIGINATOR / LEADER

Reverend Canon Sally Bingham, an Episcopal priest and canon for the environment in the diocese of California, stands at the forefront of a growing conversation about God's mandate to humans to care for creation. "Every person of faith should become aware of their moral responsibility to be a steward of creation. God put Adam in the garden to till and to keep. Every mainstream religion has a mandate to care for creation. Sometimes [followers] have not thought about it or they have not addressed it, and then they see an opportunity to really be faithful stewards of creation and they join our program," Bingham explained to us in an interview.[44]

Bingham's program is the Interfaith Power and Light (IPL) campaign.

She went on to say, "People who sit in houses of worship and say they love God and their neighbors have a particular obligation to take care of the Earth and each other. If you sit in a pew on Sunday and say you love God and you love your neighbor, how can you not be taking care of your neighbor's air and water? They are now starting to recognize that responsibility and act."[45]

Bingham's sense of responsibility to the Earth is deeply connected to her knowledge that climate change is harming the people of the world, and her faith mandates a responsibility to care for them. With a note of sadness in her voice, she tells us:

> [Climate change] affects every single aspect of life, affects every living thing—starting with the rising sea, the temperature change, the number of long heat days that are causing people to die . . . the fact that the droughts are more extreme and are disrupting crops . . . the fact that people are starving because they can't grow food in an area that has not had any rain in five years . . . that the storms that are so much more severe than they ever were and are killing people and destroying properties.[46]

She continued, "It is happening because the climate is changing. Why is the climate changing? Because we are putting too much carbon dioxide into the atmosphere."[47]

One action that she hopes congregations will take is to join the IPL campaign. The campaign has helped religious people in over 10,000 congregations, spread across nearly forty states, become aware of their responsibility to protect the climate.[48] IPL began with an episcopal church in the diocese of California asking its congregations to buy renewable energy for their electricity. That congregation served as an example to its communities, and the movement grew rapidly from there.

Prior to COP-21 in Paris, IPL asked its members to take the Paris Pledge to cut their carbon emissions in half by 2030 and be carbon neutral by 2050.[49] When a delegation from IPL traveled to Paris, they took with them an eleven-foot-long scroll with the names of 4,500 congregations and individual households who had signed the pledge. They wanted to show the world that the faith community in the United States is committed to cutting emissions, creating jobs, and saving money at the same time.[50]

Bingham knows as well as anyone that the environment has become a political issue. She told us, "It is almost universal that if you are a Democrat you are an environmentalist and if you are a Republican you are not. That unfortunately is a big stumbling block for the issue. We don't believe, in our organization, that the environment is a political issue; we see it as an issue of science but, in the big picture, it is a moral issue. Where are our values, what do we care about, what is our responsibility to the future? It's about how to leave this world to come back to our moral integrity."[51]

While Bingham does not offer solutions for solving the politicization of environmental issues, she is enthusiastic about the willingness to think differently on the issues in the religious community. The majority of the people she speaks with are in support of her initiatives, though on occasion she receives pushback. "What we have come up against occasionally," she said, "because our focus has been on climate change, is that God would never allow anything bad to happen to creation. And then we have to do some explanation about how God has given us free choice and some of our choices have been harmful to creation. Mostly we get the comment that I had never thought about like that before."[52] Religiously, she thinks people are really on board with human beings as the species put on planet to keep it safe and healthy, not only for ourselves but the people who come after us. "There are very few people who would argue with that," Reverend Bingham said.

Pope Francis, in his encyclical on the environment, said similarly that the issue is about the moral values that every person of integrity needs to have. The message was not just for Roman Catholics, but it was for all people with a conscious. Through IPL, Reverend Bingham has been teaching this idea for over fifteen years. "And now to have somebody as well known, as famous and as popular as Pope Francis to come out and say the same thing, it has been hugely helpful to our movement," said Bingham.[53] Many have received the message enthusiastically, seeing participation in the IPL program as an opportunity to look after creation as a facet of their faith.

Reverend Bingham's hope for our future comes from the fact that more and more people are involved and concerned. She believes we are almost at a critical mass and that soon things will change for the better. "We stopped smoking almost overnight when enough people were touched by disease due to cigarettes," she said. "We are close to enough people being harmed by climate change now that it can no longer be denied. People of faith are taking a leading role and once the moral and religious leaders are involved and speaking out the movement will succeed."[54]

TIM DeCHRISTOPHER: ACTIVIST / LEADER

It takes a rare person willing to risk up to ten years in a federal prison for standing up for their beliefs, but Tim DeChristopher is that person. In 2008, DeChristopher caught wind of the Bush administration's authorization to sell thousands of acres of pristine wilderness in Utah to the energy and mining industries for their use.[55] DeChristopher, founder of the climate group Peaceful Uprising, showed up at the Utah Bureau of Land Management Oil and Gas lease auction and bid $1.8 million on thirteen parcels of land, totaling 22,000 acres, with no intention to drill (or pay).[56] He received a two-year sentence for illegally stopping the sale.

Weeks after DeChristopher bid on these properties, a federal judge, in essence, agreed with him that these leases were illegal and should not have been sold, and blocked the sale of all of the parcels. Eventually, the Obama administration's Department of Interior said the overall sale was improper and pulled all of the parcels from auction—rendering DeChristopher's incarceration absurd.[57] His sentencing and trial generated massive protests across the country. Before being taken into custody, DeChristopher issued the following statement: "Until our leaders take seriously their responsibility to pass on a healthy and just world to the next generation, I will continue this fight. . . . The reality is not that I lack respect for the law; it's that I have greater respect for justice."[58]

Speaking about the controversy, scientist (and fellow hero) James Hansen said, "DeChristopher's action speaks to the question of whether it makes sense for us, humanity, to go after every last drop of oil and gas in the ground. His action also relates to the nature of the world that DeChristopher and all other young people will live in, and to their future economic well-being."[59]

In *Storms of My Grandchildren*, Hansen went on to say, "This is a gross case of intergenerational injustice. We should all strongly support DeChristopher in his case against the US government. The government cannot realistically claim that it is ignorant of the consequences of its action."[60] Essentially, the intergenerational injustice foisted on Tim DeChristopher saved future generations from the damages of carbon in our atmosphere, but at a great cost to DeChristopher's quality of life.

After DeChristopher served twenty-one months of his two-year sentence, he was released. He then enrolled in Harvard Divinity School, where he is studying to become a Unitarian Universalist minister.[61] He pleads with the religious community to do more than put on their collars and show up for acts of civil disobedience for the climate, and more than taking their investments out of fossil fuels. While these acts are important and do help the cause, DeChristopher says that waiting to be told what to do is not moral leadership. He writes, "As a veteran of the climate movement, I suggest that we don't need religious communities merely to join the climate movement. We need religious communities to lead, challenge, and deepen the climate movement."[62]

DeChristopher suggests three main areas that religious leaders need to focus on: first, speaking the hard truths about the nature of the climate challenge; second, ending corporate personhood—i.e., laws such as were established through the *Citizens United* Supreme Court case, which allows corporations to dump money into political campaigns, often producing an imbalanced effect on election results—and re-establishing democracy in the process; and third, reminding us to connect with non-consumer ways of living.[63]

DeChristopher has little sympathy for the fact that in order for climate justice to prevail the fossil-fuel industry must fail, and fail hard. The only way for extractive industries to make profits is for them to extract the remaining fossil fuels from the ground, but this cannot be done in a way that protects climate justice. At this point, the science on climate change is so solid that extracting carbons out of the earth and adding them to the atmosphere means willingly destroying people's lives—for example, the residents of the low-lying Pacific Islands whose homes are being swallowed by the sea with every tenth of a degree of global temperature increase. "What separates the climate justice movement from other climate-related players is the mission of keeping those fossil fuels in the ground *without* guaranteeing future profits to the corporations who have already profited from exploitation," DeChristopher writes.[64] In other words, DeChristopher thinks that the fossil-fuel companies have already done so much damage to humanity that they deserve to go bankrupt.

DeChristopher reminds us that the fossil-fuel industry has killed for profit throughout its history. In his home state of West Virginia, coal has cost countless lives and has left the state the least livable in the nation.[65] He says further, "Not only has the fossil fuel industry continued trading human lives for profit, but, since it is difficult to convince free people to poison their own water sources or blow up their own backyards, it has increasingly killed democracy in order to keep killing people for profit. The exploits of the Koch brothers in this area are well known and we as a nation have normalized the way that oil companies leverage our government to launch wars and overthrow governments that are not conducive to extraction."[66] Basically, in DeChristopher's eyes, the exceedingly powerful fossil-fuel corporations have one goal: do whatever it takes to extract fossil fuels, even at the expense of global peace and human health.

Which brings him to his second point, which is that corporations have no place in our democratic or political process.[67] In 2013, Senator Bernie Sanders (I-VT) introduced an amendment, endorsed by President Obama, to overturn the Supreme Court's decision on *Citizens United v. Federal Election Commission*,[68] the 2010 Supreme Court decision that paved the way for unlimited corporate and union spending in elections. Since Congress did not act on Senator Sanders's 2013 amendment, he will bring it back in the new Congress.[69] We agree completely with Sanders, Obama, and DeChristopher on this point—if a handful of millionaires and special interest groups can swing elections with egregious campaign spending, how will true citizens' rights be achieved? With nearly 70 percent of the country in favor of climate-change policy—but billionaires and huge corporations spending massive amounts of money to support candidates that call climate change a hoax—the poor (who will be most impacted) have no voice.

DeChristopher's stance further emphasizes what many of our new heroes touch on: the need for a change in how we exist in the world—what we view as essential to progress and happiness—is currently far too linked to consumption and gross domestic product. DeChristopher suggests that part of the church's role should be to remind us that we are more than consumers.[70] We are citizens, community members, and family members. He writes, "Churches are uniquely suited to develop our identities as children of God, pieces of an interdependent web of existence, or bearers of divine sparks of creativity."[71] Being nonconsumers in the world—conserving resources and producing a smaller carbon footprint—as much as possible is essential to empower the revolutionary change that the climate crisis demands.

ROBERT BULLARD: ORIGINATOR / AUTHOR

Robert Bullard, often called the "Father of Environmental Justice," uses his expertise and media savvy to garner attention for communities burdened with environmental hazards.[72] He has dedicated his career to protecting minority and low-income communities from becoming toxic pollution dump sites. Bullard sees environmental justice issues as being at the heart of everything. In his words, "The right to vote is a basic right, but if you can't breathe and your health is impaired and you can't get to the polls, then what does it matter?"[73]

"Just because you are poor," Bullard told CNN, "just because you live physically on the wrong 'side of the tracks' doesn't mean that you should be dumped on."[74] His voice was heard and the environmental justice movement began to evolve.

In 1994, President Bill Clinton invited Dr. Bullard to the White House to witness the signing of an executive order that would require the federal government to consider the environmental impact of policies on low-income communities before implementing them.[75] Later, Bullard cowrote a report titled "Toxic Wastes and Race at Twenty, 1987–2007: Grassroots Struggles to Dismantle Environmental Racism," which prompted the EPA to state, "The EPA is committed to delivering a healthy environment for all Americans and is making significant strides in addressing environmental justice concerns."[76]

Robert Bullard grew up in Elba, Alabama, a small town where residents were very aware of the civil rights movement. His parents, in fact, were activists in the movement.[77] He earned an undergraduate degree in government from Alabama A&M—a historically black university—and then a master's degree in sociology from Atlanta University.[78] Two years after completing his sociology doctorate at Iowa State University, Bullard began a study documenting environmental discrimination under the Civil Rights Act.[79] He found that, despite the fact that only 25 percent of Houston residents were African American, 100 percent of the city's solid waste sites, 75 percent of the privately owned landfills, and 75 percent of the city-owned incinerators were located in black neighborhoods.[80] Since the city of Houston did not have zoning at this time, he knew that individuals in government orchestrated these sitings. The injustice of it drew him into the cause.

Dr. Bullard worked his way through academia, holding research positions and professorships at a number of universities in Texas, Tennessee, and California, among others. He currently holds a position as Dean of the Barbara Jordan–Mickey Leland School of Public Affairs at Texas Southern University in Houston, Texas. As with many of our climate-change heroes, Bullard works in academia but is very active in political matters and in advocating for justice.

Dr. Bullard has received many well-deserved awards in his life, including being named one of *TheGrio*'s 100 Black History Makers in the Making (2010), one of *Planet Harmony*'s Ten African American Green Heroes (2010), and one of *Newsweek*'s 13 Environmental Leaders of the Century (2008).[81] In 2013, he became the first African American to win the Sierra Club John Muir Award, and in 2014 the Sierra Club named its new Environmental Justice Award after him.[82]

Bullard's book *Dumping in Dixie: Race, Class, and Environmental*

Quality[83] was the first work on environmental justice issues.[84] Bullard believes that sustainability cannot exist without justice. As he puts it, "This whole question of environment, economics, and equity is a three-legged stool. If the third leg of that stool is dealt with as an afterthought, that stool won't stand. The equity components have to be given equal weight."[85] To him, part of the solution is to pair mainstream environmental groups with environmental justice groups that have the ability to mobilize large numbers of constituents. This type of grassroots movement will get people marching and filling up courtrooms and city council meetings, kicking off conversations about the environmental movement. Bullard believes that the reality of the imbalanced impact of environmental issues—such as a climate in peril—on the poor will force collaboration between people from a diversity of socioeconomic, cultural, and racial backgrounds—we are all in this together.[86] This collaboration will bring awareness that our actions in the developed world have impacts that are not isolated to just us;[87] it is harder to turn away from the realities of our ubiquitous pollution when we are face-to-face with those shouldering the brunt of its effects.

While this is a step in the right direction, we need to expand the idea, applying the framework of environmental justice across developing countries, seeking to find actions and policies that will heal our climate's health and concurrently serve the poorest peoples in developing countries around the world.

Part Five

FINDING SOLUTIONS

Chapter 29

HOW THE CLIMATE-CHANGE WAR CAN BE WON

As inhabitants of Earth during the twenty-first century, we are faced with solving the most challenging and perplexing threat ever posed to humankind. Note that we did not say "environmental problem," since this is a problem for our *existence* on Earth. The environment will recover, no matter what damage we inflict on it—we are just tenants on this planet. We are the ones whose future is in jeopardy.

During a radio speech aimed at rallying lagging British spirits during World War II, Winston Churchill famously described Russia as a "riddle wrapped in a mystery inside an enigma."[1] Churchill's intent was not to slam the Russians since they were—at the time—allies. He just wanted to make sure the British citizens understood that, while the Russians might seem unusual to them—they were a powerful force that would be critical to defusing and defeating the scourge of Hitler's Nazi aggression.

Today, we are dealing with another type of "riddle wrapped in a mystery inside an enigma." Only, this time, we are facing a degradation of our home ecosystem, and it is a problem of our own making. But if we handle the situation well, climate change may very well be analogous to the Russia that Churchill was addressing. In fact, climate change may just turn out to be our staunchest ally as we move forward with maintaining a suitably habitable planet for humankind.

Here's what we mean: climate change sounded the alarm and provided the loud wake-up call we needed to finally realize that we were poisoning our own environment. Without the scientifically based global-warming alerts produced by climatologists and activists all over the globe, we may very well have just proceeded along our merry way until we crashed and burned forever. But Earth is sending us very clear signals—droughts in California; historic floods in Texas, South Carolina, and Louisiana; loss of permafrost; disappearing glaciers; frequent and more devastating storms of all types. These signals are causing people to wake up to the threat of climate change. Fortunately, because of the remarkable foresight and research of scientists like Svante Arrhenius, Gilbert Plass, Roger Revelle, David Keeling, Stephen Schneider, and James Hansen—we are making inroads toward understanding the enigma of climate change. We know for certain that we need to take dramatic and immediate action to counteract climate change. We have learned from the visionaries in this book how to "defuse and defeat" the world's newest "scourge"—the damage done to our planet by the burning of fossil fuels.

The good news is that we are not too late. We can still save and maintain the essential qualities of our life here on Earth. The bad news is that we have to start taking the appropriate actions *right now*. Not by 2100, not by 2050, not by 2020—but *RIGHT NOW!*

So what are the appropriate actions that we need to take to counteract climate change? This is precisely the question we intended to answer when we decided to write this book.

The first step for the majority of the world is to recognize that humans are one species in the community of a natural ecosystem. As we discussed when we talked about Naomi Klein, we need to "unlearn the myth that we have total dominion over nature."[2] Professor Thomas Lovejoy further emphasizes this: "We need to stop thinking about nature as something that can survive in little bits and pieces in human dominated landscapes, and think about human aspiration as being embedded in nature, so that it is all connected."[3]

This is why we are so careful to emphasize that climate change is not an environmental problem but rather a problem of maintaining human existence as we know it. The real lesson we learned from studying the environmental movement from "conservation to climate change," is that all of our heroes were not working to save the environment; they were working to restore our habitat as home. As wonky/hippy/tree hugger as it may sound, we cannot escape the fact that we are animals, and without being able to meet our basic needs—water, air, food, and shelter—we will not survive, no matter how fancy our technology or how high our gross domestic product. You can't eat money, and we have yet to figure out how to virtually get a good night's sleep.

Another reason we feel so strongly that climate change be labeled as a humanitarian problem and not an environmental problem is because history and media propaganda have made environmental problems seem like a bourgeois concern. A little digging around in the Gallup polls on climate change gave us the following to ponder: In 2000, the number of Americans concerned about climate change was 72 percent. At that time, we had a thriving economy, low crime rate, welfare dependency and joblessness were down, and the stock market was soaring—people had the leisure and the luxury to worry about "the polar bears." Fast forward to 2008 when, year after following year, the concern about climate change has dropped, says Geoff Feinberg, research director for the Yale Program on Climate Change Communications.[4] He attributed it to the Great Recession, saying that "environmental protection and attention is a priority often stemming from affluence."[5] In fact, nothing should be further from the truth. As Van Jones and Robert Bullard have so eloquently highlighted in chapter 4, the lack of concern about the environment often makes low-income communities in particular very sick.

It is imperative to understand climate change as a problem that will affect all of humanity, without regard to class, socioeconomic status, religious persuasion, political affiliation, age, or ethnicity. In this regard, it is a true equalizer—*we are all in this together*. Fortunately, polls taken in March 2016 show that 64 percent of adults worry a "great deal" or "fair amount" about climate change, up from 55 percent this time last year.[6] While still largely an issue that Democrats are more concerned about, Republicans and Independents are slowly becoming aware of the problem—84 percent of Democrats are concerned while 40 percent of Republicans (up from 31 percent in 2015) and 64 percent of Independents (up from 55 percent) expressed concern.[7]

A universal concern about protecting our home habitat is the grand paradigm shift we need to best move forward. But there are immediate practicalities pressing. For this, we turn our discussion of climate-change solutions to the United States. It is our belief that, the United States has the power, means and, quite frankly, the moral obligation to establish the international prototype for combating climate change.

If we proceed down the proper path, the actions we take here at home will be scrutinized and emulated by the other free and democratic nations in the world. Just as climate change must be addressed as non-partisan here in the United States, it must also be viewed as apolitical in the rest of the world. Global solutions will be forthcoming when the world accepts this principle: *Climate change is a universal crisis that will affect every country and every person on Earth unless we work together to find and implement solutions.*

Chapter 30

SOLUTIONS FOR THE UNITED STATES

INSIST THAT CLIMATE SCIENTISTS TELL IT LIKE IT IS

As environmental scientists, we are quite familiar with scientific protocol. We understand all the processes—from theorem postulation to peer review—that scientific analysis dictates. We know that scientists must be extremely cautious before coming to conclusions and making definitive statements. Just like biology and ecology, the study of climate—"climatology"—is a soft science, which means that there are no absolute right or wrong answers. There are only *statements of probability*. Because of this, climate scientists are reluctant to make definitive statements that could be taken as guarantees that something will or will not happen in the future.

Since the scientific process requires that findings be presented as probabilities of occurrence and not as 100 percent assurances, room is left for the climate change deniers to say, "Aha, so there is a measure of uncertainty about whether global warming is really occurring and is actually being caused by the activities of humans!" Of course, what they neglect to point out is that 97 percent of climatologists are saying there is a consensus on the reality of anthropogenic climate change[1] at confidence intervals of 95 percent (essentially meaning a 95 percent chance of accuracy). In the scientific world, that is as close as you can get to providing a guarantee or statement of absolute fact. But it's still not a black-or-white, definitive pronunciation about exactly how climate change will impact the future of humans on our planet.

To the climate-change deniers—as well as many of the fence-sitters—this is the window of opportunity they need to make their case that the whole process is some sort of scientific hoax. All the while, the fossil-fuel industry and their well-heeled lobbyists are taking this rhetoric right to the bank.

In her book *Merchants of Doubt*, Naomi Oreskes describes the machinations of the climate-change denial movement in detail.[2] Since the climatologists remain overly cautious about predicting the future of climate change, the doubters continue to argue that no one really knows for sure what is likely to happen in the future. And—because of this—there is no reason to worry or take any so-called dramatic and incredibly expensive actions. Just let everything play out, they say, and in good time—over the next one hundred years or so—American technology and ingenuity will allow us to adapt and maintain a livable planet.

At first blush, doing nothing may sound like a preferable approach, but the cold hard facts simply do not bear this out. As we have discussed throughout this book, climate change—or, if you prefer, the more stern but realistic *climate disruption*—is happening now, it's happening fast, and we do not have the sweet luxury of sitting back and waiting decades for things to get worse before we do anything. "Worse" is already here, and it is growing progressively more so every day. If we wait any longer, we will be on a long, slippery slope to planetary degradation for humankind, and there will be no way to apply the brakes. Of course, life on Earth will still exist in some form. But for us humans, it just will not be the life we are all currently familiar with—not even close.

To counteract this "maintenance of doubt" situation, we need the world's top climate scientists to step up to the plate, eschew protocol, stop equivocating completely—not even giving just a little bit of wiggle room—and proclaim the urgency of the situation and the need for imminent solutions. They must emulate the courage and commitment of such past environmental heroes as John Muir, Harriett Lawrence Hemenway, Marjory Stoneman Douglas, Rachel Carson, and David Brower in standing up to all-powerful corporate conglomerates and instilling a sense of urgency in the public's collective mind. They need to get us out into the streets using our voices in protest. They must state that—without any doubt whatsoever—human-caused climate change is happening now and it is getting worse literally by the hour. Professional climate scientist and blogger Dr. Joseph Romm told us in a recent interview that scientists now have a moral imperative to tell the world about the seriousness of climate change and the need to take action.[3] We agree with him one hundred percent!

This book provides a compelling compendium of solutions to inform even the most adamant skeptic. Do you have a friend who says that God will take care of the planet so we don't need to? Visit Katharine Hayhoe, Tim DeChristopher, or Sally Bingham's bios in part 4. How about a coworker who says climate-change actions will bankrupt our economy? Check out "bubble predictor" Jeremy Grantham's solutions and economic analysis, Tom Steyer's political projections, or Mark Jacobson's information about the financial incentives for renewables. What about your cousin who abhors big government and sees environmental regulations as a way to control our personal assets? Visit Bob Inglis or James Hansen's bios and read about their counterpoints and suggestions for free-market regulation of climate change—it can be done.

Does your engineering neighbor say that the technology is not there to move to one hundred percent renewables? Al Gore, Mark Ruffalo, Michael Mann, or Elon Musk can help you answer that. Do you have a friend who works in the nonprofit industry and asserts that environmental issues are only for rich yuppie-hippies who can afford to buy all organic foods and attend expensive yoga classes? Introduce her to Van Jones and Robert Bullard, who illustrate how climate-change policies create clean, affordable energy and healthy living environments for everyone—removing the dangerous chemicals, spills, leaks, and mines from the backyards of low-income areas.

But we must be honest: information—even well-researched, well-documented information—does not change people's minds. So the next step is to get involved, raise awareness, and create a paradigm shift so that climate change becomes a widely accepted truth.

JOIN/CREATE LOCAL GRASSROOTS ORGANIZATIONS: GROUNDSWELLS OF PUBLIC SUPPORT

As we conducted our research and interviews for this book, something that came up repeatedly was that climate-change action has to be initiated at the grassroots level. As the old adage goes, "all politics is local." Most commonly associated with former US Speaker of the House Tip O'Neill, this phrase means that whatever you are trying to accomplish politically usually succeeds best by working from the ground up.[4] The same certainly goes for climate change.

Grassroots movements are really nothing new. In fact—as we have discussed in previous sections of this book—they have often played an essential role in the accomplishments of many of our past environmental heroes. Ralph Waldo Emerson and Henry David Thoreau—by following their concept of transcendentalism—gained numerous followers for the ideals of living in harmony with nature and practicing passive resistance. William Temple Hornaday used his museum skills to inspire visitors to save the American bison from extinction. Harriett Lawrence Hemenway held high society teas to keep wading bird feathers out of women's hats. The "Archdruid" David Brower created media ads that rallied people against dams in national parks. Rachel Carson honed her exquisite writing skills to gain popular support for saving the bald eagle, peregrine falcon, brown pelican, and osprey. Chico Mendes became a martyr to thousands in his fight to save the Brazilian rainforest from massive clearcutting. Lois Marie Gibbs rescued more than eight hundred families who were living on top of a cancer-inducing hazmat site.

If you need some current ideas for starting your own grassroots movement, check out these organizations that have been founded by some of our contemporary climate-change heroes (as described in detail earlier in this book):

Green for All (www.greenforall.org)—an environmental justice and minority group sustainability movement started by activist and CNN journalist Van Jones.

Interfaith Power & Light (www.interfaithpowerandlight.org)—a religious response to global warming started by Episcopal minister Sally Bingham.

C40 Cities (www.c40.org)—a global network of cities aimed at taking action against climate change and led by former New York City mayor Michael Bloomberg.

Bold Nebraska (www.boldnebraska.org)—a state movement founded by activist Jane Kleeb that kept the Keystone XL pipeline out of her state.

Divest Harvard (www.divestharvard.com)—cofounded by Chloe Maxim, one of the first groups in the United States aimed at convincing colleges, universities, and other institutions to get rid of their investments in fossil-fuel companies.

But if you want one organization in particular to research as a prototype for grassroots climate organization, we suggest taking a look at 350. org (www.350.org). Founded by environmentalist and bestselling author Bill McKibben—another of our current climate-change heroes—350.org has always been at the forefront of climate-change activism. In fact, Mr. McKibben has been thinking and writing about the hazards of climate change for more than thirty years. His skills at organizing and leading rallies, marches, and demonstrations throughout the world are legendary and are a perfect model.

Also, before you think about starting a new group, check to see if there is already one in your area. The place to look is the Citizens Climate Lobby (CCL) (https://citizensclimatelobby.org). They currently have 324 active locations worldwide—with new chapters starting up every month—so it is a good bet they already have a group close to where you live. CCL holds monthly meetings, presents informative talks, organizes outreach events and rallies, and writes letters to Congress in support of climate-change legislation.

Similarly, check with the local chapters of such conservation organizations as the Audubon Society, National Wildlife Federation, Sierra Club, and Nature Conservancy. There you will find many other talented and dedicated conservation-oriented people who share your concerns about climate change and are anxious to join forces to do something about it. If you are technologically savvy, use social media (Facebook, Twitter, Snapchat, etc.) to reach out to people and build your group—whether it's one that already exists or one you start yourself.

So once you have joined or started a grassroots organization, what do you do next? Great ideas to emphasize are marches, rallies, and peaceful demonstrations—anything that will get the attention of the local media. If you can target your events toward large cities or state capitals, so much the better.

No matter what you do, always remember to emphasize the positive. You want your message to be an upbeat one—that climate change is not going to get the best of us. By working together, we will find solutions. Arrange for speakers—the more knowledgeable and optimistic, the better. Even add in musical accompaniment—local high school and college bands help to engage the community and add levity.

Remember—as our heroes in this book continually emphasize—everyone has a role to play in this transition—engineers, lawyers, doctors, mechanics, stockbrokers, bricklayers, ditch diggers, cooks, musicians, artists, teachers . . . literally everyone! We all have a set of skills we can apply toward solutions—whether you are steering your career toward climate action, putting photovoltaics on the roof of your house, caulking your windows, putting a smart thermometer in your office, growing your own organic vegetables to reduce some of your food miles and replenish soil, changing to a more efficient car like a hybrid or electric vehicle, holding a benefit concert, or going to Washington and trying to influence politics. The most important thing is to apply your unique abilities to make things happen.

In the final analysis, the ultimate objective of any grassroots movement should be getting the attention of legislators—both in state capitals and on Capitol Hill in Washington, DC. We believe that the most substantive and effective climate-change solutions will come from government-derived policies, directives, and regulations.

DIVEST AND TRANSITION TO A NEW ECONOMY: THE LONG-TERM SOLUTION

The climate system will not be fixed until our economic system is fixed. So the first message we want to discuss is extremely important, although controversial—divestment. We will be crystal clear about one thing—divestment will not hurt the fossil-fuel industries financially. Any stocks you or your company dump will most likely get snatched right back up—at least for the next few years. Nevertheless, we suggest divestment for two reasons—your own personal long-term financial gain and as a grassroots protest against companies that are funding denial and polluting our health.

If making educated and sensible investments is important to you, then being highly informed about climate science is a sharp tool in your toolbox.[5] It is well established that 80 percent of the carbon reserves must be kept in the ground to keep warming under 2.0 C above preindustrial levels. For fossil-fuel companies—the coal industry in particular—to be financially viable, they must turn these reserves into profit by extracting and burning them—the exact thing that the 2015 climate treaty in Paris declared that they cannot do. Any government regulation to limit fossil-fuel use devalues the share price of the existing carbon reserves, causing these shares to become stranded assets (obsolete or nonperforming; basically losing money for the asset holders). Therefore, an investment in fossil-fuel corporations that are still extracting fossil fuels is not a stable long-term investment. According to the treaty signed at the climate conference in Paris, the nations of the world have committed to making sure that these fossil-fuel industries fail over the next ten years.

All fossil-fuel companies are not created equal, so we are not suggesting that *all* of them are out to maximize profits, no matter the cost to our planet. It's just that there are some that have funded denial, buried scientific evidence about the harms of greenhouse-gas emissions, and knowingly killed for profit throughout their history. Divestment should aim at these companies, to highlight their transgressions and damage their social license and acceptability. The Carbon Underground ranks the top two hundred public fossil-fuel companies by the potential carbon emissions content of their reported reserves. (The list can be found at www.gofossilfree.org/top-200/.)

We would be negligent if we were to tell you to dump your fossil-fuel investments without providing direction on how to re-invest. Choose companies designed to bring down emissions and reinvigorate community and local initiatives, with the ultimate goal of maximizing transition to a new sustainable economy. In the book *Limits to Growth*, Dennis and Donella Meadows and Jorgen Randers suggest a shift in values away from increasing consumerism, infinite economic growth, and GDP, and toward the values of sustainability, beauty, efficiency, equity, and community.[6] Focus your financial investments on the sectors of our economy that embrace these values and are already low carbon—for example, caregiving, teaching, social work, the arts, and public-interest media.[7]

Some examples of business areas to focus on are wind, water, and solar technologies, sea walls, and drought-resistant crops—with the caveat, of course, that specific companies within these industries will still come with risks. For an in-depth treatment of what this new economy could look like, see Gus Speth's book *The Bridge at the Edge of the World*[8] and Richard Heinberg's *The End of Growth*[9] (along with their websites, www.thenextsystem.org and www.postcarbon.org, respectively).

Make no mistake, this is a long-term change in investment strategy. It will take time and patience. It will probably be ugly before it is beautiful. But in the long run, it will be worth the wait. Efforts in this direction will build humane, resilient communities with the strength to withstand anything that climate change sends our way. In the interim though, we will need short-term solutions.

IMPLEMENT A CARBON FEE: THE ESSENTIAL SOLUTION

Without a doubt, the primary immediate solution most commonly recommended by our climate-change heroes is implementation of a carbon fee. It should be noted here that cap-and-trade-programs—which are often discussed as another way to monetarily penalize CO_2 emissions—are no longer viewed in a favorable light because of the myriad ways these systems can be gamed for economic gain and maximizing revenues. But the carbon fee has no such problems. There is no "if this, then that" game playing involved. There is just a flat fee paid by (1) any entity that produces CO_2 emissions and/or (2) any entity that uses CO_2 emitting products.

Here is where the solutions diverge: some of our heroes suggest we should return one hundred percent of the revenues generated to the public through dividends. Others think that the government should reinvest the revenue in renewables of their choice. We fall somewhere in between. A portion of the revenue should be given back to the public, at least at first, in part to offset any increase in prices due to the fee. The government should also use the funds to establish a flourishing renewable energy market and to build the infrastructure that will be required to transmit electric power to all users and markets (thereby making any systems that burn oil or natural gas for heating, cooking, etc., obsolete). By electrifying everything, it makes it easier to match power demand with supply.[10] With a portion of the carbon fee supporting a rapid transition to wind, water, and solar (WWS), and the market mechanism in place to encourage this transition, in theory, we should be able to increasingly diminish our dependence on fossil fuels.

Because of the almost universal consensus for a carbon fee from our climate-change heroes, we recommend that all grassroots organizations should push for special congressional action that establishes a carbon fee at the federal level as soon as possible. Once implemented, this carbon fee should increase annually, as suggested by James Hansen, and be applied to habitual CO_2 generators, emitters, and users. For example, if a company is still using oil to run its buildings, it has to pay for those emissions; if a citizen is still purchasing gas for their car, there should be an extra pollution fee added to the cost; if a company is prospecting or extracting fossil fuels, all subsidies should be revoked and the company should receive a hefty fee that incorporates the externalities of fossil-fuel use that we have discussed (for example, public and environmental health). This fee (and the fund generated from eliminating federal subsidies for fossil fuels, which we will discuss later) can be amassed into a Federal Carbon Fund (FCF).

If set up properly, the resulting FCF will accomplish three key factors in one fell swoop. First, the fee will severely penalize the companies that continue to burn fossil fuels for power generation. This will provide a strong incentive for fossil-fuel companies to reduce—and eventually cease—their extraction, processing, and distribution of new fossil-fuel reserves. They will be forced to leave these unexploited resources in the ground and, hopefully, in the long term find new ways of fueling their power plants—namely by WWS renewable energy sources.

Second, the FCF will provide the financial backbone for WWS expansion throughout the United States. The FCF can allocate some of this money to new WWS energy companies as well as any existing fossil-fuel companies that see the writing on the wall and commit to adjusting their operations in order to stay in business.

FCF funding could allow for a variety of activities, including (1) retrofitting existing fossil-fuel-generating sources and infrastructure, (2) building new WWS generation plants and infrastructure, and (3) conducting wide-ranging WWS research and development. It is important to note here that the research and development would involve upgrades to existing WWS technologies and would also help develop innovative sources of clean energy, such as those proposed by the Breakthrough Energy Coalition—funded by billionaire investors like Elon Musk, Bill Gates, and Jeremy Grantham, whom we discussed in part 4.

In fact, over the long run, the expansion of WWS facilities in the United States will more than pay for itself through the economic boom that will be associated with the generation of millions of new jobs (not to mention the reduction in health problems associated with fossil-fuel emissions). Scads of new WWS work opportunities will open up in four basic employment sectors: (1) energy research and development, (2) facility construction, (3) marketing and sales, and (4) operation and maintenance. According to research by Stanford's Mark Jacobson and his colleagues, a plan that would transition all fifty states to 100 percent renewables by 2050 (with an 80–85 percent conversion by 2030) would provide "$3.9 million 40-year construction jobs and $2.0 million 40-year operation jobs for the energy facilities alone, the sum of which would outweigh the $3.9 million jobs lost in the conventional energy sector."[11] The conversion would also eliminate 62,000 premature deaths due to air pollution, and $3.3 trillion per year in 2050 global-warming costs to the world due to US emissions.[12]

And third, the FCF will provide dividends to individual companies and homeowners who have converted their operations and households from fossil-fuel use to WWS. This will provide a major financial incentive for making the move to rooftop solar panels, backyard windmills, electric cars, and, eventually, for tying into local WWS generating systems. Plus, it will set up a whirlwind feedback cycle of: more and more WWS users, which will equal fewer and fewer fossil-fuel users, which will equal more pressure on fossil-fuel companies to switch their power generation and distribution sources.

If you think this all sounds like a long shot, we want to assure you that it is not. In fact, there are already several US precedents in place for using legislative action to stand up to huge corporations, getting them to completely change how they do business. We talked about several of these when we examined the history of the US environmental movement.

First, starting in the late sixties, we had the battles with the US automobile industry over toxic exhaust pollution ("brown clouds"). In the end, the auto manufacturers were forced to retool their assembly lines to produce vehicles that ran on unleaded gasoline and used catalytic converters. Next, in 1976, came the fight with the chemical giant DuPont over the hole in the ozone layer. The final verdict here was that Freon—the product primarily responsible for producing ozone-destroying CFCs—was banned for good. Finally, we had the conflict with power plants and manufacturing facilities over the generation of acid rain, which started in the late 1980s. The US EPA took care of this problem with the Acid Rain Program (ARP) that eventually set caps on emissions of both of the responsible pollutants—NOx and SO_2.

In addition to pressuring fossil-fuel industries to stop burning fossil fuels, there are many other reasons why a carbon fee is the most important component of any plan for solving climate change. First and foremost: it will generate money . . . and we mean lots of money, especially in the short-term. According to the Center for Climate and Energy Solutions, a carbon fee has the potential to raise significant revenues for the government.[13] Depending on the carbon fee imposed, the money raised could be tens or even hundreds of billions of dollars each year. The Center for Climate and Energy Solutions states that, for example, "A carbon fee starting at about $16 per ton of CO_2 in 2014 and rising four percent over inflation would raise more than $1.1 trillion in the first ten years, and more than $2.7 trillion over a 20-year period."

ENFORCE THE CLEAN AIR ACT

As we discussed earlier, we believe the federal Clean Air Act (CAA), in its existing format (i.e., without any new amendments), provides at least part of the legal/regulatory authority Congress needs to make both Big Oil and Big Coal toe the line. In a 2007 decision, the US Supreme Court ruled that emissions of CO_2 and other GHG are public health hazards that merit regulation under the CAA.[14]

Moreover, according to information posted on the EPA's website, "On December 7, 2009, the [EPA] Administrator signed two distinct findings regarding greenhouse gases under section 202(a) of the Clean Air Act:

- **Endangerment Finding:** The Administrator finds that the current and projected concentrations of the six key well-mixed greenhouse gases; carbon dioxide (CO_2), methane (CH_4), nitrous oxide (N_2O), hydrofluorocarbons (HFCs), perfluorocarbons (PFCs), and sulfur hexafluoride (SF_6) in the atmosphere threaten the public health and welfare of current and future generations.
- **Cause or Contribute Finding:** The Administrator finds that the combined emissions of these well-mixed greenhouse gases from new motor vehicles and new motor vehicle engines contribute to the greenhouse gas pollution which threatens public health and welfare."[15]

Despite the fact that some of the regulatory clout to counteract climate change already exists under the CAA, either a new federal act specifically addressing the causes of and solutions to climate change (e.g., the "Climate Change Control Act") or additional targeted amendments to the CAA are critically needed. In fact, it would make the most sense to move the adjudication and enforcement of climate-change laws and regulations from the US EPA to our proposed new federal agency known as the Department of Energy, Resiliency, and Sustainability (DOERS) (see below).

This move would place all activities associated with climate change—from designing and siting new WWS facilities to issuing violations and levying fines—in the hands of the people working together under the same federal banner. When it comes to tackling the world's most powerful industry—Big Oil—and its teams of high-powered lobbyists, climate-change heroes need all the help they can get.

EXPEDITE RENEWABLE FACILITY CONSTRUCTION

To use a favorite federal government term, the conversion to renewable energy resources—wind, water, and solar (WWS)—must be "fast-tracked." In other words, the permitting process for researching, designing, upgrading, and constructing facilities associated with all WWS power generation needs should be mostly exempted from federal regulatory review processes, which often delay putting new projects in the ground for several years. This means that—beyond ensuring that each project will not cause some sort of calamitous environmental impacts (e.g., wiping out an entire ecosystem or decimating an already endangered species)—proposed WWS projects will not have to go through any sort of detailed Environmental Impact Statement (EIS) assessment, review, or public hearing process under the National Environmental Policy Act (NEPA) or any of the associated federal regulations (i.e., CAA, Clean Water Act [CWA], Endangered Species Act [ESA], etc.).

We realize that such an expedited strategy for renewables is going to raise the hackles of many of the NGO conservation organizations throughout the United States and around the world. As such, there will likely be considerable resistance to exempting a whole category of energy resource development (i.e., WWS) from federal regulatory review.

To counter this opposition before it starts, we suggest considering the following two points. First, there is a significant precedent for exempting a specific type of energy development from the aforementioned federal statutes. As we mentioned previously, these exemptions were granted in 2005 to the fracking industry.

In order to accelerate the location and development of natural gas—considered to be a "cleaner source of fossil fuels"—fracking wells were allowed to be drilled indiscriminately all over the landscape, typically without any significant federal review and compliance documentation being required. And we have to believe that the proven hazards of fracking—contaminated drinking-water wells, exploding water faucets, thousands of acres of desecrated wildlife habitats, and billions of gallons of toxic wastewater[16]—are far worse than any environmental impacts that will be generated by constructing solar panel arrays and wind turbine farms, no matter how large they are.

Next, if you want to think about the essential reason to decide to fast-track WWS development, know that we are really now in a situation where we have to choose between the lesser of two evils. In order to most effectively and efficiently ward off the long-term, life-altering impacts of climate change on the human condition, we have to take some risks. And one of these risks will involve believing that fast-tracking WWS facilities on a nationwide basis will not result in any long-term environmental damage from which we cannot recover. On the other hand, if we do not fast-track WWS and cease burning fossil fuels, we now know—beyond the shadow of a doubt—that there will be long-term environmental damage from which we cannot recover; in the form of melting glaciers, rising sea levels, and increasing storm severity.

ELIMINATE FEDERAL SUBSIDIES ON FOSSIL FUELS

According to a November 2015 report compiled by the group Oil Change International (OCI) and the UK-based think tank Overseas Development Institute (ODI), national subsidies to oil, gas, and coal producers amount to $20.5 billion annually in the United States, with almost all of those being received in the form of tax or royalty breaks.[17] According to Avaneesh Pandey, writing about the OCI/ODI Report in the *International Business Times*, "Federal subsidies amount to $17.2 billion annually, while subsidies in a number of oil-, gas- and coal-producing states average $3.3 billion annually." This report also notes that the United States is set apart from the world's other G20 countries—a group representing the world's largest economies—by the sheer variety of tax exemptions for fossil-fuel producers.[18]

In the United States, as Pandey again documents, "Deductions for cleaning up oil spills allows companies to claim the cost as a standard business expense."[19] The BP oil-spill disaster (the *Deepwater Horizon* accident) allowed BP to claim $9.9 billion in tax deductions in 2010 for the $32.2 billion in clean-up costs they reportedly incurred, by way of this subsidy provision.[20]

Furthermore, BP's final settlement with the US government and five state governments—for regulatory penalties and damages incurred—totaled more than $20 billion. But only $5.5 billion of this was in the form of a non-tax-deductible penalty. Inexplicably, BP was able to write the remainder off.[21]

In our opinion this is absolute absurdity. How does an oil company that royally screws up one of its major operations deserve to get partially bailed out by the US government? BP should not have received even a penny in tax deductions for their clean-up and damage reparation costs in the Gulf of Mexico. On a broader scale, how much sense does it make for the US government to be subsidizing the operations of fossil-fuel companies at the same time that we're supposed to be leading the world in reducing CO_2 emissions? US subsidies to the fossil-fuel industries must stop—and do so soon.

INITIATE FEDERAL SUBSIDIES ON RENEWABLES

Finally, we should be prepared to strongly support federal subsidies—to the tune of billions of dollars—to get the WWS industry fully up to speed and rolling along as quickly as possible. However, the way we see it, none of this money is coming out of the pocketbooks of individual Americans.

As we discussed earlier in this section, the primary source of funding for WWS design and construction will be provided indirectly by the fossil-fuel giants themselves. Big Oil will make these contributions in two ways. First, they will pay carbon fees if they continue to generate CO_2/GHG emissions. They will also be charged a substantial tax-added penalty for any profits they generate from selling fossil-fuel products. Second, some of the funds recuperated from eliminating fossil-fuel subsidies can be moved to support renewables. An additional source of "revenue" may be the settlements from the lawsuits popping up against some of the oil giants such as Exxon/Mobil for knowing the damages of climate change and burying their research. The first of these has already occurred—the Conservation Law Foundation (CLF), based in Boston, announced at a press conference on May 17, 2016, that it had "served formal notice of a lawsuit against ExxonMobil for its decades-long campaign to discredit climate change and knowingly endanger people and communities."[22] This is the first lawsuit against a fossil-fuel company, but we are certain it is not the last, as more people learn about the science they have squelched.

COORDINATE/PROMOTE MEDIA EVENTS

One of the biggest drawbacks we continually see with the present climate-change movement is the lack of attention from the mainstream media. This was glaringly apparent even during the most recent UN Climate Change Conference (COP-21)—which we both attended—held on November 29 through December 11, 2015, at Le Bourget Conference Center on the outskirts of Paris, France.

The primary coverage COP-21 was receiving was buried behind the front pages or in secondary sections of major newspapers and relegated to the second segment on the primary television networks. Oh to be sure, climate change is often featured these days in conservation and environmental magazines—like *Audubon*, the *Sierra Club Bulletin*, and *National Wildlife*—as well as on public television and national public radio. But this is generally only "preaching to the choir." The vast majority—probably in excess of 95 percent—of people who read conservation publications and watch public media are already climate-change believers. They don't need further convincing. They already realize climate change is a serious problem, and they want to take action as soon as possible.

To effectively convince people of the seriousness of climate change and rally their support for positive resolutions, we have to reach out to the masses—take our message to the skeptics and even to the deniers.

And the best way to do this is by getting increased coverage from mainstream US television networks—NBC, MSNBC, FOX, CBS, ABC, and CNN—as well as international networks like the BBC, Russian Television (RT), Abu Dhabi TV, and Al Jazeera, since climate change is affecting the whole world and everybody, everywhere needs to be part of the solution.

Special climate-change programming could be kicked off with a series of concerts broadcast worldwide using a catchy moniker. For a prototype, think of "We Are the World"—the charity single originally recorded by scores of musical artists and other celebrities in 1985, who came together as the supergroup USA for Africa. Michael Jackson and Lionel Ritchie wrote the song, and the album sold more than 20 million copies worldwide.[23] "We Are the World" was performed live a number of times (not always with the same performers), with the most famous occasion being at the Live Aid concert.

Held on July 13, 1985, the Live Aid concert raised money for humanitarian aid for those affected by the famine in Ethiopia. Prince Charles and Princess Diana officially opened the concert at Wembley Stadium in London, and—nearly two hours later—it began at JFK Stadium in Philadelphia as well. The two-location "superconcert" went on for sixteen hours, linked by satellite to more than a billion viewers around the world. Smaller concerts took place in many countries at the same time, with the world truly coming together for a cause. Demonstrating a triumphant blending of technology and good will, Live Aid raised more than $125 million for famine relief in Africa.[24]

An event like this aimed at climate change must be placed in the hands of skilled Hollywood "power people"—like our climate-change heroes James Cameron, Leonardo DiCaprio, and Mark Ruffalo—with proven abilities for directing and pulling off shows like this. To really make this work, musicians from every venue—rock, blues, jazz, country, folk, hip-hop, heavy metal, even classical—should be involved. The concert should provide something for everyone to latch onto and want to watch. Celebrity speakers also should be recruited, representing all facets of the current political spectrum—Republicans, Democrats, evangelicals, and Independents. Their messages may vary in tone, but they should all address the same bottom line—finding solutions to the climate-change crisis.

If you think putting on an event like this would be very expensive, you're absolutely right! Big name talent is always going to require big-time money. While we may be able to appeal to the environmental sensitivities of many performers in terms of their fees, a lot of money will still be required to pull off the type of extravaganzas that will be needed to reach a vast audience.

This is where our financial supporters and investors will be critical. They must be willing to front high production costs. However, if these events work as they should, they will attract the wealthy—ideally many who will want to be associated with the climate-change movement—and the productions will, in fact, bring in a lot of income. Revenue generated by sales of twenty million albums is, as they say, nothing to sneeze at.

DESIGN AND IMPLEMENT CLEAN ENERGY PLANS (CEPs)

As we see it, based on research and interviews with our climate-change heroes, the bulk of the actual shift to clean energy will occur at the local and state levels. The basis for establishing state (or, as appropriate, local) Clean Energy Plans (CEPs) will be Stanford professor Mark Jacobson's 50 States-50 Plans Initiative, which is featured on the website for the Solutions Project (TSP). As we talked about in chapter 23, Professor Jacobson is another of our climate-change heroes, as is one of his TSP founding partners—actor Mark Ruffalo. (TSP has also provided basic CEPs for 139 countries, although international implementation has a lot more uncertainty because of the highly variable finances and politics of each country involved.)

If you look at the TSP website, you may be surprised to learn that the capacity for providing sufficient WWS to run all fifty US states already exists. It is just a matter of harnessing these renewable energy sources and then building the wide-ranging infrastructures needed to delivery this power to consumers.

So the models for designing CEPs for each state are already in place. And this is where our proposed Department of Energy, Resiliency, and Sustainability (DOERS)—which we will talk about next—steps in and provides the funds from our recommended federal carbon fund (FCF) that we discussed earlier in this chapter. Ideally, a portion of FCF-generated monies will be used for both building new power plants and upgrading and/or constructing the infrastructure needed to bring the CEPs to reality.

ESTABLISH A DEPARTMENT OF ENERGY, RESILIENCY, AND SUSTAINABILITY (DOERS)

After grassroots organizations have worked to take the steps we've discussed toward solving climate change, we will need a new federal agency to manage everything. Government involvement at the federal level is essential to coordinating and implementing an effective and efficient rollout of climate-change solutions.

Of course, we recognize that there will be a lot of opposition to this proposal. We understand that there is a sector of the population with a deeply held belief that big government can't solve problems because big government is the problem.

But here's the thing: throughout US history, the federal government has typically been key in solving the national and international crises we have faced. Abraham Lincoln brought an end to the horrors of slavery when he signed the Emancipation Proclamation. Franklin Delano Roosevelt's alphabet soup of social programs led our nation through the Great Depression. The US Soil Conservation Service worked with Midwestern farmers to end the Dust Bowl.

A multitude of federal environmental laws passed in the seventies now provide critical protection of our air, land, water, and wildlife—and they grew out of the environmental upheaval of the sixties. The Wall Street blowup of 2008 was resolved by federal bailouts. Can you imagine any of these things being accomplished if the federal government had not been involved? And so it is with climate change. The federal government has to be at the helm—the captain steering the ship—if you will.

This is why we believe a new federal agency—potentially called the Department of Energy, Resiliency, and Sustainability (DOERS)—needs to be established. We do realize, of course, that this could also be accomplished—perhaps more expeditiously—by simply renaming and revitalizing the existing Department of Energy (DOE). However, we believe that a fresh start all the way around—from hiring new experts charged with a new mission to implementing a new management structure—would be the best way to proceed.

But either way, the primary rationale for the existence of DOERS would be this: the agency will provide a national focal point—the engine of the train, so to speak—for responding to grassroots actions that are taking place at the local and state levels. DOERS will also—as necessary and appropriate—take the lead in transitioning state and local activities into national laws and policies.

First and foremost, DOERS will be responsible for managing revenues generated by the federal carbon fee (i.e., what we have been calling the federal carbon fund, or FCF). Since DOERS will have the best handle on what is going on with regard to nationwide energy issues, Congress will annually allocate the generated money to the agency for distribution to qualifying companies and homeowners. DOERS will also have oversight into efficacy and integrity with regard to how the distributed FCF monies are used.

The key here is that the whole process should be self-perpetuating. The wind, water, and solar companies that receive money will want to use the funds to continue expanding their clean energy facilities so they can make more money. Individual homeowners will want to continue switching their overall lifestyles away from fossil fuels so that they can also receive more money each year.

DOERS will also provide oversight and clearinghouse functions for the design and implementation of state and local Clean Energy Plans (CEPs). As we noted in the above section, CEPs are another key component for making climate-change solutions a reality.

In summary, DOERS should be responsible for the following:

- Enforcing applicable provisions of the Clean Air Act.
- Eliminating subsidies for fossil fuels.
- Initiating subsidies for WWS facilities.

- Fast-tracking construction of WWS facilities
- Supporting/coordinating CEPs.

Accomplishing DOERS' monumental goal of solving the climate-change crisis will require a very special person, possessing a one-of-a-kind skill set. The secretary of energy, resilience, and sustainability must be a proven world leader, with recent, first-hand knowledge of our planet's major political and social systems. He/she must also be politically connected to the world, including having direct relationships with world leaders. He/she needs, as well, to have a keen understanding of the science and ramifications of climate change.

From the professional side, the DOERS secretary of energy, resilience, and sustainability should be extremely intelligent, articulate, and profoundly adept at public speaking. He/she must also possess the charisma and character to reach out to all peoples of the world and rally unyielding support and dedication to the cause. Finally, he/she must have a deep commitment for seeing the climate-change crisis resolved as quickly and efficiently as possible. Many of the heroes we discuss throughout this book would be excellent candidates for Secretary and/or other executive level positions within DOERS.

Chapter 31

SOLUTIONS FOR THE WORLD

SCHEDULE AND CONDUCT PRODUCTIVE CLIMATE-CHANGE CONFERENCES EVERY YEAR

As we discussed in chapter 21, the twenty-first Conference of the Parties (COP-21) International UN Climate Change Conference, which was held November 29 through December 11, 2015, in Paris, France, received a generally mixed review. On one hand, COP-21 was precedent setting in that all parties in attendance—a total of more than 190 countries—agreed to sign a treaty (the Paris Agreement). In some ways, this is exceptional. For the first time in the history of the COP meetings, the entire world came to an agreement—fossil fuels are not the way forward.

Since COP-21, there has been a perceptible shift away from the idea that green energy is just for tree huggers toward a view of renewables as the key to savvy new technology that gives us a bright future—a technological edge. Fossil fuels have become, not just dirty but old-fashioned, clumsy and futureless. Reliance on them signals a country that will be left behind by the rest of the world—both technologically and economically—foregoing the potential to be considered a future global leader.

On the other hand, most activists, and even delegates from many attending nations, believe that what was signed really did not amount to much. For one thing, most countries did not use a lot of specifics when defining their emission-reduction targets—for example, naming exactly when and how these targets would be met. We are hopeful that the "stock-taking" (assessing climate action every five years) and the "ratcheting" or "ramping up" (increasing levels of climate action commitment every five years) portions of this agreement will keep countries on track, but there is no guarantee.[1]

So what does the Paris Agreement really mean for the ongoing efforts of the world to stave off climate-change impacts? Well, the attendees did all agree on a long-term goal of trying to limit the rise in global temperature to "well below" 2.0 C (3.6 F) higher than preindustrial levels.[2] But many activists viewed that as not good enough, since most climate scientists believe that a global rise of more than 1.5 C (2.7 F) will mean dramatic negative impacts for millions of people living on planet Earth.

Given the stronger consensus and commitment shown by the world at COP-21, it is essential to continue scheduling COPs every year. These global meetings need to continue to be held at revolving locations throughout the world. Also, just as President Obama did as a prelude to COP-21, the United States must take the lead in rallying all participating countries and getting them on board ahead of time with their climate action goals. The message to participating countries must be loud and clear each year: we are all in this together and there is a deep need for real accomplishments and tangible, achievable goals.

Next, it is critical that the international climate-change treaties produced each year are living documents that are revisited and revised by each country—not every five years, as was decided at COP-21, but annually. Each year, each participating country must account for their climate action in a global stocktake. For example, according to the Paris Agreement, beginning in 2018, and every five years thereafter, the signing nations will get together to look at the progress of their combined efforts on climate-change action.[3] "Stocktake" was a big buzzword at the Paris climate summit, and for good reason, as it forms the base of the "ambition mechanism," the cycle of action in which the world's progress is continually analyzed and strengthened.[4] Constantly taking stock of where we are and where we need to go ensures that our climate action goals just keep getting better. The idea of an annual stocktake is to keep the constant focus of Earth's nations on greenhouse gas emission targets—if we wait five years to do a stocktake, it may be too late to move the needle.

The ambition mechanism does not mean only that countries pay attention to their progress, however, but also that they regularly tighten their GHG emission targets. Countries at COP-21 agreed to submit new climate plans, called nationally determined contributions (NDCs), with ramped up actions on emissions, every five years.[5] Again, this ratcheting process should be aggressive and annual and done in a conscientiously honest and fair manner. We recognize that these are new, untested approaches, and time will determine whether they are effective, though they hold great promise on paper. There is some precedent for success, as the approaches resemble the type of peer pressure and imitation method used in the 1961 Peace Corps Initiative and the 1975 Helsinki Accord.[6]

Additionally, each country should be required to present a suite of solutions demonstrating how the strengthening emission-reduction targets will be accomplished. This, in turn, will allow all of the participating nations to take into account each other's input, sharing information about what works and doesn't work and what technology holds promise for the future.

Finally, the collective target for these NDC emission-reduction plans must be aimed at keeping the global temperature rise to a maximum of 1.5 C (2.7 F)—not 2.0 C (3.6 F) or even higher.

INITIATE LARGE-SCALE RESTORATION OF NATURAL ECOSYSTEMS

Everything every one of us can do to protect and preserve natural landscapes is important to reducing the total amount of CO_2 that re-enters our atmosphere and contributes to the thickness of the greenhouse blanket that is warming our Earth. Thomas Lovejoy celebrates ecological restoration as a way for everyone to get involved with healing the planet.[7] Anyone can seed grasses to restore a coastal wetland or plant a tree. Participation in a solution is accessible, which makes climate change feel less overwhelming as a problem to tackle.

The greatest effect, however, will come from restoring huge swaths of land, especially if they support natural forests. When trees are cut down, they release stored carbon to the atmosphere, contributing to global warming directly, and are no longer available as a collector, or sink, for carbon. According to the World Carfree Network (WCN), the destruction of forests contributes at least 1 percent more greenhouse-gas pollution than all the cars and planes together.[8] That is why one of the Nature Conservancy's (TNC) primary goals is buying, protecting, and—if necessary—restoring large tracts of undeveloped land all over the world.

As an example, TNC is coordinating with the hunter-gatherer Hadza people of Tanzania to secure their traditional homelands. First, the Hadza earn carbon-offset credits by keeping their carbon-sequestering woodlands intact. Then—supported through Carbon Tanzania—they sell their certified credits on international markets.[9] Similarly, TNC has helped the Yurok tribe in Northern California to generate income by selling carbon credits while concurrently protecting thousands of acres of natural forests.[10]

TNC is also working locally to restore oyster bars, coral reefs, sea grass beds, and mangrove swamps in places as diverse as metropolitan New York City, the Gulf of Mexico, and the island nations of the Pacific and Indian Oceans, providing an organizational model for other groups to follow. In addition to adding more natural landscapes to the planet, restoring these features also improves the resilience of coastal communities to sea-level rise and violent storms.[11]

INVESTIGATE FEASIBILITY OF ALTERNATIVE TECHNOLOGIES

Based on everything we've discussed in our book, we steadfastly believe that we can stop the runaway train of climate-change effects before they become irreversible by simply doing what makes the most sense. In a nutshell, that is achieving a total switch in the way the world is run on fossil fuels to renewable energy sources—WWS—by 2050 at the latest.

But we also realize that if at some point that goal does not appear attainable, we may need to consider some alternative technologies designed to provide the world with at least temporary relief from global warming. An important clarification to make is that geoengineering solutions such as the one we are about to present are incredibly controversial. As Joseph Romm writes in *Climate Change: What Everyone Needs to Know* "geoengineering is akin to a risky, never tested, course of chemotherapy prescribed to treat a condition curable through diet and exercise—or, in this case, GHG emissions reductions."[12]

Engineering, by definition, is about managing a system that is understood, and we do not fully understand the climate system, therefore we cannot geoengineer it.[13] Furthermore, since most geoengineering-type solutions involve doing things that treat the symptom and not the cause, the minute we stop doing them we will be back where we would have been anyway. Since these ideas are global in scale, whatever downside they have will also be global in scale.[14] Again, we are confident in our ability to switch to 100 percent wind, water, and solar by 2050, and therefore do not anticipate having to employ any such alternative techniques; because of this, we will only mention and briefly describe one geoengineering concept here.

CREATE A SOLAR SHIELD

Many geoengineering techniques for reducing the impact of climate change have been bandied about since Dr. James Hansen warned the world about the problem in 1988. The most commonly mentioned of these methods involves creating a solar shield by artificially injecting sulfur particles (sulfates) into the upper atmosphere to create, in effect, a giant umbrella over Earth.

Climate-change hero Professor Paul Crutzen was one of the first scientists to recommend this procedure in 2006. But in a 2016 interview, Dr. Crutzen told us that creating a solar shield is "very risky and should be considered only as an *ultima ratio* [last resort] if more conservative methods fail."[15]

The potential hazards of creating a solar shield over Earth range from significant alteration of the global water cycle to accelerated ozone depletion associated with the increased sulfur that is injected into the atmosphere.[16] Also, since these shields would only be temporary fixes, abrupt and significant rises in global temperatures—including severe droughts—could occur when they are removed.

DEMONSTRATE HUMAN RESILIENCY: ALASKA

In the future, some locations may have to accept the fact that major changes are on the horizon and that lifestyles must be adapted accordingly. For example, in Alaska, Claude Garoutte—a former Colorado lumberjack—came up with the prototype design for an "eco-house." Traditionally, houses in the far north have depended on the firmness of the frozen ground for their foundation, but the eco-houses are designed to compensate for the melting permafrost. "We know you can't hold back Mother Nature," Garoutte says. And, even though the new houses were intended for use on the Alaskan tundra, Garoutte believes they will "set a standard for the world to follow."[17]

As Madeline Ostrander, writing for *Audubon* magazine, describes the houses,

> Encased in thick foam insulation and powered partly by solar panels, the eco-houses use 80 to 90 percent less energy than the typical North Slope dwelling. Instead of a sewer hookup, the buildings are connected to an external black tank that uses bacteria to compost human waste. Each house features an iron foundation that resembles a giant bed frame, with six legs. If the ground below the house were to shift or sink, a handheld jack could be used to shorten or lengthen the foundation legs and make the house level again.[18]

Ostrander also describes how the bottom edges of the rear part of the foundation frame of each eco-house curve upward like the blades of a toboggan. According to Garoutte, this makes it possible to move the house—like a sled—to another location, if the underlying tundra simply becomes too difficult to deal with.[19]

While the eco-houses are still a work in progress, Garoutte firmly believes in their design principles. Ostrander writes, "Any vision of the future in which people use their homes as escape pods, attempting to flee collapsing terrain, is necessarily apocalyptic—but there's something utopian about it, too."[20]

DEMONSTRATE HUMAN RESILIENCY: FLORIDA

In the February 2015 issue of *National Geographic*, Laura Parker describes how a development company called Dutch Docklands is investigating ways to work with climate change, to go with the flow.[21] Seeing a way to profit from the climate changes brought by global warming, the company has proposed a floating village in Florida. Located in a flooded quarry near Miami, the village would consist of "29 private, artificial islands, each with a sleek, four-bedroom villa, a sandy beach, a pool, palm trees, and a dock long enough to accommodate an 80-foot yacht."[22] Each floating island mansion would cost $12.5 million.

Dutch Docklands is aiming the project toward the wealthy who are concerned about how climate change and rising sea levels might affect their property values. The islands will be anchored to the bottom of the lake to keep them in place, but rising sea levels will never be a concern.

While only the richest would be able to buy their own private floating island, Dutch Docklands believes that the concept will open up endless possibilities for places around the world that will be hardest hit by sea-level rise caused by climate change: "Floating communities with floating parks and floating schools. A floating hospital."[23] "People only see the negative effects of flooding," a company employee told Parker. "We need to show people there is a way to make money out of this. For the government, there are tax dollars. For developers, their investment is secured for the next 50 years. There is a lot of money involved in this climate change. It will be a whole new industry."[24]

CONCLUSION

In the overall history of human life here on Earth, we have never faced a more broad-based and imminent environmental peril than that posed by climate change and global warming. On a geologic time scale, we are accelerating toward our own oblivion at laser-focused warp speed. Right now—every day—the world is adding another spike of atmospheric pollution to the shroud that may eventually doom our own species to extinction.

Climate change is not something that might become a problem in the future—maybe by 2030 or 2050 or 2100. It is a problem right now, getting worse every day that we sit by and pretend that it is not really happening. Just ask the subsistence farming families in the Bangladesh river deltas who are being forced by rising sea levels to leave their homes and flee to the sweltering, teeming ghettos of Dhaka. Or the citizens of Venice, Italy, who are spending billions of dollars on mechanical seawalls that will keep the ocean from swallowing the artistic, architectural, and historical treasures of their magical city. Or the native islanders of the South Pacific archipelagos who are watching their homelands disappear forever right before their eyes. Or the city managers in Miami Beach, Florida, who are already $100 million into fighting the city's incessant flooding. We could go on and on, but we think—hope—you understand our point about the immediacy of the climate-change peril.

But here is the good news we hope you gleaned from our words throughout this book. Climate change does not have to remain a problem. In fact, if we focus and work together, climate change can be well on its way toward full resolution within as little as fifteen years—maybe even sooner.

If we act with temerity, we can use the perpetual, inextinguishable energy of Earth—the sun's glorious rays, the wind's constant breezes, and the water's endless waves—to work for us all. And, in the process, we'll leave the polluting fossil fuels right where they belong—buried in the ground, never to see the light of day.

Think about it: Renewable energy here on Earth is abundant and omnipresent. Every time you go outside, you see and feel it everywhere. It's like an endless symphony written by a master composer and played by a world-class orchestra. The golden rays of streaming sunlight are the strings—always there, maintaining the basic rhythm of the interwoven movements. The wind provides the percussion—rising from gentle whispering breezes of the snare drum to bold resounding gusts of the tympani. Then moving water blends in with the woodwinds and the brass—transitioning from gently lapping melodic notes of the flute to lazy ripples of an oboe's dulcet tones and concluding with rolling waves of trumpet blasts. We can use this symphony to enrich our lives and create an elegant, sustainable future.

We are right on the cusp of what we can call the "Renewable Revolution," providing a mighty parallel to the Industrial Revolution. As we discussed earlier in the book, the Industrial Revolution resulted in the transformation of our nation from a rural agrarian society to an urban, manufacturing society. Now we are about to totally transform ourselves again, from a hard-edged, fossil-fuel driven economy to a softer-sided renewable energy world community.

Though, as Bill McKibben said, it will take a political, concerted effort analogous to our initiatives in World War II,[1] the transformation from fossil fuels to renewable energy is already possible. The Solutions Project lays out plans for converting each of our states—plus many countries—from fossil fuels to renewable resources. And we can accomplish this at the same time as we create numerous new industries in the wind, water, and solar power sector.

Along with this industrial boom will come millions of new jobs, leading to increased financial security for everyone. Now that's a win-win scenario we can all live with. Our children, grandchildren, and all future generations will look back and be forever grateful to us for being proactive and resolving the climate-change dilemma.

Summon the Heroes . . . by working together we can all get this done!
—Budd Titlow and Mariah Tinger

ACKNOWLEDGMENTS

Though all of the material in this book grew from our perspective and interpretation of the challenges and solutions of climate change, our deepest gratitude goes to the visionaries who spent precious time with us, sharing their exceptional and inspiring knowledge over the past year. We are particularly grateful to Thomas Lovejoy, who beguiled us with delightful stories alongside his critical climate insights, and his wonderful assistant Carmen Thorndike. Sarah Shanley Hope left us breathless after presenting everything that the Solutions Project is working on (and even skipped lunch for us!). Reverend Canon Sally Bingham may not realize this, but she was our first interview, kicking off a wave of hope and excitement for this project. Bill McKibben prioritized a meeting with us, despite being buried in world-altering grass-roots actions in Paris—and was always quick with a word of inspiration, support, and encouragement. Jeremy Grantham and Ramsay Ravenel devoted an entire Friday afternoon to bouncing around ideas for heroes and insight on the economic aspects of climate change. Gus Speth easily endeared himself to us during our chat as the kind of guy you wish you could join every night for a dinner conversation. Father John Chryssavgis provided invaluable advice and counsel. Janine Benyus took time away from her own writing retreat in a cabin in Montana, and, despite the chilly air, spoke with us from a spot by the lake for a full hour. Naomi Oreskes welcomed us and a camera crew into her office (patiently sitting through the sound-checks) and then stunned us with scintillating information. James Balog squeezed in an interview despite a trip to Mt. Kilimanjaro the next morning. Budd had a particularly fun time connecting with Joe Romm, as both share a passion for science and writing, and grew up with fathers who ran newspapers. Katharine Hayhoe, in true college professor fashion, offered some very constructive criticism on our initial set of interview questions (greatly improving them as a result) and then gave us a rock-star interview in Paris, joined by her friend Jonathan Patz. Bob Inglis made us laugh and broadened our horizons and perspective. Mark Jacobson's time was short, but his information was rich—his remarkable research offers an enormous light at the end of the (often dark) climate-change tunnel. Paul Hawken provided us with first-hand input on his remarkably ambitious Project Drawdown research and book. Paul Crutzen updated us on his viewpoints on geoengineering solutions to climate change. Wally Broecker provided us with his detailed description of how carbon capture and burial could work. And, finally, Amory Lovins gave us a detailed look inside some landmark climate-change studies being conducted at his Rocky Mountain Institute.

And we are grateful to Debby Titlow (Budd's wife and Mariah's mom), for being our secret third author, despite her own illustrious writing and teaching career. We offered her the position, but she just laughed and then corrected our sentence structure in the offer email. She has been a rock for both of us, reviewing our words, brainstorming titles, acting as a voice of reason for ideas that were too "out there" and a muse for our writing.

To Lorraine and John Tinger, for whisking away Mariah's kids (Budd's grandkids) for adventures so that she could write, always taking a deep interest in our book, and always being the first to support any social networking initiatives (including our Kickstarter project—thank you!).

To Merisa Titlow (Budd's daughter and Mariah's sister), for deep insight into many key players and for being a powerful sounding board for suggestions and approach. Additionally, Merisa and her husband, Sam Eisenstein, supported our Kickstarter project.

To David Denny (Budd's son and Mariah's brother), for challenging Mariah's views in particular, as her devil's advocate, and pushing all of her buttons as a good older brother should. And for constantly providing "you can get this done" support to us every step we took along the way.

To Dr. Tim Weiskel, Mariah's professor at Harvard University, for his continued counsel, inspiration, promotion, and advice, in person, and through the content on his website www.EcoEthics.net.

To the good sisters, especially Sister Mary Frances, at St. Scholastica Priory in Petersham, Massachusetts; and Preston Browning at the Wellspring House. Both provided a quiet and motivational place to write.

A big thanks to Jason LaChapelle, who provided counsel and videography of our heroes (for a potential future project!).

To all of our Kickstarter supporters, who believed in us, believed in our book, and sponsored our research trip to the climate change conference (COP-21) in Paris. The firsthand experiences we had there added important depth and understanding to the book, and allowed us to be part of a historical moment. First and foremost among the supporters are Ryan and Rekha (Madan) Hayden, and the rest of the Madan and Hayden clans, who are like family to Mariah. They continually delight us with their curiosity, passion, and support for climate-change solutions. To our fabulous, kind neighbors Abby and George—their four beautiful, vibrant children remind us regularly of the importance of the work we are doing for future generations. Urvi Mujumdar was the very first person to support our project, and she thoughtfully connected me with her uncle, Jayant Sathaye, recipient of the 2007 Nobel Peace Prize for climate change alongside Al Gore. We are grateful to Urvi for her support and Jayant for offering his valuable insight. Vanessa Parker-Geisman, David Morin, Nicole Tateosian, and Sara Radkiewicz supported our Kickstarter, but also spent hours discussing the book with us on hikes and over dinners; We thank Jacob Holzberg-Pill and Grace Oedel, who are working climate miracles with Dig-In Farm; Molly Benson, who was so inspired by the book idea that she supported our Kickstarter in honor of her friends David Scoglio, Amanda Kelly, and Kristen and Brian Burdt; Reeta and Prabhu Rao, Chris and Kristen Buchanan, Tom Breider, and Lauren Kyser Newsom, who are all deeply involved in climate-change solutions themselves; dear friends Cindy Koenig, Florentien de Ruiter, and Tom Bock, Julia and Dave Long, Brad and Katy Langhorst, Katie and Ron Cooper, Wendy Palto, Andy and Lisa Schneider, Jana Eisenstein, Kath Hardcastle, Kurt and Kelly Anderson, Jeremy and Jennifer Schley Johnson, Ken and Kelly Kozloff, Tim and Erin Morin, Brendan Kearney, Gregory Soutiea, Jinju and Justin Fong, Alex White aka "A", Eric Weeks and Kelly Mac, Michelle and Saul Farber, Kara Brickman, and Hao Chen and Michael Winder—thank you all so much! Tina Tobey and Chris Mack, supported our Kickstarter, and Tina is helping us to spread the message in exciting ways. A big thanks to Dana Zemack, who deftly guided me through optimizing the Kickstarter process; and last, but not least, we thank George Clark, who not only supported our Kickstarter, but dispensed invaluable advice as Harvard's environmental librarian, including some suggestions for amazing titles!

Budd would like to thank Andrew Miller, his next-door neighbor and retired professional aquatic ecologist, for providing valuable sources of material and serving as a continuing sounding board (over dark beers) for our book's progress. Also his thanks goes to the Board of Directors (BOD) of the Apalachee Audubon Society—especially Dr. Sean McGlynn—who graciously allowed him to defer his BOD presidency for a year so that he could write this book.

Mariah extends a big shout out to her support team. She had several babysitters who lovingly cared for her children, but notable were Paige Coyne and Danielle Bertaux, who always delighted the children, and the staff at the Beverly Athletic Center, Emily, Kelly, Brianna, Nina and Mel, who watched her kids while she cleared her mind and sorted her thoughts over a five-mile run.

Mariah thoroughly enjoyed sharing conversations about the book with her dear and cherished elders—her Great-Aunt Betty Igleheart and her grandmother, Mama Ches Gearing. Mama Ches thoughtfully contributed some research articles, too!

Mariah offers her deepest love and gratitude to her incredible family. Every time she holds her two young children, Harrison and Sierra, she prays they will grow up in a world that is clean, healthy, equitable, and full of nature's beauty. They are her every-day motivators. Her husband, Brian, an accountant by trade, has developed a secondary passion for all things environmental. He has gone deep into the process of making their home carbon neutral, even when the investment did not offer the return on investment he would hope for, acknowledging that it was just "the right thing to do" for the planet. Most importantly, she treasures his endless hours of patient support and encouragement. He is the crimson columbine to her hummingbird energy—rooted at home, deeply supportive, providing a scaffolding for her creative colors to shine.

Mariah thanks her father, Budd. Writing this book with him has been a deep treasure—an experience that she will never forget. He is the first person who taught her to appreciate nature, inspiring a life and career devoted to Earth's breathtaking beauty. She is honored to connect with him in this joint celebration of hope through our writing. And Budd would be remiss for not thanking Mariah for providing the initial inspiration and continued enthusiasm for writing this book. Her remarkable people skills were also the glue that held everything together—especially throughout the interviewing process.

Finally, we owe a deep debt of gratitude to Steven L. Mitchell, Sheila Stewart (in abundance!), Hanna Etu, and the rest of the team at Prometheus Books for giving us the opportunity to bring *Protecting the Planet* to life. Everyone was incredibly helpful, but Sheila Stewart, in particular, offered profound guidance and editorial skills to help us produce a book that we hope will make a difference to the health and welfare of future generations of people on Earth.

NOTES

INTRODUCTION

1. Joseph Romm, *Climate Change: What Everyone Needs to Know* (New York: Oxford University Press, 2016), p. 71.

2. Carbon Dioxide Information Analysis Center, "800,000-Year Ice-Core Records of Atmospheric Carbon Dioxide (CO_2)," Carbon Dioxide Information Analysis Center, http://cdiac.ornl.gov/trends/co2/ice_core_co2.html (accessed February 28, 2016).

3. Andrea Thompson and Brian Kahn, "What Passing a Key CO_2 Mark Means to Climate Scientists," *Climate Central*, November 20, 2015, www.climatecentral.org/news/co2-400-ppm-scientists-meaning-19713 (accessed February 28, 2016).

4. James Hansen et al., "Target Atmospheric CO_2: Where Should Humanity Aim?" *Open Atmospheric Science Journal*, October 15, 2008, arxiv.org/abs/0804.1126 (accessed February 28, 2016).

5. Brian Kahn, "CO_2 on Path to Cross 400 ppm Threshold for a Month," *Climate Central*, March 18, 2014, www.climatecentral.org/news/co2-on-path-to-cross-400-ppm-threshold-for-a-month-17189 (accessed February 28, 2016); Adam Vaughn, "Global Carbon Dioxide Levels Break 400 ppm Milestone," *Guardian*, May 6, 2015.

6. Thompson and Kahn, "What Passing a Key CO_2 Means."

7. Naomi Klein, *This Changes Everything: Capitalism vs. the Climate* (New York: Simon & Schuster, 2015).

CHAPTER 1: WHAT EXACTLY IS CLIMATE CHANGE?

1. Tania Lombrozo, "Global Warming Explained, in About a Minute," NPR, December 16, 2013, http://www.npr.org/sections/13.7/2013/12/15/251437395/global-warming-explained-in-about-a-minute (accessed February 12, 2016).

2. Joseph Romm, *Climate Change: What Everyone Needs to Know* (New York: Oxford University Press, 2016), pp. 1–2.

3. Ibid., p. 70.

CHAPTER 2: WHAT IS THE HISTORY OF CLIMATE CHANGE?

1. James R. Fleming, "Joseph Fourier, the 'Greenhouse Effect,' and the Quest for a Universal Theory of Terrestrial Temperatures," *Endeavour* 23, no. 2 (1999): 72–75.

2. John Mason, "The History of Climate Science," *Skeptical Science*, April 7, 2013, https://www.skepticalscience.com/print.php?n=1473 (accessed August 4, 2015).

3. Spencer Weart, *The Discovery of Global Warming: Revised and Expanded Edition* (Cambridge, MA: Harvard University Press, 2008), p. 6.

4. Spencer Weart, "The Carbon Dioxide Greenhouse Effect," *The Discovery of Global Warming*, March 2015, American Institute of Physics, https://www.aip.org/history/climate/CO2.htm#s1 (accessed September 13, 2016).

5. Weart, *The Discovery of Global Warming*, p. 6.

6. "1931–1965: Hurlburt to Keeling," CO_2.Earth, https://www.co2.earth/193101965-hurlburt-to-keeling (accessed September 8, 2016).

7. Weart, *Discovery of Global Warming*, p. 19.

8. Gilbert Plass, James Rodger Fleming, Gavin Schmidt, "Carbon Dioxide and the Climate," *American Scientist* 98, no. 1 (January-February 2010): 58, http://www.americanscientist.org/issues/feature/2010/1/carbon-dioxide-and-the-climate/99999 (accessed July 29, 2015).

9. Ibid.

10. Ibid.

11. Gilbert Plass, James Rodger Fleming, Gavin Schmidt, "Carbon Dioxide and the Climate," overview, *American Scientist* 98, no. 1 (January-February 2010), http://www.americanscientist.org/issues/feature/2010/1/carbon-dioxide-and-the-climate (accessed July 29, 2015).

12. Roger Revelle and Rhodes Fairbridge, "Carbonates and Carbon Dioxide," in *Treatise on Marine Ecology and Paleoecology*, ed. Joel W. Hedgpeth, The Geological Society of America Memoir 67, vol. 1 (December 30, 1957): 239–95.

13. David Malakoff, "Global Warming: From Theory to Fact," *Causes*, NPR, May 21, 2007, http://www.npr.org/templates/story/story.php?storyId=10307560 (accessed August 4, 2015).

14. Ibid.

15. Ibid.

16. QUEST Staff, "The Keeling Curve Explained," *QUEST*, KQED: Science, December 12, 2014, http://ww2.kqed.org/quest/2014/12/12/the-keeling-curve-explained/ (accessed July 26, 2016).

17. Malakoff, "Global Warming: From Theory to Fact."

18. S. M. Enzler, "History of the Greenhouse Effect and Global Warming," Lenntech, 2004, http://www.lenntech.com/greenhouse-effect/global-warming-history.htm (accessed August 4, 2015).

19. Lyndon B. Johnson, "Special Message to the Congress on Conservation and Restoration of Natural Beauty, February 8, 1965," in *Public Papers of the Presidents of the*

United States: Lyndon B. Johnson, 1965, vol. 1, entry 54 (Washington, DC: Government Printing Office, 1966), pp. 155–65, found online at LBJ Presidential Library, http://www.lbjlib.utexas.edu/johnson/archives.hom/speeches.hom/650208.asp.

20. Adam Frank, "The Forgotten History of Climate Change Science," *13.7: Cosmos and Culture*, NPR, May 13, 2014, http://www.npr.org/sections/13.7/2014/05/13/312128173/the-forgotten-history-of-climate-change-science (accessed February 13, 2016).

21. Ibid.

CHAPTER 3: HOW DOES CLIMATE CHANGE OCCUR?

1. Associated Press, "Carbon Dioxide Emissions Rise to 2.4 Million Pounds Per Second," CBS News, December 2, 2012, http://www.cbsnews.com/news/carbon-dioxide-emissions-rise-to-24-million-pounds-per-second/ (accessed March 11, 2016).

2. Tapio Schneider, "How We Know Global Warming Is Real," *Skeptic* 14, no. 1 (2008): 31–37.

3. Ibid.

4. Ibid.

CHAPTER 4: WHAT IS THE PROOF FOR CLIMATE CHANGE?

1. National Centers for Environmental Information, "State of the Climate: Global Analysis—Annual 2015," NOAA (National Oceanic and Atmospheric Administration), January 2016, https://www.ncdc.noaa.gov/sotc/global/201513 (accessed February 13, 2016).

2. "Attack on the Clean Air Act," Union of Concerned Scientists, http://www.ucsusa.org/global_warming/solutions/reduce-emissions/clean-air-act.html#.VwIfGvkrKUl (accessed November 13, 2015).

3. Joseph Romm, *Climate Change: What Everyone Needs to Know.* (New York: Oxford University Press, 2016), p. 10.

4. Tapio Schneider, "How We Know Global Warming Is Real," *Skeptic* 14, no. 1 (2008): 31–37.

5. Ibid.

6. British Antarctic Survey, "Ice Cores and Climate Change," Natural Environment Research Council, May 18, 2015, https://www.bas.ac.uk/data/our-data/publication/ice-cores-and-climate-change/ (accessed March 11, 2016).

7. James Hansen et al. "Target Atmospheric CO_2: Where Should Humanity Aim?" *Open Atmospheric Science Journal* 2 (2008): 217–31, http://www.columbia.edu/~jeh1/2008/TargetCO2_20080407.pdf (accessed July 26, 2016).

CHAPTER 5: WHAT ARE THE CURRENT IMPACTS AND WHERE ARE THEY OCCURRING?

1. *Chasing Ice*, directed by Jeff Orlowski (Los Angeles: Exposure Labs, 2012).

2. Earth Science Communications Team, "Climate Change: How Do We Know?" *Global Climate Change: Vital Signs of the Planet*, ed. Holly Shaftel, NASA's Jet Propulsion Laboratory, http://climate.nasa.gov/evidence (accessed February 14, 2016).

3. Ibid.

4. Ibid.

5. "Global Warming Is Happening Now," National Wildlife Federation, https://www.nwf.org/Wildlife/Threats-to-Wildlife/Global-Warming/Global-Warming-is-Happening-Now.aspx (accessed July 9, 2015).

6. Ibid.

7. "Polar Bear," National Wildlife Federation, https://www.nwf.org/Wildlife/Wildlife-Library/Mammals/Polar-Bear.aspx (accessed February 2016).

8. *Chasing Ice*, directed by Jeff Orlowski.

9. "Global Warming Is Happening Now."

10. "Sea Level Rise: Oceans are Getting Higher—Can We Do Anything about It?" *National Geographic*, http://ocean.nationalgeographic.com/ocean/critical-issues-sea-level-rise/ (accessed July 9, 2015).

11. Laura Parker, "Climate Change Economics: Treading Water," *National Geographic*, February 2015, http://ngm.nationalgeographic.com/2015/02/climate-change-economics/parker-text (accessed August 4, 2015).

12. Ibid.

13. Amy Sherman, "Fact-Checking Rick Scott on the Environment and Sea-Level Rise," *PolitiFact Florida*, March 11, 2015, http://www.politifact.com/florida/article/2015/mar/11/fact-checking-rick-scott-environment-and-sea-level/ (accessed August 2015).

14. Laura Parker, "Climate Change Economics: Treading Water."

15. Ibid.

16. Ibid.

17. Keenan Orfalea, "IPCC Report Examines Climate Change's Effects on Mississippi River Delta and Strategies for Adaptation," Restore the Mississippi River Delta, August 5, 2014, http://www.mississippiriverdelta.org/blog/2014/08/05/ipcc-report-examines-climate-changes-effects-on-mississippi-river-delta-and-strategies-for-adaptation/ (accessed February 14, 2016).

18. Ibid.

19. Ibid.

20. Ibid.

21. Ibid.

22. Thomas R. Karl, Jerry M. Melillo, and Thomas C. Peterson, eds., *Global Climate*

Change Impacts in the United States (New York: Cambridge University Press, 2009).

23. "Mississippi Delta, LA, USA," *Climate Hot Map: Global Warming Effects around the World*, Union of Concerned Scientists, 2011, http://www.climatehotmap.org/global-warming-locations/mississippi-delta-la-usa.html (accessed August 2015).

24. Dan Vergano, "Climate Change Softens Up Already Vulnerable Louisiana," *USA Today*, August 6, 2013, http://www.usatoday.com/story/news/nation/2013/08/06/climate-hurricanes-wetlands-global-warming/2595657/ (accessed August 2015).

25. Ibid.

26. Ibid.

27. Ibid.

28. "MOSE Project, Venice, Venetian Lagoon, Italy," Water-Technology.net, http://www.water-technology.net/projects/mose-project/ (accessed March 11, 2016).

29. Ibid.

30. Shanta Barley, "Climate Change Could Swamp Venice's Flood Defence," *New Scientist*, August 24, 2009, https://www.newscientist.com/article/dn17668-climate-change-could-swamp-venices-flood-defence/ (accessed February 14, 2016).

31. "Climate Change in the Pacific Islands," US Fish and Wildlife Service, November 2, 2011, https://www.fws.gov/pacific/climatechange/changepi.html (accessed July 2015).

32. Karl Mathiesen, "Climate Change in the Marshall Islands and Kiribati, Before and After—Interactive," *Guardian*, March 11, 2015, http://www.theguardian.com/environment/ng-interactive/2015/mar/11/climate-change-in-the-marshall-islands-and-kiribati-before-and-after-interactive (accessed July 10, 2015).

33. Ibid.

34. Joanna Foster, "Epic King Tides Offer Glimpse of Climate Change in Marshall Islands," *ThinkProgress: Climate*, March 6, 2014 http://thinkprogress.org/climate/2014/03/06/3372301/marshall-islands-flood-king-tide/ (accessed July 2015).

35. Christopher Jorebon Loeak, "A Clarion Call from the Climate Change Frontline," *Huffington Post*, September 18, 2014, http://www.huffingtonpost.com/christopher-jorebon-loeak/a-clarion-call-from-the-c_b_5833180.html (accessed July 10, 2015).

36. "Republic of Maldives," *Climate Hot Map: Global Warming Effects around the World*, Union of Concerned Scientists, 2011, http://www.climatehotmap.org/global-warming-locations/republic-of-maldives.html (accessed August 2015).

37. "Maldives Cabinet Makes a Splash," BBC News, October 17, 2009, http://news.bbc.co.uk/2/hi/8311838.stm (accessed March 2016).

38. Ibid.

39. "Republic of Maldives," *Climate Hot Map.*

40. "Global Warming Science," Union of Concerned Scientists, http://www.ucsusa.org/our-work/global-warming/science-and-impacts/global-warming-science#.VwbYs_krLIV (accessed July 19, 2015).

41. Joel Bach and David Gelber, "A Dangerous Future," *Years of Living Dangerously*, season 1, ep. 8, originally aired on Showtime on June 2, 2014.

42. Ibid.

43. Ibid.

44. M. Sophia Newman, "Will Climate Change Spark Conflict in Bangladesh?" *The Diplomat*, June 27, 2014, http://thediplomat.com/2014/06/will-climate-change-spark-conflict-in-bangladesh/ (accessed July 2015).

45. Bach and Gelber, "A Dangerous Future."

46. Ibid.

47. Ibid.

48. Newman, "Will Climate Change Spark Conflict in Bangladesh?"

49. Ibid.

50. Service Assessment Team, *Hurricane/Post-Tropical Cyclone Sandy, October 22–29, 2012*, NOAA (National Oceanic and Atmospheric Administration), May 2013, p. 10, http://www.nws.noaa.gov/os/assessments/pdfs/Sandy13.pdf (accessed July 2015).

51. Todd Gutner, "Hurricane Sandy Grows to Largest Atlantic Tropical Storm Ever," CBS Boston, October 28, 2012, http://boston.cbslocal.com/2012/10/28/hurricane-sandy-grows-to-largest-atlantic-tropical-storm-ever/ (accessed February 2016).

52. Edward Mason, "Hello Again, Climate Change," *Harvard Gazette*, November 6, 2012, http://news.harvard.edu/gazette/story/2012/11/hello-again-climate-change/ (accessed February 2016).

53. Jeff Masters, "Why Did Hurricane Sandy Take Such an Unusual Track into New Jersey?" *WunderBlog*, Weather Underground, October 31, 2012, https://www.wunderground.com/blog/JeffMasters/comment.html?entrynum=2283 (accessed February 16, 2016).

54. Mark Fischetti, "Climate Change Hastened Syria's Civil War," *Scientific American*, March 2, 2015, http://www.scientificamerican.com/article/climate-change-hastened-the-syrian-war/ (accessed July 2015).

55. "What Was the Largest Hurricane to Hit the United States?" *Geology.com: Geoscience News and Information*, http://geology.com/hurricanes/largest-hurricane/ (accessed March 2016).

56. "Hurricane Katrina—A Look Back 10 Years Later," National Weather Service Weather Forecast Office: New Orleans/Baton Rouge, August 17, 2015, http://www.srh.noaa.gov/lix/?n=katrina_anniversary (accessed March 2016).

57. Allison Plyer, "Facts for Features: Katrina Impact," Jesuit Social Research Institute, Loyola University, New Orleans, August 28, 2015, http://www.loyno.edu/jsri/news/facts-features-katrina-impact (accessed July 2016).

58. Marijn Rijken et al., "D1.1 Domain Analysis, Scope, and Requirement: Grant Agreement No. 313308," COBACORE (Community Based Comprehensive Recovery), March 31, 2014, http://www.cobacore.eu/wp-content/uploads/2014/05/D1.1-Scope-and-Requirements.pdf (accessed March 26, 2016).

59. Daniel Stone, "Rising Temperatures May Cause More Katrinas," *National Geographic*, March 19, 2013, http://news.nationalgeographic.com/news/2013/03/130319

-hurricane-climate-change-katrina-science-global-warming/ (accessed July 2015).

60. Ibid.

61. Richard Harris, "Why Typhoon Haiyan Caused So Much Damage," NPR, November 11, 2013, http://www.npr.org/2013/11/11/244572227/why-typhoon-haiyan-caused-so-much-damage (accessed July 2015).

62. Eduardo D. Del Rosario, "Final Report Re Effects of Typhoon 'Yolanda' (Haiyan)," National Disaster Risk Reduction and Management Council, 2014, http://www.ndrrmc.gov.ph/attachments/article/1329/FINAL_REPORT_re_Effects_of_Typhoon_YOLANDA_(HAIYAN)_06-09NOV2013.pdf (accessed March 30, 2016).

63. John Vidal and Damian Carrington, "Is Climate Change to Blame for Typhoon Haiyan?" *Guardian*, November 13, 2013, http://www.theguardian.com/world/2013/nov/12/typhoon-haiyan-climate-change-blame-philippines (accessed July 2015).

64. Chris Mooney, "What the Massive Snowfall in Boston Tells Us about Global Warming," *Washington Post*, February 10, 2015, https://www.washingtonpost.com/news/energy-environment/wp/2015/02/10/what-the-massive-snowfall-in-boston-tells-us-about-global-warming/ (accessed July 2015).

65. Ibid.

66. Joel Bach and David Gelber, "End of the Woods," *Years of Living Dangerously*, season 1, ep. 2, originally aired on Showtime on April 20, 2014.

67. National Centers for Environmental Information, "National Overview—October 2015," NOAA (National Oceanic and Atmospheric Administration), November 2015, https://www.ncdc.noaa.gov/sotc/national/201510 (accessed July 2016).

68. Cole Mellino, "9 States Report Record Low Snowpack Amid Epic Drought," EcoWatch, April 14, 2015, http://www.ecowatch.com/9-states-report-record-low-snowpack-amid-epic-drought-1882032630.html (accessed July 2016).

69. Sarah Zhang, "Alaska's on Fire and It May Make Climate Change Even Worse," *Wired*, July 8, 2015, http://www.wired.com/2015/07/alaskas-fire-may-make-climate-change-even-worse (accessed July 2015).

70. "Thawing of Permafrost Expected To Cause Significant Additional Global Warming, Not Yet Accounted For In Climate Predictions," United Nations Environment Programme, November 27, 2012, http://www.unep.org/newscentre/default.aspx?DocumentID=2698&ArticleID=9338 (accessed July 2016).

71. Ibid.

72. Ibid.

73. Ibid.

74. Joseph Romm, *Climate Change: What Everyone Needs to Know* (New York: Oxford University Press, 2016), pp. 81–84, 124–27, 141–44.

75. Libby Blanchard, "The Permafrost Problem: The IPCC's Fifth Assessment and the Missing Climate Heavyweight," *Huffington Post*, January 10, 2015, http://www.huffingtonpost.co.uk/libby-blanchard/ipcc-report_b_6135756.html (accessed July 2016).

76. Blanchard, "Permafrost Problem."

77. Zhang, "Alaska's on Fire."

78. Ibid.

79. Ibid.

80. Romm, *Climate Change*.

81. Zhang, "Alaska's on Fire."

82. Romm, *Climate Change*, p. 80.

83. Ibid., p. 84.

84. Andrew Nikiforuk, "Bark Beetles, Aided by Climate Change, Are Devastating U.S. Pine Forests," *Washington Post*, December 5, 2011, https://www.washingtonpost.com/national/health-science/bark-beetles-aided-by-climate-change-are-devastating-us-pine-forests/2011/11/08/gIQAoBoCWO_story.html (accessed July 2015).

85. Michael D. Lemonick, "Why Bark Beetles Are Chewing through U.S. Forests," *Climate Central*, January 7, 2013, http://www.climatecentral.org/news/why-bark-beetles-are-chewing-their-way-through-americas-forests-15429 (accessed July 2015).

86. Teresa Chapman, Thomas T. Veblen, and Tania Schoennagel, "Spatiotemporal Patterns of Mountain Pine Beetle Activity in the Southern Rocky Mountains," *Ecology* 93, no. 10 (October 2012): 2175–85, http://www.ncbi.nlm.nih.gov/pubmed/23185879 (accessed July 2015).

87. Kenneth F. Raffa, Erinn N. Powell, and Philip A. Townsend, "Temperature Driven Range Expansion of An Irruptive Insect Heightened By Weakly Coevolved Plant Defenses," *Proceedings of the National Academy of Sciences* 110, no. 6 (February 5, 2013): 2193–98.

88. Terrell Johnson, "California Isn't Alone: Historic Droughts Are Happening around the World," Weather Channel, July 28, 2015, https://weather.com/science/environment/news/california-historic-drought-world-brazil-africa-korea (accessed February 2016); Tom Moore, "California Snowpack and Drought Show Improvement over Last Year," Weather Channel, May 6, 2016, https://weather.com/news/climate/news/snowpack-sierra-drought-california (accessed September 4, 2016).

89. Johnson, "California Isn't Alone."

90. Joel Bach and David Gelber, "Dry Season," *Years of Living Dangerously*, season 1, ep. 1, originally aired on Showtime on April 13, 2014.

91. Thomas Friedman, "Without Water, Revolution," *New York Times: Sunday Review*, May 18, 2013, http://www.nytimes.com/2013/05/19/opinion/sunday/friedman-without-water-revolution.html? _r=0 (accessed August 17, 2015).

92. Ibid.

93. Greg Botelho and Elise Labott, "Israel Says Syria Used Chemical Weapons; Russia Warns of 'Iraqi Scenario,'" CNN, April 24, 2013, http://www.cnn.com/2013/04/23/world/meast/syria-civil-war/ (accessed July 2015).

94. Ibid.

95. Ibid.

96. Ibid.

97. Ibid.

98. John D. Banusiewicz, "Hagel to Address 'Threat Multiplier' of Climate Change," US Department of Defense, October 13, 2014, http://www.defense.gov/News/Article/Article/603440 (accessed March 2016).

99. Ibid.

100. National Wildlife Federation Staff, "Global Warming Impacts on Estuaries and Coastal Wetlands," National Wildlife Federation, 1996–2016, https://www.nwf.org/Wildlife/Threats-to-Wildlife/Global-Warming/Effects-on-Wildlife-and-Habitat/Estuaries-and-Coastal-Wetlands.aspx (accessed February 2016).

101. Ibid.

102. David Gessner, "The Birds of British Petroleum," *Audubon*, July-August 2015, http://www.audubon.org/magazine/july-august-2015/the-birds-british-petroleum (accessed June 2016).

103. National Wildlife Federation Staff, "Global Warming Impacts on Estuaries and Coastal Wetlands," National Wildlife Federation, 1996–2016, https://www.nwf.org/Wildlife/Threats-to-Wildlife/Global-Warming/Effects-on-Wildlife-and-Habitat/Estuaries-and-Coastal-Wetlands.aspx (accessed February 2016).

104. Delaware Department of Natural Resources and Environmental Control, "Climate Change: Wetlands and Intertidal Habitat Loss," State of Delaware, http://www.dnrec.delaware.gov/ClimateChange/Pages/climatechangewetlandsloss.aspx (accessed February 2016).

105. Stephen Davis, "Everglades Restoration and Sea Level Rise," Everglades Foundation, June 9, 2014, http://www.evergladesfoundation.org/2014/06/09/everglades-restoration-and-sea-level-rise/ (accessed February 2016).

106. Ibid.

107. Ibid.

108. Ocean Portal Team, "Ocean Acidification," *Ocean Portal: Find Your Blue*, 2015, http://ocean.si.edu/ocean-acidification (accessed July 2015).

109. "Endangered Oceans," Center for Biological Diversity, http://www.biologicaldiversity.org/campaigns/endangered_oceans/ (accessed July 2015).

110. Ocean Portal Team, "Ocean Acidification."

111. Ibid.

112. Ibid.

113. Ibid.

114. Ibid.

115. "Endangered Oceans."

116. Ibid.

117. Ibid.

118. "Sea Level," *Climate Hot Map: Global Warming Effects around the World*, Union of Concerned Scientists, 2011, http://www.climatehotmap.org/global-warming-effects/sea-level.html (accessed June 2016).

119. Ibid.

120. "Climate Change and Sea-Level Rise in Florida: An Update of the Effects of Climate Change on Florida's Ocean and Coastal Resources," The Florida Oceans and Coastal Council of Tallahassee, Florida, December 2010, p. 13, http://seagrant.noaa.gov/Portals/0/Documents/what_we_do/climate/Florida%20Report%20on%20Climate%20Change%20and%20SLR.pdf (accessed July 2016).

121. Bach and Gelber, "Dry Season."

122. Dennis Pillion, "Florida Oysters in Crisis: 'Our Industry Needs to be Shut Down,' Apalachicola Seafood Rep Says," *AL.com*, September 11, 2014, http://www.al.com/news/beaches/index.ssf/2014/09/florida_oysters_in_crisis_our.html (accessed July 2015).

123. Bach and Gelber, "Dry Season."

124. Jennifer Portman, "Bay May Need to be Closed to Oyster Harvesting," *Tallahassee Democrat*, September 11, 2014, http://www.tallahassee.com/story/news/2014/09/10/bay-may-need-closed-oyster-harvesting/15395795/ (accessed June 2016).

125. Benjamin Sovacool, "The Costs of Failure: A Preliminary Assessment of Major Energy Accidents, 1907–2007," *Energy Policy* 36, no. 5 (May 2008): 1802–20, https://ideas.repec.org/a/eee/enepol/v36y2008i5p1802-1820.html (accessed July 2015).

126. Amy Louviere, "MSHA News Release," Mine Safety and Health Administration, US Department of Labor, December 6, 2011, http://arlweb.msha.gov/Media/PRESS/2011/NR111206.asp (accessed July 2015).

127. United States Mine Rescue Association, "Finley Coal Company Explosion: Nos. 15 and 16 Mine Explosion," *Mine Disasters in the United States*, http://usminedisasters.com/saxsewell/finley.htm (accessed July 2015).

128. Louviere, "MSHA News Release."

129. "Une derniere marche au centre-ville pour la residents de Lac Megantic," ICI Radio-Canada, December 6, 2014, http://ici.radio-canada.ca/regions/estrie/2014/12/06/001-megantic-marche-ville.shtml (accessed August 2015).

130. Ari Phillips, "Pipeline Ruptures in California, Spilling Thousands of Gallons of Oil Into the Ocean," *ThinkProgress: Climate*, May 20, 2015, http://thinkprogress.org/climate/2015/05/20/3660974/santa-barbara-pipeline-spill/ (accessed August 2015).

131. Ibid.

132. "Alberta's Oil Sands 2006," Alberta Department of Energy, December 2007, http://env.chass.utoronto.ca/env200y/ESSAY09/OilSands06.pdf.

133. Tim Dickinson, "Inside the Koch Brothers' Toxic Empire," *Rolling Stone*, September 24, 2014, http://www.rollingstone.com/politics/news/inside-the-koch-brothers-toxic-empire-20140924 (accessed August 2015).

CHAPTER 7: IN THE BEGINNING: AN AMERICAN MELODRAMA

1. "Bering Land Bridge," *National Park Service*, https://www.nps.gov/bela/learn/historyculture/Bering-Land-Bridge.htm (accessed July 2016).

2. Scott Armstrong Elias, "First Americans Lived on Bering Land Bridge for Thousands of Years," *Scientific American*, March 4, 2014, http://www.scientificamerican.com/article/first-americans-lived-on-bering-land-bridge-for-thousands-of-years/ (accessed July 2016).

3. "Bering Land Bridge," *National Park Service*.

4. Paul Harrison, "North American Indians: The Spirituality of Nature," *World Pantheism*, December 12, 1996, http://www.pantheism.net/paul/history/native-americans.htm (accessed June 2015).

5. Ibid.

6. Benjamin Kline, *First Along the River: A Brief History of the US Environmental Movement*, 4th ed. (Lanham, MD: Rowman & Littlefield, 2011).

7. Joshua Johns, "A Brief History of Nature and the American Consciousness," *Nature and the American Identity*, April 5, 1996, http://xroads.virginia.edu/~cap/nature/cap2.html (accessed June 2015).

8. Ibid.

CHAPTER 8: MANIFEST DESTINY: EXPLORE AND CONQUER

1. "Louisiana Purchase," *History*, A&E Networks, 2009, http://www.history.com/topics/louisiana-purchase (accessed July 2015).

2. "Westward Expansion," *History*, A&E Networks, 2009, http://www.history.com/topics/westward-expansion (accessed July 2015).

3. Ibid.

4. William Drown and Solomon Drown, *Compendium of Agriculture, or the Farmer's Guide, in the Most Essential Parts of Husbandry and Gardening; Compiled from the Best American and European Publications, and the Unwritten Opinions of Experienced Cultivators*, (Providence: Field & Macy, 1824).

5. Ibid.

6. John Mason, "The History of Climate Science," Skeptical Science, April 7, 2013, http://www.skepticalscience.com/history-climate-science.html (accessed July 2016).

7. "John James Audubon: The American Woodsman: Our Namesake and Inspiration," National Audubon Society, http://www.audubon.org/content/john-james-audubon (accessed July 2015).

8. Ibid.

9. "John James Audubon," *Encyclopedia of World Biography*, 2016, http://www

.notablebiographies.com/An-Ba/Audubon-John-James.html (accessed July 2016).

10. "John James Audubon," *Encyclopedia*.

11. Ibid.

12. John James Audubon, *The Birds of America* (London: Metro Books, 2012).

13. Ron Flynn, "Is Your Print an Original?" Audubon: Free Information and Articles for J. J. Audubon Print Collectors, 2008, http://www.auduboninfo.net/authenticate/authenticate.htm (accessed August 2016).

14. John James Audubon, *Ornithological Biography* (Neuilly sur Seine, France: Ulan Press, 2012).

15. Ken Chowder, "John James Audubon: Drawn from Nature," *American Masters*, PBS, July 25, 2007, http://www.pbs.org/wnet/americanmasters/john-james-audubon-drawn-from-nature/106/.

16. "John James Audubon: The American Woodsman."

17. Ibid.

18. Ibid.

19. "Manifest Destiny," *History*, A&E Networks, 2010, http://www.history.com/topics/manifest-destiny (accessed July 2015).

20. Michelle Facos, *An Introduction to Nineteenth Century Art* (London: Routledge, 2011).

21. Kathy Weiser, "California Legends: The California Gold Rush," *Legends of America*, June 2015, http://www.legendsofamerica.com/ca-goldrush.html (accessed March 30, 2016).

22. Ibid.

CHAPTER 9: A MULTIPLICITY OF HEROES AND THE CIVIL WAR

1. Ralph Waldo Emerson, *Nature* (New York: Duffield, 1909).

2. Ibid.

3. Henry David Thoreau, *Civil Disobedience*, 1849, http://xroads.virginia.edu/~hyper/walden/Essays/civil.html (accessed March 30, 2016).

4. Henry David Thoreau, *Walden* (New York: CE Merrill, 1910).

5. Ibid.

6. Ann Woodlief, "Henry David Thoreau: Biography," *American Transcendentalism Web*, http://transcendentalism-legacy.tamu.edu/authors/thoreau/ (accessed July 2015).

7. Ibid.

8. Ibid.

9. Elizabeth Witherell and Elizabeth Dubrulle, "Life and Times of Henry David Thoreau," *The Writings of Henry D. Thoreau*, 1995, http://thoreau.library.ucsb.edu/thoreau_life.html (accessed March, 2016).

10. Ibid.

11. Ibid.

12. Thoreau, *Walden*.

13. Ibid.

14. Richard Smith, "Guest Post: Thoreau Leaves Walden," *American Literary Blog*, September 6, 2010, http://americanliteraryblog.blogspot.com/2010/09/guest-blog-thoreau-leaves-walden.html (accessed July 2016).

15. Charlotte Zoë Walker, *Sharp Eyes: John Burroughs and American Nature Writing* (Syracuse, NY: Syracuse University Press, 2000), p. 292.

16. Clara Barrus, "Teacher and Student," chapter 10 in *John Burroughs: Boy and Man* (New York: Doubleday, Page, 1920), http://www.catskillarchive.com/jb/bm-10.htm (accessed July 2016).

17. Ibid.

18. John Burroughs, *Notes on Walt Whitman as Poet and Person* (New York: Haskell House, 1971).

19. *Encyclopedia of World Biography*, s.v. "John Burroughs," http://www.encyclopedia.com/topic/John_Burroughs.aspx (accessed March 30, 2016).

20. *Encyclopedia Brittanica Online*, s.v. "John Burroughs," http://www.britannica.com/biography/John-Burroughs (accessed March 30, 2016).

21. David Schuyler, *Sanctified Landscape: Writers, Artists, and the Hudson River Valley, 1820–1909* (Ithaca: Cornell University Press, 2012), p. 133.

22. Ibid., p. 140.

23. John Burroughs, *John Burroughs' America: Selections from the Writings of the Naturalist* (Mineola, NY: Dover, 1997), p. 3.

24. Richard Green, *Te Ata: Chickasaw Storyteller, American Treasure* (Norman, OK: University of Oklahoma Press, 2006), p.115.

25. Zosia Zaks, "Appreciation, John Burroughs: Forgotten Naturalist," *EcoTopia*, http://ecotopia.org/ecology-hall-of-fame/john-burroughs/appreciation/ (accessed July 2016).

26. "Frederick Law Olmsted Biography," *Biography.com*, http://www.biography.com/people/frederick-law-olmsted-9428434 (accessed July 2016).

27. Ibid.

28. "Frederick Law Olmsted Biography."

29. Michelle Werts, "Father of Landscape Architecture and Urban Parks," *American Forests*, April 26, 2012, https://www.americanforests.org/blog/father-of-landscape-architecture-and-urban-parks/ (accessed July 2015).

30. Frederick Law Olmsted, *The Cotton Kingdom* (Boston: De Capo, 1996).

31. Charles E. Beveridge, "The Olmsted Firm—An Introduction," National Association for Olmsted Parks, http://www.olmsted.org/the-olmsted-legacy/the-olmsted-firm/an-introduction (accessed July 2016).

32. Frederick Law Olmsted, http://www.fredericklawolmsted.com/ (accessed July 2015).

33. Charles Darwin, *The Origin of Species* (New York: United Holdings Group, 2011).

34. T. H. Huxley, "Criticisms on 'The Origin of Species,'" in *Darwiniana*, vol. 2 of *Collected Essays*, ed. Charles Blinderman and David Joyce, in *The Huxley File*, 2003, http://alepho.clarku.edu/huxley/CE2/CritoS.html (accessed July 2016); Francis Darwin, *Charles Darwin: His Life Told in an Autobiographical Chapter, and in a Selected Series of His Published Letters* (London: John Murray, 1892), http://darwin-online.org.uk/converted/published/1892_letter_F1461/1892_letter_F1461.html (accessed July 2016); "Skeptics of Darwinian Theory," Access Research Network, http://www.arn.org/quotes/critics.html (accessed July 2016).

35. Asa Gray, "Darwin on the Origin of Species: A Book Review," *Atlantic*, July 1860, http://www.theatlantic.com/magazine/archive/1860/07/darwin-on-the-origin-of-species/304152/ (accessed July 2016).

36. Ibid.

37. "Charles Darwin Biography," *Biography.com*, http://www.biography.com/people/charles-darwin-9266433 (accessed July 2015).

38. "HMS Beagle Voyage," *AboutDarwin.com*, February 10, 2008, http://www.aboutdarwin.com/voyage/voyage03.html (accessed July 2016).

39. Ibid.

40. Ibid.

41. Ibid.

42. Gordon Chancellor and Randal Keynes, "Darwin's Field Notes on the Galapagos: 'A Little World within Itself,'" *Darwin Online*, October 2006, http://darwin-online.org.uk/EditorialIntroductions/Chancellor_Keynes_Galapagos.html (accessed July 2015).

43. Ibid.

44. Charles Darwin, *The Voyage of the Beagle* (Minneapolis: Zenith, 2015).

45. Dennis O'Neil, "Darwin and Natural Selection," *Early Theories of Evolution: 17th–19th Century Discoveries That Led to the Acceptance of Biological Evolution*, http://anthro.palomar.edu/evolve/evolve_2.htm (accessed July 2015).

46. Heather Scoville, "8 People Who Influenced Charles Darwin," *About Education*, 2016, http://evolution.about.com/od/Pre-Darwin/ss/8-People-Who-Influenced-Charles-Darwin.htm (accessed July 2016).

47. Ibid.

48. Ibid.

49. O'Neil, "Darwin and Natural Selection."

50. Ibid.

51. John van Wyhe, "Mind the Gap: Did Darwin Avoid Publishing His Theory for Many Years?" *Royal Society Journal of the History of Science* 61, no. 2 (May 22, 2007), http://rsnr.royalsocietypublishing.org/content/61/2/177 (accessed July 2016).

52. Ibid.

53. Adrian J. Desmond, "Charles Darwin: British Naturalist," *Encyclopedia Britannica*, June 10, 2016, https://www.britannica.com/biography/Charles-Darwin (accessed July 2016).

54. Ibid.

55. Nick Matzke, "Survival of the Pithiest," *Panda's Thumb*, September 3, 2009. http://pandasthumb.org/archives/2009/09/survival-of-the-1.html (accessed March 30, 2016).

56. Forrest M. Mims III, "Measuring the Earth's Water Vapor Blanket," *My NASA Data*, June 24, 2016, http://mynasadata.larc.nasa.gov/science_projects/measuring-the-earths-water-vapor-blanket-2/ (accessed July 2016).

57. George Perkins Marsh, *Man and Nature* (Seattle: University of Washington Press, 2003).

58. Paul Waldman, "American War Dead, By the Numbers," *The American Prospect*, May 26, 2014, http://prospect.org/article/american-war-dead-numbers (accessed July 2015).

59. Wayne D. Rasmussen, "Lincoln's Agricultural Legacy," National Agriculture Library, United States Department of Agriculture, https://www.nal.usda.gov/lincolns-agricultural-legacy (accessed July 2015).

60. Marsh, *Man and Nature*.

61. George Perkins Marsh, *The Earth as Modified by Human Action* (New York: Scribner, Armstrong, 1874).

62. Cutler J. Cleveland, "Marsh, George Perkins," *The Encyclopedia of Earth*, December 20, 2010, http://www.eoearth.org/view/article/154491/ (accessed July 2015) [site discontinued].

63. "George Perkins Marsh: Renaissance Vermonter," George Perkins Marsh Institute, Clark University, http://www2.clarku.edu/departments/marsh/about/ (accessed July 2015).

64. Ibid.

65. Cleveland, "Marsh, George Perkins."

66. Ibid.

CHAPTER 10: THE RISE OF INDUSTRIAL AMERICA: PROGRESS OR PERISH

1. "The Second Industrial Revolution," *Boundless*, https://www.boundless.com/u-s-history/textbooks/boundless-u-s-history-textbook/the-gilded-age-1870-1900-20/the-second-industrial-revolution-144/the-second-industrial-revolution-761-6982/ (accessed January 14, 2016).

2. Richard White, *Railroaded: The Transcontinentals and the Making of America* (New York: W. W. Norton, 2012).

3. "History Times: The Industrial Revolution," Gilder Lehrman Institute of American History, https://www.gilderlehrman.org/history-by-era/gilded-age/essays/history-times-industrial-revolution (accessed July 2015).

4. Ibid.

5. Ibid.

6. Ibid.

7. "First National Park Resulted from the Yellowstone Expedition," *About.com: Education*, January 31, 2016, http://history1800s.about.com/od/thegildedage/a/first-national-park.htm (accessed July 2016).

8. "The World's First Hydroelectric Power Plant Began Operation," *America's Library: America's Story*, http://www.americaslibrary.gov/jb/gilded/jb_gilded_hydro_2.html (accessed July 2015).

9. "William Temple Hornaday: Saving the American Bison," *Stories from the Smithsonian*, http://siarchives.si.edu/history/exhibits/stories/william-temple-hornaday-saving-american-bison (accessed July 2015).

10. Stefan Bechtel, *Mr. Hornaday's War: How a Peculiar Victorian Zookeeper Waged a Lonely Crusade for Wildlife That Changed the World* (Boston: Beacon Press, 2012), pp. 5–6.

11. "William Temple Hornaday: Saving the American Bison."

12. William Temple Hornaday, *The Extermination of the American Bison, With a Sketch of Its Discovery and Life History* (Washington: Smithsonian Institution, 1889).

13. "William Temple Hornaday," National Wildlife Federation, https://www.nwf.org/Who-We-Are/History-and-Heritage/Conservation-Hall-of-Fame/Hornaday.aspx (accessed July 2015).

14. Ibid.

15. "William Temple Hornaday: Saving the American Bison."

16. Andrew Glass, "June 30, 1864, Lincoln Creates Yosemite Park," *Politico*, June 30, 2009, http://www.politico.com/story/2009/06/june-30-1864-lincoln-creates-yosemite-park-024332 (accessed July 2016).

17. "This Day in History, 1890: Yosemite National Park Established," *History*, A&E Networks, October 1, 2016, http://www.history.com/this-day-in-history/yosemite-national-park-established (accessed July 2016).

18. Ibid.

19. Michael P. Cohen, *The History of the Sierra Club: 1892–1970* (Oakland, CA: Sierra Club, 1988), chapter three, http://vault.sierraclub.org/history/origins/ (accessed July 2016).

20. Ibid.

21. "Explore the Adirondacks," *Visit Adirondacks*, Adirondack Regional Tourism Council, http://visitadirondacks.com/about (accessed March 30, 2016).

22. Ibid.

23. Paul R. Ehrlich, David S. Dobkin, and Darryl Wheye, "Plume Trade," Stanford University, 1988, https://web.stanford.edu/group/stanfordbirds/text/essays/Plume_Trade.html (accessed July 2015).

24. Ibid.

25. Ibid.

26. Kate Kelly, "Harriet Lawrence Hemenway (1858–1960): Saving Birds One Hat at a Time," *America Comes Alive*, http://americacomesalive.com/2014/04/08/harriet-lawrence

-hemenway-1858-1960-saving-birds-one-hat-at-a-time/#.V6UA1_krLDd (accessed July 2015).

27. Ibid.

28. Ibid.

29. Ibid.

30. "History of Audubon and Science-Based Bird Conservation," *Audubon*, http://www.audubon.org/content/history-audubon-and-waterbird-conservation (accessed July 2016).

31. Kelly, "Harriet Lawrence Hemenway."

32. Ibid.

33. "Migratory Bird Treaty Centennial, 1916–2016: Timeline," US Fish and Wildlife Service, January 28, 2016, https://www.fws.gov/birds/mbtreaty100/timeline.php (accessed July 2016).

34. "Digest of Federal Resource Laws of Interest to the US Fish and Wildlife Service: Migratory Bird Treaty Act of 1918," US Fish and Wildlife Service, https://www.fws.gov/laws/lawsdigest/migtrea.html (accessed July 2016).

35. Ibid.

36. "Migratory Bird Treaty Act: Birds Protected," US Fish and Wildlife Service, September 16, 2015, https://www.fws.gov/birds/policies-and-regulations/laws-legislations/migratory-bird-treaty-act.php (accessed July 2016).

37. "Birding with Jerry Hall: Egrets and Eagles Were Once Shot by the Thousands All in the Name of Fashion," *San Marcos Daily Record*, March 10, 2013, http://www.sanmarcosrecord.com/birding-jerry-hall-egrets-and-eagles-were-once-shot-thousands-all-name-fashion (accessed July 2015).

CHAPTER 11: WELCOME TO THE PROGRESSIVE ERA

1. Benjamin Kline, *First Along the River: A Brief History of the US Environmental Movement*. (Lanham, MD: Rowman & Littlefield, 2011), p. 59.

2. *Encyclopedia Brittanica Online*, s.v. "Social Darwinism," https://www.britannica.com/topic/social-Darwinism (accessed July 2016).

3. "The Progressive Era (1890–1920)," *The Eleanor Roosevelt Papers Project*, https://www.gwu.edu/~erpapers/teachinger/glossary/progressive-era.cfm (accessed July 2015).

4. Howard Ehrlich, "Theodore Roosevelt: A Life of Public Service," Theodore Roosevelt Association, http://www.theodoreroosevelt.org/site/c.elKSIdOWIiJ8H/b.9298051/k.FEAC/Theodore_Roosevelt_A_Life_of_Public_Service.htm (accessed July 2015).

5. Ibid.

6. "The Conservationist," Theodore Roosevelt Association, http://www.theodoreroosevelt.org/site/c.elKSIdOWIiJ8H/b.8344385/k.114A/The_Conservationist.htm (accessed July 2015).

7. Kline, *First Along the River*, pp. 62–67.

8. Ibid.

9. "Conservationist," Theodore Roosevelt Association.

10. Ibid.

11. "Pelican Island National Wildlife Refuge, Florida: History," US Fish and Wildlife Services, http://www.fws.gov/refuge/pelican_island/about/history.html (accessed March 30, 2016).

12. Kline, *First Along the River*, p. 62.

13. Theodore Roosevelt, *The Winning of the West* (Lincoln: University of Nebraska Press, 1995); "Theodore Roosevelt Biography," Biography.com, http://www.biography.com/people/theodore-roosevelt-9463424#early-life (accessed July 2015).

14. Ibid.

15. Ibid.

16. "U.S. Forest Service History: Gifford Pinchot (1865–1946)," *The Forest History Society*, May 1, 2015, http://www.foresthistory.org/ASPNET/People/Pinchot/Pinchot.aspx (accessed July 2015).

17. Kline, *First Along the River*, p. 64.

18. "Gifford Pinchot (1865–1946)," Grey Towers National Historic Site, US Department of Agriculture, Forest Service, http://www.fs.usda.gov/detail/greytowers/aboutgreytowers/history/?cid=stelprd3824502 (accessed March 30, 2016).

19. Gifford Pinchot, *Breaking New Ground* (Washington, DC: Island Press, 1947), p. 27.

20. Frank Graham, *Man's Dominion: The Story of Conservation in America* (New York: M Evans, 1971).

21. Ibid.

22. "U.S. Forest Service History: Gifford Pinchot (1865–1946)."

23. "Gifford Pinchot (1865–1946)," Grey Towers National Historic Site.

24. "U.S. Forest Service History: Gifford Pinchot (1865–1946)."

25. Ibid.

26. Ibid.

27. Ibid.

28. James G. Bradley, "The Mystery of Gifford Pinchot and Laura Houghteling," *Pennsylvania History: A Journal of Mid-Atlantic Studies* 66, no. 2 (Spring 1999): 199–214, https://journals.psu.edu/phj/article/viewFile/25565/25334 (accessed July 2015).

29. "The Dunbar Childhood of John Muir," BBC, October 14, 2009, http://news.bbc.co.uk/local/edinburghandeastscotland/hi/people_and_places/nature/newsid_8282000/8282633.stm (accessed July 2015).

30. Ibid.

31. "Fountain Lake Farm: John Muir's Boyhood Home," Sierra Club, 2016, http://vault.sierraclub.org/john_muir_exhibit/geography/wisconsin/fountain_lake_farm.aspx (accessed July 2016).

32. Mark R. Stoll, "God and John Muir: A Psychological Interpretation of John Muir's

Life and Religion," Sierra Club, 1993, http://vault.sierraclub.org/john_muir_exhibit/life/god_john_muir_mark_stoll.aspx (accessed July 2016).

33. "Dunbar Childhood," BBC.

34. Stoll, "God and John Muir."

35. "Dunbar Childhood," BBC.

36. Stoll, "God and John Muir."

37. "John Muir: A Brief Biography," Sierra Club, 2016, http://vault.sierraclub.org/john_muir_exhibit/life/muir_biography.aspx (accessed July 2016).

38. Ibid.

39. Ibid.

40. Ken Burns, "People: John Muir (1838–1914)," *The National Parks: America's Best Idea*, PBS, http://www.pbs.org/nationalparks/people/historical/muir/ (accessed March 30, 2016).

41. Ibid.

42. "John Muir: A Brief Biography," Sierra Club.

43. Ibid.

44. Richard J. Hartesveldt, "Roosevelt and Muir in Yosemite: The First Conservationists," *Undiscovered Yosemite*, http://www.undiscovered-yosemite.com/Roosevelt-and-Muir.html (accessed July 2015).

Since both men loved to talk, there's little doubt that their ruminations by the light of a campfire laid the groundwork for many new national parks. Many of these protected iconic landscapes throughout the American West, including Grand Canyon, Arizona; Chaco Canyon, New Mexico; Devils Tower, Wyoming; El Morro, New Mexico; Gila Cliff Dwellings, New Mexico; Natural Bridges, Utah; Jewel Cave, South Dakota; Montezuma Castle, Arizona; Navajo, Arizona; Pinnacles, California; Tonto, Arizona; Muir Woods, California; Tumacacori, Arizona, Petrified Forest, Arizona; and Lassen Peak and Cinder Cone, California (now Lassen Volcanic National Park).

45. "The Camping Trip That Changed the Nation," *MacGillivray Freeman's National Parks Adventure*, 2015, http://nationalparksadventure.com/the-camping-trip-that-changed-the-nation/ (accessed March 2016).

46. "Theodore Roosevelt," *The John Muir Exhibit*, Sierra Club, http://vault.sierraclub.org/john_muir_exhibit/people/roosevelt.aspx.

47. Tim Palmer, *Endangered Rivers and the Conservation Movement* (Lanham, MD: Rowman & Littlefield, 2004), p. 51.

48. "John Muir: A Brief Biography," Sierra Club.

49. John Muir, *Our National Parks* (San Francisco: Sierra Club Books, 1991), p. 42.

50. John Muir Trust, "Welcome: John Muir (1838–1914)," *Discover John Muir*, https://discoverjohnmuir.com/ (accessed March 2016).

51. "Oldest National Parks in the United States," *World Atlas*, 2016, http://www.worldatlas.com/articles/oldest-national-parks-in-the-united-states.html (accessed August 16, 2016).

52. "About Us: History," National Park Service, https://www.nps.gov/aboutus/history.htm (accessed July 2015).

53. Ibid.

54. "Who We Are: Anna Botsford Comstock," National Wildlife Federation, https://www.nwf.org/Who-We-Are/History-and-Heritage/Conservation-Hall-of-Fame/Comstock.aspx (accessed July 2015).

55. Anna Botsford Comstock, *Handbook of Nature Study* (Ithaca, NY: Comstock Publishing/Cornell University Press, 1939).

56. "Anna Botsford Comstock Facts," *Your Dictionary*, 2010, http://biography.yourdictionary.com/anna-botsford-comstock (accessed August 17, 2016).

57. Ibid.

58. "Who We Are: Anna Botsford Comstock."

59. Ibid.

60. Ibid.

61. Ibid.

62. "Anna Botsford Comstock Facts," *Your Dictionary*.

63. Donald E. Hurlbert, "360 Degree View of Martha, the Last Passenger Pigeon," Smithsonian National Museum of Natural History, June 23, 2014, http://vertebrates.si.edu/birds/Martha/index.html (accessed January 14, 2016).

64. Naomi Klein, *This Changes Everything: Capitalism vs. the Climate* (New York: Simon and Schuster, 2014).

CHAPTER 12: WATCH OUT: HERE COME THE ROARING TWENTIES

1. Izaak Walton, *The Compleat Angler* (1676; New York: Oxford University Press, 2014).

2. "About Us," Izaak Walton League of America, 2016, http://www.iwla.org/about-us.

3. "History: Ansel Adams," Sierra Club, 2016, http://vault.sierraclub.org/history/ansel-adams/ (accessed March 30, 2016).

4. Ibid.

5. Ibid.

6. "Ansel Adams Biography," *Biography.com*, http://www.biography.com/people/ansel-adams-9175697 (accessed July 2015).

7. Ansel Adams, *Sierra Nevada: The John Muir Trail* (New York: Little, Brown, 2006).

8. "People & Events: Ansel Adams (1902–1984)," *American Experience*, PBS, http://www.pbs.org/wgbh/amex/ansel/peopleevents/p_aadams.html (accessed August 17, 2016).

9. The Ansel Adams Gallery, http://anseladams.com/ (accessed August 17, 2016).

10. "History: Ansel Adams," Sierra Club.

11. Ibid.

12. William Turnage, "Ansel Adams Biography," The Ansel Adams Gallery, http://

www.anseladams.com/ansel-adams-information/ansel-adams-biography/.

13. "Bob Marshall," Wilderness.net, http://www.wilderness.net/NWPS/Marshall (accessed July 2015).

14. Ibid.

15. "Robert Marshall," The Wilderness Society, http://wilderness.org/bios/founders/robert-marshall (accessed January 15, 2016).

16. Robert Marshall, "The Problem of the Wilderness," *Scientific Monthly* 30, no. 2 (February 1930): 141–48.

17. Ibid.

18. James Glover, *A Wilderness Original: The Life of Bob Marshall* (Seattle, WA: Mountaineers Books, 1986).

19. Robert Marshall, *Arctic Village: A 1930s Portrait of Wiseman, Alaska*, rep. ed. (Fairbanks, AK: University of Alaska Press, 1991).

20. "Bob Marshall," Wilderness.net.

21. Ibid.

22. "Mission and Impact," The Wilderness Society, http://wilderness.org/article/mission-and-impact (accessed August 12, 2015).

23. Robert Marshall, "The Universe of Wilderness Is Vanishing," *Trends* 9, no. 1 (Jan/Feb/Mar 1972): 11–14, http://npshistory.com/newsletters/park_practice/trends/v9n1.pdf (accessed August 12, 2015).

24. "The Bob Marshall Wilderness Country," *Ultimate Montana*, 2010, http://www.ultimatemontana.com/sectionpages/Section13/attractions/thebob.html (accessed March 30, 2016).

CHAPTER 13: WOE IS US: THE GREAT DEPRESSION AND THE DUST BOWL

1. "1928 Presidential Campaign Slogans: Herbert Hoover," US Presidents, http://www.presidentsusa.net/1928slogan.html (accessed July 2015).

2. "The Great Depression," *History*, A&E Networks, 2009, http://www.history.com/topics/great-depression (accessed July 2015).

3. Rick Szostak, "Great Depression," *Encyclopedia.com: Dictionary of American History*, 2003, http://www.encyclopedia.com/topic/Great_Depression.aspx (accessed July 2015).

4. William DeGregorio, "Franklin D. Roosevelt," *The Complete Book of U.S. Presidents*, 7th ed. (Fort Lee, NJ: Barricade Books, 2009), available at http://presidentialham.com/u-s-presidents/franklin-d-roosevelt-with-ham/ (accessed July 2015).

5. Benjamin Kline, *First Along the River: A Brief History of the US Environmental Movement*, 4th ed. (Lanham, MD: Rowman & Littlefield, 2011), p. 75.

6. Ibid., p. 73.

7. Ibid., p. 74.

8. Ibid., p. 75.

9. National Wildlife Refuge System, US Fish and Wildlife Service, https://www.fws.gov/refuges/ (accessed June 2016).

10. "FDR Biography," FDR Presidential Library and Museum, http://www.fdrlibrary.org/fdr-biography (accessed July 2015).

11. Ibid.

12. Franklin D. Roosevelt, "A Message to Congress on the Use of Our Natural Resources," in *The Public Papers and Addresses of Franklin D. Roosevelt*, vol. 4 (1935; New York City: Random House, 1938), p. 59, http://newdeal.feri.org/speeches/1935c.htm (accessed August 18, 2016).

13. "Surviving the Dust Bowl," transcript, *American Experience*, 2007, http://www.pbs.org/wgbh/americanexperience/features/transcript/dustbowl-transcript/ (accessed July 2016).

14. "Surviving the Dust Bowl: The Drought," *American Experience*, 2007, http://www.pbs.org/wgbh/americanexperience/features/general-article/dustbowl-drought/ (accessed March 2016).

15. Eric Foner and John Garraty, "Dust Bowl," *The Reader's Companion to American History*, 1991, http://www.history.com/topics/dust-bowl (accessed March 2016).

16. "Surviving the Dust Bowl," transcript.

17. Ibid.

18. Ibid.

19. Ibid.

20. John Steinbeck, *The Grapes of Wrath* (New York: Penguin Books, 1999), p. 233.

21. Ibid.

22. John Steinbeck, *Cannery Row* (New York: Penguin Group, 1945).

23. John Steinbeck, *Sweet Thursday* (New York: Penguin Classics, 2008).

24. Ibid. p. 138.

25. John Steinbeck, *America and Americans* (New York: Penguin Classics, 2003).

26. Ibid.

27. "John Steinbeck Biography," National Steinbeck Center, http://www.steinbeck.org/pages/john-steinbeck-biography (accessed July 2015).

28. Ibid.

29. Ibid.

30. John Steinbeck, *The Pastures of Heaven* (London: Penguin Classics, 1995).

31. Clifford Eric Gladstein and Mimi Reisel Gladstein, "Revisiting the *Sea of Cortez* with a 'Green' Perspective," in *Steinbeck and the Environment: Interdisciplinary Approaches*, ed. Susan F. Beegel, Susan Shillinglaw, and Wesley N. Tiffney Jr. (Tuscaloosa, AL: University of Alabama Press, 2011), pp. 162–63.

32. Lorelei Cederstrom, "The 'Great Mother' in *The Grapes of Wrath*," in *Steinbeck and the Environment*, p. 76.

33. John Steinbeck, *Burning Bright* (London: Penguin Books, 2006).

34. John Steinbeck, *East of Eden* (London: Penguin Classics, 1992).

35. John Steinbeck, *The Winter of Our Discontent* (London: Penguin Classics, 2008).

36. John Steinbeck, *Travels with Charley: In Search of America* (London: Penguin Books, 1980).

37. "John Steinbeck Biography," National Steinbeck Center.

38. J. I. Merritt, "Roger Tory Peterson Is Rewriting America's Birding Bible Chapter and Verse," *People*, March 17, 1980, http://www.people.com/people/archive/article/0,20076045,00.html (accessed March 30, 2016).

39. "Biography," Roger Tory Peterson Institute of Natural History, http://rtpi.org/roger-tory-peterson/roger-tory-peterson-biography/ (accessed August 17, 2016).

40. Merritt, "Roger Tory Peterson Rewriting."

41. Roger Tory Peterson, *A Field Guide to the Birds* (Boston: Houghton Mifflin, 1947).

42. Roger Tory Peterson, *Guide to the Eastern Birds* (Boston, Houghton Mifflin, 1980).

43. Roger Tory Peterson, *Guide to the Western Birds* (Boston, Houghton Mifflin, 1969).

44. Matthew C. Perry, "Roger Tory Peterson," Washington Biologists' Field Club, 2007, http://www.pwrc.usgs.gov/resshow/perry/bios/petersonroger.htm (accessed March 30, 2016).

45. Roger Tory Peterson, *Birds Over America* (New York: Dodd, Mead, 1948).

46. Douglas Carlson, *Roger Tory Peterson: A Biography* (Austin: University of Texas Press, 2007).

47. Perry, "Roger Tory Peterson."

48. "Biography," Roger Tory Peterson Institute.

49. Ibid.

50. Perry, "Roger Tory Peterson."

51. "Biography," Roger Tory Peterson Institute.

52. Ibid.

53. "Jay Norwood 'Ding' Darling," US Fish and Wildlife Service, 2008, https://www.fws.gov/dingdarling/about/dingdarling.html (accessed March 2016).

54. "Jay Norwood 'Ding' Darling (1876–1962)," The Meadowlark Gallery, http://www.meadowlarkgallery.com/DarlingJayNorwoodDing.html (accessed July 2015).

55. "'Ding' Darling: Cartoonist and Conservationist," The Robinson Library, March 12, 2015, http://www.robinsonlibrary.com/finearts/drawing/humor/darling.htm (accessed July 2015).

56. Ibid.

57. "Who Are We: Jay Norwood 'Ding' Darling," National Wildlife Federation, http://www.nwf.org/Who-We-Are/History-and-Heritage/Conservation-Hall-of-Fame/Darling.aspx (accessed July 2015).

58. Ibid.

59. "Jay Norwood 'Ding' Darling," US Fish and Wildlife Service.

60. "'Ding' Darling: Cartoonist and Conservationist," Robinson Library.

61. "Jay Norwood 'Ding' Darling," Meadowlark Gallery.

62. "Jay N. 'Ding' Darling (1876–1962): Our Namesake," "Ding" Darling Wildlife Society, http://dingdarlingsociety.org/articles/our-namesake.

63. John Mason, "The History of Climate Science," *Skeptical Science*, April 7, 2013, https://www.skepticalscience.com/print.php?n=1473 (accessed March 30, 2016).

CHAPTER 14: ENVIRONMENTAL CONCERNS MOVE TO THE BACK BURNER

1. Donald Worster, *Nature's Economy: The Roots of Ecology* (New York: Random House, 1982).

2. Josef Essberger, "I am become death, the destroyer of worlds," *EnglishClub*, https://www.englishclub.com/ref/esl/Quotes/Death/I_am_become_death_the_destroyer_of_worlds._2620.htm (accessed March 30, 2016).

3. "Faceless Body Belonged to My Sister: Hiroshima, Nagasaki Nuke Survivors Recall Horrors 70 Years On," RT, August 5, 2015, https://www.rt.com/news/311608-hiroshima-nagasaki-nuclear-anniversary/ (accessed July 2016).

4. Ibid.

5. Benjamin Kline, *First Along the River: A Brief History of the US Environmental Movement* (Lanham, MD: Rowman & Littlefield, 2011).

6. Ibid.

7. Ruthanne Vogel, "Everglades Biographies: Marjory Stoneman Douglas, 1890–1998," *Reclaiming the Everglades: South Florida's Natural History, 1884 to 1934*, http://everglades.fiu.edu/reclaim/bios/douglas.htm (accessed July 2015).

8. Rosalie E. Leposky, "Marjory Stoneman Douglas: Defender of the Everglades," Friends of the Everglades, 1997, http://www.everglades.org/about-marjory-stoneman-douglas/ (accessed July 2015).

9. "Marjory Stoneman Douglas," National Women's Hall of Fame, 2016, https://www.womenofthehall.org/inductee/marjory-stoneman-douglas/ (accessed July 2016).

10. Leposky, "Marjory Stoneman Douglas."

11. T. D. Allman, "Beyond Disney," *National Geographic Magazine*, March 2007, http://ngm.nationalgeographic.com/print/2007/03/orlando/allman-text (accessed August 17, 2016).

12. Carl Hiaasen, *Team Rodent: How Disney Devours the World* (New York: Ballantine Books, 1998), p. 4.

13. "Carl Hiaasen Quotes," *BrainyQuote*, http://www.brainyquote.com/quotes/authors/c/carl_hiaasen.html (accessed July 2015).

14. Marjory Stoneman Douglas, with John Rothchild, *Voice of the River* (Sarasota, FL: Pineapple Press, 1987), p. 14.

15. Marjory Stoneman Douglas, *The Everglades: River of Grass* (New York: Rinehart, 1947).

16. Ken Burns, "People: Marjory Stoneman Douglas (1890–1998)," *National Parks: America's Best Idea*, http://www.pbs.org/nationalparks/people/behindtheparks/douglas/ (accessed March 30, 2016).

17. Ibid.

18. Rupert Cornwell, "Obituary: Marjory Stoneman Douglas," *Independent*, London, May 25, 1998, p. 16.

19. Burns, "People: Marjory Stoneman Douglas."

20. "Marjory Stoneman Douglas Quotes," *BrainyQuote*, http://www.brainyquote.com/quotes/authors/m/marjory_stoneman_douglas.html (accessed July 2015).

21. Ken Chowder, "Wild by Law: The Rise of Environmentalism and the Creation of the Wilderness Act," directed by Lawrence R. Hott and Diane Garey, *American Experience*, PBS, 1991.

22. Aldo Leopold, *A Sand County Almanac, with Essays on Conservation from Round River* (New York: Ballantine, 1970), pp. 138–39.

23. Curt D. Meine, *Aldo Leopold: His Life and Work* (Madison, WI: University of Wisconsin Press, 1988).

24. Chowder, "Wild by Law."

25. "Aldo Leopold," Wilderness.net, http://www.wilderness.net/NWPS/Leopold (accessed July 2015).

26. Leopold, *Sand County Almanac.*

27. "Aldo Leopold," Wilderness.net.

28. Chowder, "Wild by Law."

29. "The Leopold Legacy: Aldo Leopold," The Aldo Leopold Foundation, http://www.aldoleopold.org/AldoLeopold/leopold_bio.shtml (accessed March 30, 2016).

30. Leopold, *Sand County Almanac*, p. xvii.

31. Ibid., p. 190.

32. Kline, *First Along the River*, p. 81.

33. Elizabeth Nix, "7 Things You Might Not Know about the Hoover Dam," *History*, September 14, 2015, A&E Networks, http://www.history.com/news/history-lists/7-things-you-might-not-know-about-the-hoover-dam (accessed July 2016).

34. Brandon Loomis, "The Dam Not Built: Echo Park at Dinosaur National Monument," *Arizona Republic*, October 14, 2013, http://archive.azcentral.com/travel/articles/20131012echo-park-dinosaur-national-monument.html (accessed March 30, 2016).

35. Ibid.

36. Ibid.

37. Ibid.

38. Ibid.

39. Ibid.

40. Eliot Porter, *The Place No One Knew: Glen Canyon on the Colorado* (San Francisco: Sierra Club, 1963).

41. Loomis, "Dam Not Built."

42. Mark W. T. Harvey, *A Symbol of Wilderness: Echo Park and the American Conservationist Movement* (Albuquerque: University of New Mexico Press, 1994).

43. Loomis, "Dam Not Built."

44. "David Brower Legacy," Earth Island Institute, http://www.earthisland.org/index.php/aboutUS/legacy/ (accessed August 2015).

45. Ibid.

46. Ibid.

47. Ibid.

48. "David Brower," Wilderness.net, http://www.wilderness.net/NWPS/Brower (accessed August 16, 2016).

49. "The Wilderness Within: Remembering David Brower," Rewilding Institute, 2012, http://rewilding.org/rewildit/support-rewilding/books-for-sale-from-the-rewilding-institute-2/the-wildness-within-remembering-david-brower/.

50. Ibid.

51. Glen Canyon Institute, "David Brower," Glen Canyon Institute, http://www.glencanyon.org/about/david-brower (accessed August 2015).

52. John McPhee, *Encounters With the Archdruid* (New York: Farrar, Straus and Giroux, 1971).

53. Kenneth Brower, "A Voice for the Wilderness: An Excerpt from *The Wildness Within*," *Earth Island Journal*, http://www.earthisland.org/journal/index.php/eij/article/wildness/ (accessed August 16, 2016).

54. "David Brower Legacy," Earth Island Institute.

55. Gavin Schmidt, "The Carbon Dioxide Theory of Gilbert Plass," *Real Climate: Science from Climate Scientists*, January 4, 2010, http://www.realclimate.org/index.php/archives/2010/01/the-carbon-dioxide-theory-of-gilbert-plass/ (accessed August 17, 2016).

56. Spencer Weart, "Roger Revelle's Discovery," *The Discovery of Global Warming*, July 2007, American Institute of Physics, https://www.aip.org/history/climate/Revelle.htm (accessed March 30, 2016).

57. David Malakoff, "Global Warming: From Theory to Fact," *Causes*, NPR, May 21, 2007, www.npr.org/templates/story/story.php?storyid=10307560 (accessed March 2016).

CHAPTER 15: FULL SPEED AHEAD: PREPARING FOR THE ENVIRONMENTAL YEARS

1. Rachel Carson, *Silent Spring* (Boston: Houghton Mifflin Company, 1962).

2. Linda Lear, "Rachel Carson's Silence," *Pittsburgh Post-Gazette*, April 13, 2014, http://www.post-gazette.com/opinion/Op-Ed/2014/04/13/THE-NEXT-PAGE-Rachel-Carsons-silence/stories/201404130058 (accessed August 22, 2016).

3. Mark Hamilton Lytle, *The Gentle Subversive: Rachel Carson, Silent Spring, and the Rise of the Environmental Movement* (New York: Oxford University Press, 2007), p. 38.

4. Ibid.

5. Linda Lear, "Rachel Carson's Biography," *The Life and Legacy of Rachel Carson*, 1998, http://www.rachelcarson.org/Bio.aspx#.Vbug5kXOeHk (accessed August 2015).

6. Rachel Carson, *The Sea Around Us* (New York: Oxford University Press, 1951).

7. Carson, *Silent Spring*, p. 13.

8. Bryan Walsh, "How Silent Spring Became the First Shot in the War over the Environment," *Time*, September 25, 2012, http://science.time.com/2012/09/25/how-silent-spring-became-the-first-shot-in-the-war-over-the-environment/ (accessed August 22, 2016).

9. Tim Lambert, "The Unending War on Rachel Carson," *ScienceBlogs*, May 19, 2007, http://scienceblogs.com/deltoid/2007/05/19/the-unending-war-on-rachel-car/ (accessed August 21, 2016).

10. Mark Stoll, "Rachel Carson's Silent Spring, A Book That Changed the World," *Environment and Society Portal*, 2012, http://www.environmentandsociety.org/exhibitions/silent-spring/personal-attacks-rachel-carson (accessed August 21, 2016).

11. Daniel Botkin and Edward Keller, *Environmental Science: Earth as a Living Planet*, 7th ed. (Hoboken, NJ: John Wiley, 2009), pp. 231 and 609.

12. Walsh, "How Silent Spring Became the First Shot in the War over the Environment."

13. Lewis Herber [Murray Bookchin], *Our Synthetic Environment* (New York: Knopf, 1962).

14. Naomi Klein, *This Changes Everything: Capitalism vs. the Climate* (New York: Simon & Schuster, 2014).

15. Herber [Bookchin], *Our Synthetic Environment*, pp. 237–45.

16. Ibid. See also, Klein, *This Changes Everything*, pp. 451–55.

17. "A Quick History of the Land and Water Conservation Fund Program: 1964 and All That," National Park Service: Land and Water Conservation Fund, September 19, 2008, https://www.nps.gov/ncrc/programs/lwcf/history.html (accessed August 2015).

18. Bridget Hunter, "Stewart Udall: A Legacy of Conservation," *IIP Digital*, March 24, 2010, http://iipdigital.usembassy.gov/st/english/article/2010/03/20100324145231abretnuh0.8080408.html#axzz45LyYIIix (accessed August 2015).

19. Ibid.

20. Ibid.

21. Keith Schneider and Cornelia Dean, "Stewart L. Udall, Conservationist in Kennedy and Johnson Cabinets, Dies at 90," *New York Times*, March 20, 2010, http://www.nytimes.com/2010/03/21/nyregion/21udall.html?_r=0 (accessed August 2015).

22. Ibid.

23. Hunter, "Stewart Udall."

24. Stewart Udall, *The Quiet Crisis* (New York: Holt, Rinehart and Winston, 1963), p. viii.

25. Ibid., p. 202.

26. Anne Ferrer and Miles Weiss, "Obituary: Stewart Udall / U.S. Interior Secretary Under JFK and LBJ," *Pittsburgh Post-Gazette*, March 21, 2010, http://www.post-gazette.com/news/obituaries/2010/03/21/Obituary-Stewart-Udall-U-S-Interior-secretary-under-JFK-and-LBJ/stories/201003210264 (accessed August 15, 2015).

27. Hunter, "Stewart Udall."

28. Stewart and Lee Udall, "A Message to Our Grandchildren," *High Country News*, March 31, 2008, https://www.hcn.org/issues/367/17613.

29. "The Scenic Hudson Decision," Marist Environmental History Project, http://library.marist.edu/archives/mehp/scenicdecision.html (accessed August 2015).

30. Ibid.

31. Ibid.

32. Ibid.

33. Benjamin Kline, *First Along the River: A Brief History of the US Environmental Movement*, 4th ed. (Lanham, MD: Rowman & Littlefield, 2011), p. 104.

34. Ibid.

35. Ibid.

36. Lyndon B. Johnson, "Special Message to the Congress on Conservation and Restoration of Natural Beauty," February 8, 1965, *The American Presidency Project*, ed. Gerhard Peters and John T. Woolley, http://www.presidency.ucsb.edu/ws/?pid=27285 (accessed August 2015).

37. Jone Johnson Lewis, "Lady Bird Johnson Quotes," *About.com: Education*, March 31, 2016, http://womenshistory.about.com/od/quotes/a/ladybirdjohnson.htm (accessed June 2016).

38. "Ms. Lyndon Baines Johnson (Claudia Alta Taylor), 1912–2007," University of Texas, LBJ Presidential Library, http://www.lbjlib.utexas.edu/johnson/archives.hom/biographys.hom/ladybird_bio.asp (accessed August 20, 2016).

39. Ibid.

40. Holly Shok, "Zahniser, Howard Clinton," Pennsylvania Center for the Book, Fall 2009, http://pabook2.libraries.psu.edu/palitmap/bios/Zahniser__Howard.html (accessed August 2015).

41. Ibid.

42. Ibid.

43. Ibid.

44. "Remarks by Ed Zahniser at the Wilderness Rendezvous," The Murie Center, October 12, 2014, http://www.muriecenter.org/jh-wilderness-remarks/ (accessed July 2015).

45. "The Beginnings of the National Wilderness Preservation System," Wilderness.net, September 21, 2015, http://www.wilderness.net/NWPS/fastfacts (accessed July 2016).

46. Don Hopey, "Savior of Wild Places: Franklin Native Howard Zahniser Worked Tirelessly for the Passage of '64 Wilderness Act," *Pittsburgh Post-Gazette*, August 13, 2001, http://old.post-gazette.com/healthscience/20010813zahniser0813p2.asp (accessed July 13, 2015).

47. Ibid.

48. Ibid.

49. "Howard Zahniser: Author of the Wilderness Act," Wilderness.net, http://www.wilderness.net/nwps/zahniser (accessed August 2015).

50. "E. O. Wilson," E. O. Wilson Biodiversity Foundation, 2016, http://eowilsonfoundation.org/e-o-wilson/ (accessed July 2016).

51. E. O. Wilson, "My Wish: Build the Encyclopedia of Life," *TED Talks*, March 2007, https://www.ted.com/talks/e_o_wilson_on_saving_life_on_earth/transcript?language=en (accessed July 2016).

52. Ibid.

53. "Edward O. Wilson, PhD: Biography," *Academy of Achievement*, June 3, 2013, http://www.achievement.org/autodoc/page/wil2bio-1.

54. "E. O. Wilson," E. O. Wilson Biodiversity Foundation.

55. Edward O. Wilson, *Biophilia* (Cambridge: Harvard University Press, 1984).

56. Edward O. Wilson, ed., *Biodiversity* (Washington, DC: National Academy Press, 1988).

57. Michael M. Gunter Jr., *Building the Next Ark: How NGOs Work to Protect Biodiversity* (Hanover, NH: Dartmouth College Press, 2004), p. 11.

58. Edward O. Wilson, *Sociobiology: The New Synthesis* (Cambridge, MA: Belknap Press of Harvard University Press, 1975).

59. "Edward O. Wilson, Ph.D.: Biography," *Academy of Achievement*.

60. Ibid.

61. Edward O. Wilson, *On Human Nature* (Cambridge: Harvard University Press, 1978).

62. "Edward O. Wilson, Ph.D.: Biography," *Academy of Achievement*.

63. Bert Hölldobler and Edward O. Wilson, *The Ants* (Cambridge, MA: Belknap Press of Harvard University Press, 1990).

64. E. O. Wilson, *Anthill: A Novel* (New York: W. W. Norton, 2010).

65. Ibid.

66. Edward O. Wilson, *Letters to a Young Scientist*. (New York: Liveright, 2013).

67. "E. O. Wilson Quotes," *BrainyQuote*, http://www.brainyquote.com/quotes/authors/e/e_o_wilson.html (accessed November 08, 2015).

68. Michael Ruse, "Edward O. Wilson: American Biologist," *Encyclopedia Brittanica Online*, July 9, 2014, https://www.britannica.com/biography/Edward-O-Wilson (accessed August 22, 2016).

69. Lisa Hymas. "E. O. Wilson Wants to Know Why You're Not Protesting in the Streets," *Grist*, April 30, 2012, http://grist.org/article/e-o-wilson-wants-to-know-why-youre-not-protesting-in-the-streets/.

70. Paul Ehrlich, *The Population Bomb* (New York: Ballantine, 1968).

CHAPTER 16: THE ENVIRONMENTAL HEYDAYS

1. "Keep America Beautiful – (Crying Indian) – 70s PSA Commercial," YouTube video, 1:06, posted by "Justin Engelhaupt," December 31, 2010, https://www.youtube.com/watch?v=8Suu84khNGY (accessed July 2016).

2. "David Brower (1912–2000): Grand Canyon Battle Ads," Sierra Club, 2016, http://content.sierraclub.org/brower/grand-canyon-ads (accessed July 2016).

3. *A Fierce Green Fire: The Battle for a Living Planet*, DVD, directed by Mark Kitchell (New York: First Run Features, 2013).

4. Ibid.

5. Ibid.

6. Ibid.

7. Jennifer Latson, "The Burning River That Sparked a Revolution," *Time*, June 22, 2015, http://time.com/3921976/cuyahoga-fire/ (accessed July 2016).

8. Ibid.

9. Ibid.

10. James Brooke, "Denver Seeing the Light Past Its 'Brown Cloud,'" *New York Times*, April 21, 1998, http://www.nytimes.com/1998/04/21/us/denver-seeing-the-light-past-its-brown-cloud.html (accessed July 2016).

11. "Legislation: A Look at US Air Pollution Laws and Their Amendments," American Meteorological Society, https://www.ametsoc.org/sloan/cleanair/cleanairlegisl.html (accessed July 2016).

12. "Emission Facts: The History of Reducing Tailpipe Emissions," US Environmental Protection Agency, May 1999, https://www3.epa.gov/otaq/consumer/f99017.pdf (accessed July 2016).

13. "Milestones in Mobile Source Air Pollution Control and Regulations," US Environmental Protection Agency, March 23, 2016, https://www3.epa.gov/otaq/consumer/milestones.htm (accessed July 2016).

14. "History of the Catalytic Converter," Catalytic Converters, 2015, https://www.catalyticconverters.com/history/ (accessed July 2016).

15. Ibid.

16. "Milestones in Mobile Source Air Pollution Control and Regulations."

17. "S. 622 (94th): Energy Policy and Conservation Act," GovTrack.us, December 22, 1975, https://www.govtrack.us/congress/bills/94/s622/text (accessed July 2016).

18. James B. Treece, "10 Ways the 1973 Oil Embargo Changed the Industry," *Automotive News*, October 14, 2013, http://www.autonews.com/article/20131014/GLOBAL/131019959/10-ways-the-1973-oil-embargo-changed-the-industry (accessed July 2016).

19. Marc Davis, "How the US Automobile Industry Has Changed," *Investopedia*, http://www.investopedia.com/articles/pf/12/auto-industry.asp (accessed July 2016).

20. Treece, "10 Ways the 1973 Oil Embargo Changed the Industry."

21. Victor B. Scheffer, *The Shaping of Environmentalism in America* (Seattle: University of Washington Press, 1991).

22. Benjamin Kline, *First Along the River: A Brief History of the US Environmental Movement* (Lanham, MD: Rowman & Littlefield, 2011).

23. Ibid.

24. Chris Williams, "Hothouse Earth: Capitalism, Climate Change, and the Fate of Humanity," *International Socialist Review* 62 (November-December 2008), http://www.isreview.org/issues/62/feat-hothouseearth.shtml (accessed August 17, 2015).

25. "National Environmental Policy Act," *NEPA Informational Guide*, National Marine Fisheries Service, https://www.greateratlantic.fisheries.noaa.gov/prot_res/atgtrp/osm/NEPA%20Overview.pdf (accessed August 20, 2016).

26. Michael Gorn, "Russel E. Train: Oral History Interview," US Environmental Protection Agency, May 5, 1992, https://www.epa.gov/aboutepa/russell-e-train-oral-history-interview (accessed July 2016).

27. Ibid.

28. Ibid.

29. Ibid.

30. Keith Schneider, "Russell E. Train, Conservationist Who Helped Create the E.P.A., Dies at 92," *New York Times*, September 17, 2012 http://www.nytimes.com/2012/09/18/science/earth/russell-e-train-92-dies-helped-create-the-epa.html?_r=0 (accessed March 30, 2016).

31. Steve Ertel, "Statement on the Passing of WWF Founder and Chairman Emeritus Russell E. Train (1920–2012)," World Wildlife Fund, September 17, 2012. http://www.worldwildlife.org/press-releases/statement-on-the-passing-of-wwf-founder-and-chairman-emeritus-russell-e-train-1920-2012 (accessed August 12, 2015).

32. Schneider, "Russell E. Train."

33. Ibid.

34. Ibid.

35. Russell E. Train, *Politics, Pollution, and Pandas: An Environmental Memoir* (Washington, DC: Island Press, 2003).

36. Ibid., p. 102.

37. Kline, *First Along the River*, p. 105.

38. Claudia Copeland, "Clean Water Act: A Summary of the Law," *Congressional Research Service*, April 3, 2010, http://www.in.gov/idem/files/rules_erb_20130213_cwa_summary.pdf (accessed November 9, 2015).

39. Ibid.

40. "Coastal Zone Management Act," National Oceanic and Atmospheric Administration: Office for Coastal Management, https://coast.noaa.gov/czm/act/ (accessed July 2016).

41. Ibid.

42. Daniel Lewis, "Scientist, Candidate and Planet Earth's Lifeguard," *New York Times*, October 1, 2012, http://www.nytimes.com/2012/10/02/us/barry-commoner-dies-at-95.html (accessed March 30, 2016).

43. Barry Commoner, *The Closing Circle: Nature, Man, and Technology* (New York: Random House, 1971).

44. Peter Dreier, "Remembering Barry Commoner," *Nation*, October 1, 2012, http://www.thenation.com/article/remembering-barry-commoner/ (accessed March 30, 2016).

45. "Dioxins and Their Effects on Human Health," World Health Organization, June 2014, http://www.who.int/mediacentre/factsheets/fs225/en/ (accessed August 22, 2016).

46. Ibid.

47. Ibid.

48. Ibid.

49. Paul Ehrlich, *The Population Bomb* (New York: Ballantine, 1968).

50. Dreier, "Remembering Barry Commoner."

51. Commoner, *Closing Circle*, pp. 33–48.

52. Dreier, "Remembering Barry Commoner."

53. Matt Schudel, "Barry Commoner, Scientist and Influential Environmentalist, Dies at 95," *Washington Post*, October 2, 2012, https://www.washingtonpost.com/local/obituaries/barry-commoner-scientist-and-influential-environmentalist-dies-at-95/2012/10/02/889c8c7c-0caa-11e2-a310-2363842b7057_story.html.

54. Commoner, *Closing Circle*.

55. Donella H. Meadows, Dennis L. Meadows, Jørgen Randers, and William W. Behrens III, *Limits to Growth: A Report for the Club of Rome's Project on the Predicament of Mankind* (New York: Universe Books, 1972).

56. Christian Parenti, "'The Limits to Growth': A Book That Launched a Movement," *Nation*, December 5, 2012, https://www.thenation.com/article/limits-growth-book-launched-movement/ (accessed July 2016).

57. Leon Kolankiewicz, "Donella Meadows—A Tribute," *Social Contract Journal* 11, no. 4 (Summer 2001), http://www.thesocialcontract.com/artman2/publish/tsc1104/article_985.shtml (accessed March 30, 2016).

58. Ibid.

59. Ibid.

60. Ibid.

61. Ibid.

62. Herman Daly, "Review of *Collision Course (Endless Growth on a Finite Planet)*," *Daly News*, April 2015, http://steadystate.org/review-of-collision-course-endless-growth-on-a-finite-planet/ (accessed July 2016).

63. Ibid.

64. Kolankiewicz, "Donella Meadows—A Tribute."

65. "About Donella 'Dana' Meadows," Donella Meadows Institute, http://donellameadows.org/donella-meadows-legacy/donella-dana-meadows/ (accessed July 2016).

66. Ibid.

67. Ibid.

68. Kolankiewicz, "Donella Meadows—A Tribute."

69. "About the Donella Meadows Institute," Donella Meadows Institute, http://donellameadows.org/about-dmi/ (accessed July 2016).

70. Kolankiewicz, "Donella Meadows—A Tribute."

71. Ibid.

72. Edward Abbey, *The Monkey Wrench Gang* (Philadelphia, Lippincott: 1975).

73. Ibid.

74. Edward Abbey, *Desert Solitaire: A Season in the Wilderness* (New York: Ballantine, 1971).

75. Edwin Way Teale, "Making the Wild Scene," *New York Times*, January 28, 1968, https://www.nytimes.com/books/97/09/28/reviews/desert.html (accessed July 2016).

76. Ibid.

77. Abbey, *Desert Solitaire*.

78. Doug Peacock, "Notes from the Arroyo," forward to *The Best of Edward Abbey* (San Francisco: Sierra Club, 2005), http://www.dougpeacock.net/the-best-of-ed-abbey-foreward.html (accessed July 2016).

79. "Edward Abbey: Freedom Begins Between the Ears," Wilderness.net, www.wilderness.net/NWPS/Abbey (accessed August 21, 2016).

80. Natalie Kyriacou, "Anti-Sealing Campaign," *MyGreenWorld*, April 2, 2014, http://www.mygreenworld.org/anti-sealing-campaign/ (accessed August 21, 2016).

81. Robert Hunter, *The* Greenpeace *to Amchitka: An Environmental Odyssey* (Vancouver: Arsenal Pulp, 2004), pp. 15–17.

82. "Amchitka: The Founding Voyage," Greenpeace International, http://www.greenpeace.org/international/en/about/history/amchitka-hunter/ (accessed July 2016).

83. Ibid.

84. "About," Greenpeace (USA), http://www.greenpeace.org/usa/about/ (accessed July 2016).

85. Ibid.

86. *Fierce Green Fire*, DVD.

87. Ibid.

88. "Oil Embargo, 1973–1974," US Department of State: Office of the Historian, https://history.state.gov/milestones/1969-1976/oil-embargo (accessed March 30, 2016).

89. Gerald R. Ford, "Remarks at Dedication Ceremonies for the National Environmental Research Center, Cincinnati, Ohio," *American Presidency Project*, ed. Gerhard Peters and John T. Woolley, http://www.presidency.ucsb.edu/ws/index.php?pid=5042 (accessed August 23, 2016).

90. "Gerald Ford on Environment," *On The Issues*, http://www.ontheissues.org/Celeb/Gerald_Ford_Environment.htm (accessed August 12, 2015).

91. Ibid.

92. Ibid.

93. Wallace S. Broecker, "Climatic Change: Are We on the Brink of a Pronounced Global Warming?" *Science* 189, no. 4201 (August 8, 1975): 460–63.

94. "Wallace S. Broecker," Earth Institute of Columbia University, http://www.earth.columbia.edu/articles/view/2246 (accessed July 2016).

95. Syukuro Manabe and Richard T. Wetherald, "On the Distribution of Climate Change Resulting from an Increase in CO_2 Content of the Atmosphere," *Journal of the Atmospheric Sciences* 37, no. 1 (January 1980): 99–118.

96. "Ozone Depletion," *National Geographic*, http://environment.nationalgeographic

.com/environment/global-warming/ozone-depletion-overview/ (accessed August 13, 2015).

97. Ibid.

98. Ibid.

99. Ibid.

100. Ibid.

101. Charles Welch, "CFCs," *The Ozone Hole*, http://www.theozonehole.com/cfc.htm (accessed August 21, 2016).

102. Ibid.

103. "Basic Ozone Layer Science," US Environmental Protection Agency, http://www.epa.gov/ozone-layer-protection/basic-ozone-layer-science.

104. Ibid.

105. "Susan Solomon," Earth, Atmospheric, and Planetary Sciences, https://eapsweb.mit.edu/people/solos (accessed July 2016).

106. "Basic Ozone Layer Science."

107. Ibid.

108. "International Day for the Preservation of the Ozone Layer," United Nations, September 16, 2015, http://www.un.org/en/events/ozoneday/ (accessed July 2016).

109. Ibid.

110. "The Montreal Protocol on Substances That Deplete the Ozone Layer," Ozone Secretariat, United Nations Environment Programme, http://ozone.unep.org/en/treaties-and-decisions/montreal-protocol-substances-deplete-ozone-layer (accessed March 30, 2016).

111. "Executive Summary: Scientific Assessment of Ozone Depletion: 2006," Earth System Research Laboratory: Chemical Sciences Division, National Oceanic and Atmospheric Administration, http://www.esrl.noaa.gov/csd/assessments/ozone/2006/executivesummary.html (accessed July 2016).

112. "Montreal Protocol on Substances That Deplete the Ozone Layer."

113. Reiner Grundmann, "Climate Change and Knowledge Politics," *Environmental Politics* 16, no. 3 (June 2007): 414–32, http://stsclimate.soc.ku.dk/papers/grundmannclimatechangeandknowledgepolitics.pdf (accessed August 21, 2016).

114. Ibid.

115. "Summary of the Resource Conservation and Recovery Act," US Environmental Protection Agency, December 29, 2015, https://www.epa.gov/laws-regulations/summary-resource-conservation-and-recovery-act (accessed August 21, 2016).

116. Ridgway M. Hall Jr., Robert C. Davis Jr., Richard E. Schwartz, Nancy S. Bryson, Timothy R. McCrum, *RCRA Hazardous Wastes Handbook* (Lanham, MD: Rowman & Littlefield, 2001), section 1, p. 2.

117. Lois Marie Gibbs, "History: Love Canal: The Start of a Movement," *Lessons from Love Canal: A Public Health Resource*, 2002, https://www.bu.edu/lovecanal/canal/ (accessed July 2016).

118. "Dioxins and PCBs," European Food Safety Authority, https://www.efsa.europa.eu/en/topics/topic/dioxins (accessed July 2016).

119. Gibbs, "History: Love Canal."

120. Ibid.

121. Michael H. Brown, "Love Canal and the Poisoning of America," *Atlantic*, December 1979, http://www.theatlantic.com/magazine/archive/1979/12/love-canal-and-the-poisoning-of-america/376297/ (accessed August 21, 2016).

122. "History: Chronology—Key Dates and Events at Love Canal," Center for Health, Environment, and Justice, 2002, http://www.bu.edu/lovecanal/canal/date.html (accessed July 2016).

123. Ibid.

124. Ibid.

125. "Superfund: CERCLA Overview," US Environmental Protection Agency, September 30, 2015, https://www.epa.gov/superfund/superfund-cercla-overview (accessed July 2016).

126. Kevin Konrad, "Lois Gibbs: Grassroots Organizer and Environmental Health Advocate," *American Journal of Public Health* 101, no. 9 (September 2011): 1558–59, http://www.ncbi.nlm.nih.gov/pmc/articles/PMC3154230/ (accessed July 2016).

127. "Summary of the Comprehensive Environmental Response, Compensation, and Liability Act (Superfund)," US Environmental Protection Agency, February 8, 2016, https://www.epa.gov/laws-regulations/summary-comprehensive-environmental-response-compensation-and-liability-act (accessed July 2016).

128. Ibid.

129. Konrad, "Lois Gibbs."

130. "Founder: Lois Gibbs," Center for Health, Environment, and Justice, http://chej.org/lois/ (accessed August 21, 2016).

131. Tony Schwartz, "TV: Lois Gibbs Fights the Battle of Love Canal," *New York Times*, February 17, 1982, http://www.nytimes.com/1982/02/17/arts/tv-lois-gibbs-fights-the-battle-of-love-canal.html (accessed August 21, 2016).

132. "Founder: Lois Gibbs," Center for Health, Environment, and Justice.

133. James Hansen, "Wanning Workshop + Beijing Charts + Year-End Comments," Dr. James E. Hansen: Communications, Columbia University: Climate Science, Awareness and Solutions Program, December 29, 2015, http://www.columbia.edu/~jeh1/mailings/2015/20151229_Sleepless.pdf (accessed July 2016).

134. Ibid.

135. Ibid.

136. Mark Jacobson, "Mark Jacobson to James Hansen: Nukes Are Not Needed to Solve World's Climate Crisis," *EcoWatch*, January 4, 2016, http://www.ecowatch.com/mark-jacobson-to-james-hansen-nukes-are-not-needed-to-solve-worlds-cli-1882141790.html (accessed July 2016).

137. Ibid.

138. Navid Chowdhury, "Is Nuclear Energy Renewable Energy?" Stanford University: Introduction to Nuclear Energy, March 22, 2012, http://large.stanford.edu/courses/2012/ph241/chowdhury2/ (accessed March 30, 2016).

139. "Backgrounder on Radioactive Waste," United States Nuclear Regulatory Commission, April 3, 2015, http://www.nrc.gov/reading-rm/doc-collections/fact-sheets/radwaste.html (accessed July 2016).

140. "Why Is Uranium-235 Ideal for Nuclear Power?" *How Stuff Works: Science*, 2016, http://science.howstuffworks.com/innovation/science-questions/why-is-uranium-235-ideal-for-nuclear-power.htm (accessed July 2016).

141. "What Is Uranium? How Does It Work?" World Nuclear Association, March 2014 http://www.world-nuclear.org/information-library/nuclear-fuel-cycle/introduction/what-is-uranium-how-does-it-work.aspx (accessed July 2016).

142. "Backgrounder on the Three Mile Island Accident," United States Nuclear Regulatory Commission, December 12, 2014, http://www.nrc.gov/reading-rm/doc-collections/fact-sheets/3mile-isle.html (accessed March 30, 2016).

143. "US Nuclear Power Policy," World Nuclear Association, July 2016, http://www.world-nuclear.org/information-library/country-profiles/countries-t-z/usa-nuclear-power-policy.aspx.

144. Jimmy Carter, "Conservationist of the Year Award Remarks on Accepting the Award from the National Wildlife Federation," March 20, 1979, at *The American Presidency Project*, ed. Gerhard Peters and John T. Woolley, http://www.presidency.ucsb.edu/ws/?pid=32067 (accessed August 15, 2015).

145. Ibid.

146. Arthur Allen, "Prodigal Sun," *Mother Jones*, March/April 2000, http://www.motherjones.com/politics/2000/03/prodigal-sun (accessed August 21, 2016).

147. "About NREL," National Renewable Energy Laboratory, 2015, http://www.nrel.gov/about/ (accessed August 21, 2016).

CHAPTER 17: THE REAGAN YEARS: BIG TROUBLE FOR THE ENVIRONMENT

1. Benjamin Kline, *First Along the River: A Brief History of the US Environmental Movement* (Lanham, MD: Rowman & Littlefield, 2011), p. 114.

2. Joseph Romm, "Who Got Us into This Energy Mess? Start with Ronald Reagan," *ThinkProgress*, July 8, 2008, http://thinkprogress.org/climate/2008/07/08/202854/who-got-us-in-this-energy-mess-start-with-ronald-reagan/ (accessed July 2016).

3. Paul Boyer, *The Enduring Vision: A History of the American People* (Belmont, CA: Wadsworth Publishing, 2013).

4. David Biello, "Where Did the Carter White House Solar Panels Go?" *Scientific American*, August 6, 2010, http://www.scientificamerican.com/article/carter-white-house-solar-panel-array/ (accessed March 30, 2016).

5. Ibid.

6. "A Look Back at Reagan's Environmental Record," *Grist*, June 11, 2004, http://grist.org/politics/griscom-reagan/ (accessed December 31, 2015).

7. "James G. Watt: Biography," FamPeople.com, 2012, http://www.fampeople.com/cat-james-g-watt (accessed July 2016).

8. "A Look Back at Reagan's Environmental Record."

9. Ibid.

10. "James G. Watt," FamPeople.com.

11. Ibid.

12. Ibid.

13. "The Legacy of James Watt," *Time*, October 24, 1983.

14. Kline, *First Along the River*, pp. 114–15.

15. Ibid., p.115.

16. Ibid.

17. Phil Wisman, "EPA History (1970–1985)," US Environmental Protection Agency, November 1985, https://www.epa.gov/aboutepa/epa-history-1970-1985 (accessed July 2016).

18. Ibid. Wisman, "EPA History."

19. "Solar One and Two (Now Defunct)," *Solaripedia*, 2009, http://www.solaripedia.com/13/31/solar_one_and_two_(now_defunct).html (accessed December 31, 2015).

20. Ibid.

21. Ibid.

22. "Wind Energy Center Alumni," University of Massachusetts: Wind Energy Center, 2014, http://www.umass.edu/windenergy/about/history/alumni (accessed December 31, 2015).

23. Mike Mooiman, "Windfall?—Wind Energy in New Hampshire," *Energy in New Hampshire* (blog), June 3, 2013, http://nhenergy.blogspot.com/2013/06/windfall-wind-energy-in-new-hampshire.html (accessed August 22, 2016).

24. "About RMI: Vision and Mission," Rocky Mountain Institute, http://www.rmi.org/Vision%20and%20Mission (accessed July 2016).

25. "About RMI," Rocky Mountain Institute, http://www.rmi.org/About%20RMI (accessed July 2016).

26. "Abundance by Design," Rocky Mountain Institute, http://www.rmi.org/abundance_by_design (accessed August 22, 2016).

27. "Natural Capitalism," Rocky Mountain Institute, http://www.rmi.org/Natural++Capitalism (August 22, 2016).

28. Ibid.

29. "About WRI," World Resources Institute, http://www.wri.org/about (accessed December 31, 2015).

30. "Angels by the River: A Memoir," Chelsea Green Publishing, http://www.chelseagreen.com/angels-by-the-river (accessed August 25, 2016).

31. James Gustave Speth, *Angels by the River: A Memoir* (White River Junction, VT: Chelsea Green Publishing, 2014), book jacket.

32. "Introduction and Mission," Ocean Arks International, http://oceanarksint.org/index.php?id=mission (accessed June 13, 2015).

33. Paul J. Crutzen and John W. Birks, "The Atmosphere After a Nuclear War: Twilight at Noon," *Ambio: A Journal of the Human Environment* 11, no. 2–3 (1982): 114–25, https://www.researchgate.net/publication/236687098_The_Atmosphere_After_a_Nuclear_War_Twilight_at_Noon (accessed March 30, 2016).

34. Pervaze A. Sheikh, "Debt-for-Nature Initiatives and the Tropical Forest Conservation Act: Status and Implementation," *Congressional Research Service*, March 30, 2010, https://www.cbd.int/financial/debtnature/g-inventory2010.pdf (accessed December 31, 2015).

35. Catherine Kilbane Gockel and Leslie C. Gray, "Debt-for-Nature Swaps in Action: Two Case Studies in Peru," *Ecology and Society* 16, no. 3 (2011): 13, http://www.ecologyandsociety.org/vol16/iss3/art13/ (accessed August 22, 2016).

36. Sheikh, "Debt-for-Nature Initiatives."

37. Ibid.

38. Ibid.

39. Marc Lallanilla, "Chernobyl: Facts about the Nuclear Disaster," *LiveScience*, September 25, 2013, http://www.livescience.com/39961-chernobyl.html (accessed July 2016).

40. Ibid.

41. Ibid.

42. "Laws That Protect Our Oceans," US Environmental Protection Agency, January 20, 2016, https://www.epa.gov/beach-tech/laws-protect-our-oceans (accessed July 2016).

43. "Ocean Dumping Ban Act of 1988," US Environmental Protection Agency, November 21, 1988, https://www.epa.gov/aboutepa/ocean-dumping-ban-act-1988 (accessed August 23, 2016).

44. "Learn About Ocean Dumping," US Environmental Protection Agency, January 29, 2016, https://www.epa.gov/ocean-dumping/learn-about-ocean-dumping (accessed August 23, 2016).

45. Marlise Simons, "Brazilian Who Fought to Protect Amazon Is Killed," *New York Times*, December 24, 1988, http://www.nytimes.com/1988/12/24/world/brazilian-who-fought-to-protect-amazon-is-killed.html (accessed July 2016).

46. Jan Rocha and Jonathan Watts, "Brazil Salutes Chico Mendes 25 Years After His Murder," *Guardian*, December 20, 2013, http://www.theguardian.com/world/2013/dec/20/brazil-salutes-chico-mendes-25-years-after-murder (accessed March 30, 2016).

47. Simons, "Brazilian Who Fought to Protect Amazon Is Killed."

48. "Chico Mendes," *Encyclopedia.com*, 2004, http://www.encyclopedia.com/topic/Chico_Mendes.aspx (accessed July 2016).

49. Ibid.

50. Rocha and Watts, "Brazil Salutes Chico Mendes."

51. "Chico Mendes," *Encyclopedia.com.*

52. Ethan Spaner, "Chico Vive: Chico Mendes and Grassroots Environmentalism,"

National Wildlife Federation's Blog, April 3, 2014, http://blog.nwf.org/2014/04/chico-vive-chico-mendes-and-grassroots-environmentalism/ (accessed August 23, 2016).

53. Ibid.

54. Rocha and Watts, "Brazil Salutes Chico Mendes."

55. Ibid.

56. Ibid.

CHAPTER 18: THE CLIMATE-CHANGE DEBATE TAKES OFF

1. S. M. Enzler, "History of the Greenhouse Effect and Global Warming," Lenntech, http://www.lenntech.com/greenhouse-effect/global-warming-history.htm (accessed August 2015).

2. Ibid.

3. Ibid.

4. Ibid.

5. "First Steps to a Safer Future: Introducing the United Nations Framework Convention on Climate Change," United Nations Framework Convention on Climate Change, 2014, http://unfccc.int/essential_background/convention/items/6036.php (accessed March 27, 2016).

6. "The IPCC: Who Are They and Why Do Their Climate Reports Matter?" Union of Concerned Scientists, http://www.ucsusa.org/global_warming/science_and_impacts/science/ipcc-backgrounder.html#.V6Te37XDGxU (accessed July 2016).

7. Intergovernmental Panel Climate Change (IPCC), *Climate Change: The IPCC Scientific Assessment* (Cambridge: Cambridge University Press, 1990), p. 410.

8. Ibid.

9. Naomi Klein, *This Changes Everything: Capitalism vs. the Climate* (New York: Simon and Schuster, 2014).

10. Naomi Oreskes and Eric Conway, *Merchants of Doubt: How a Handful of Scientists Obscured the Truth on Issues from Tobacco Smoke to Global Warming* (New York: Bloomsbury, 2010), p. 233.

11. Ibid., p. 234.

12. "Koch Industries: Secretly Funding the Climate Denial Machine," Greenpeace, http://www.greenpeace.org/usa/global-warming/climate-deniers/koch-industries/ (accessed August 24, 2016).

13. Rob Hopkins, "Naomi Oreskes on the Roots of Climate Change Denial," *Transition Network: Transition Culture*, July 7, 2104, https://transitionnetwork.org/blogs/rob-hopkins/2014-07/naomi-oreskes-roots-climate-change-denial (accessed August 24, 2016).

14. "Stephen H. Schneider, Climatologist," *National Geographic*, 2016, http://www.nationalgeographic.com/explorers/bios/stephen-schneider/ (accessed March 27, 2016).

15. T. Rees Shapiro, "Stephen H. Schneider, Climate Change Expert, Dies at 65," *Washington Post*, July 20, 2010, http://www.washingtonpost.com/wp-dyn/content/article/2010/07/19/AR2010071905108.html (accessed March 27, 2016).

16. Ibid.

17. Ibid.

18. Philip Shabecoff, "Global Warming Has Begun, Expert Tells Senate," *New York Times*, June 24, 1988, http://www.nytimes.com/1988/06/24/us/global-warming-has-begun-expert-tells-senate.html?pagewanted=all (accessed July 2016).

19. Andrew Revkin, "Industry Ignored Its Scientists on Climate," *New York Times*, April 23, 2009, http://www.nytimes.com/2009/04/24/science/earth/24deny.html?_r=0 (accessed July 2016).

20. Ibid.

21. "Acid Rain," National Atmospheric Deposition Program, 2014, http://nadp.sws.uiuc.edu/educ/acidrain.aspx (accessed August 17, 2015).

22. Ibid.

23. Ibid.

24. Gene Likens, "Acid Rain," *The Encyclopedia of Earth*, October 17, 2011, http://www.eoearth.org/view/article/149814/ (accessed August 17, 2015) [site discontinued].

25. "Acid Rain Program," US Environmental Protection Agency, https://www.epa.gov/airmarkets/acid-rain-program.

26. "Acid Rain Pollution Solved Using Economics," Environmental Defense Fund, 2016, https://www.edf.org/approach/markets/acid-rain?utm_source=g (accessed March 27, 2016).

27. "Acid Rain Program."

28. Ibid.

29. National Response Team, "The *Exxon Valdez* Oil Spill: A Report to the President (Executive Summary)," US Environmental Protection Agency, May 1989, https://www.epa.gov/aboutepa/exxon-valdez-oil-spill-report-president-executive-summary (accessed March 27, 2016).

30. Brandon Keim, "The *Exxon Valdez* Spill Is All Around Us," *Wired*, March 24, 2009, http://www.wired.com/2009/03/valdezlegacy/ (accessed December 31, 2015).

31. Christine Dell'Amore, "*Exxon Valdez* Anniversary: 20 Years Later, Oil Remains," National Geographic, March 23, 2009, http://news.nationalgeographic.com/news/2009/03/090323-exxon-anniversary.html (accessed August 24, 2016).

32. Bryan Walsh, "Still Digging Up *Exxon Valdez* Oil, 20 Years Later," *Time*, June 4, 2009, http://content.time.com/time/health/article/0,8599,1902333,00.html (accessed August 24, 2016).

33. National Response Team, "*Exxon Valdez* Oil Spill."

34. "About Us," Environmental Media Association, http://www.green4ema.org/about-us/ (accessed February 24, 2016).

35. "Summary of the Clean Air Act," US Environmental Protection Agency, https://www.epa.gov/laws-regulations/summary-clean-air-act (accessed July 2016).

36. "1990 Clean Air Act Amendment Summary," US Environmental Protection Agency, October 27, 2015, https://www.epa.gov/clean-air-act-overview/1990-clean-air-act-amendment-summary (accessed August 24, 2016).

37. "The Clean Air Act," Union of Concerned Scientists, http://www.ucsusa.org/global_warming/solutions/reduce-emissions/the-clean-air-act.html#.VkX30cpGwnQ (accessed August 17, 2015).

38. Ibid.

39. "Massachusetts et al. v. Environmental Protection Agency et al.," Supreme Court, October Term 2006, http://www.supremecourt.gov/opinions/06pdf/05-1120.pdf (accessed July 2016).

40. "Clean Air Act," Union of Concerned Scientists.

41. Joe Mendelson and Catherine Bowes, "The Clean Air Act: 40 Years of Success Protecting Public Health & Environment," National Wildlife Federation, https://www.nwf.org/pdf/Policy-Solutions/CleanAirActFactSheet.pdf (accessed August 17, 2015).

42. "Attack on the Clean Air Act," Union of Concerned Scientists, February 2011, http://www.ucsusa.org/global_warming/solutions/reduce-emissions/clean-air-act.html#.VwpGL_krLIV (accessed November 13, 2015).

43. Mendelson and Bowes, "Clean Air Act."

44. Enzler, "History of the Greenhouse Effect and Global Warming."

45. "10 Warmest Years on Record Globally," *Climate Central*, January 6, 2015, http://www.climatecentral.org/gallery/graphics/10-warmest-years-globally (accessed November 13, 2015).

46. Ida Kubiszewski, "United Nations Conference on Environment and Development (UNCED), Rio de Janeiro, Brazil," Rio Earth Summit 1992, June 7, 2012, http://pdf.amazingdiscoveries.org/References/RtR/Lec4/Lec4-UNCED_Rio_de_Janeiro_Brazil.pdf (accessed November 13, 2015).

47. Ibid.

48. Ibid.

49. "First Steps to a Safer Future."

50. William J. Clinton, "Remarks on Earth Day," April 21, 1993, *The American Presidency Project*, ed. Gerhard Peters and John T. Woolley, http://www.presidency.ucsb.edu/ws/?pid=46460 (accessed July 2016).

51. Benjamin Kline, *First Along the River: A Brief History of the US Environmental Movement* (Lanham, MD: Rowman & Littlefield, 2011), p. 145.

52. Paul Wapner, "Clinton's Environmental Legacy," *Tikkun*, March/April 2001, http://www.tikkun.org/nextgen/clintons-environmental-legacy (accessed March 27, 2016).

53. Ibid.

54. Ibid.

55. Ibid.

56. Ibid.

57. Ibid.

58. Ibid.

59. Al Gore, *Earth in the Balance: Ecology and the Human Spirit* (New York: Penguin, 1992).

60. Al Gore, *An Inconvenient Truth: The Planetary Emergence of Global Warming and What We Can Do about It* (New York: Rodale, 2006).

61. "North American Free Trade Agreement (NAFTA)," Office of the United States Trade Representative, https://ustr.gov/trade-agreements/free-trade-agreements/north-american-free-trade-agreement-nafta (accessed July 2016).

62. Ibid.

63. Oliver Tickell, "Naomi Klein: A Crisis This Big Changes Everything," *Ecologist*, January 21, 2015, http://www.theecologist.org/Interviews/2710065/naomi_klein_a_crisis_this_big_changes_everything.html (accessed August 25, 2016).

64. Quentin Karpilow et al., "NAFTA: 20 Years of Costs to Communities and the Environment," Sierra Club, March 2014, https://content.sierraclub.org/creative-archive/sites/content.sierraclub.org.creative-archive/files/pdfs/0642-NAFTA%20Report_05_low.pdf (accessed July 2016).

65. Kline, *First Along the River*, p. 132.

66. Renee Skelton and Vernice Miller, "The Environmental Justice Movement," National Resources Defense Council, March 17, 2016, http://www.nrdc.org/ej/history/hej.asp (accessed March 27, 2016).

67. Ibid.

68. Robert Bullard, *Dumping in Dixie: Race, Class, and Environmental Quality* (Boulder, CO: Westview Press, 1990).

69. Ibid.

70. Eddy F. Carder, "The American Environmental Justice Movement," *Internet Encyclopedia of Philosophy*, http://www.iep.utm.edu/enviro-j/ (accessed August 24, 2016).

71. "IPCC Second Assessment: Climate Change 1995," Intergovernmental Panel on Climate Change, December 1995, https://www.ipcc.ch/pdf/climate-changes-1995/ipcc-2nd-assessment/2nd-assessment-en.pdf (accessed March 31, 2016).

72. Joseph Stromberg, "What Is the Anthropocene and Are We Really In It?" *Smithsonian Magazine*, January 2013, http://www.smithsonianmag.com/ist/?next=/science-nature/what-is-the-anthropocene-and-are-we-in-it-164801414/ (accessed March 31, 2016).

73. Ibid.

74. Paul Hawken and Amory Lovins, *Natural Capitalism: Creating the Next Industrial Revolution* (Boston: Little, Brown, 1999).

75. Ibid.

76. Joel Makower, "Inside Paul Hawken's Audacious Plan to 'Drawdown' Climate Change," *Two Steps Forward* (blog), GreenBiz, October 22, 2014, https://www.greenbiz.com/blog/2014/10/22/inside-paul-hawkens-audacious-plan-drawdown-climate-change (accessed March 31, 2016).

77. Ibid.

78. "Kyoto Protocol," United Nations Framework Convention on Climate Change, 2014, http://unfccc.int/kyoto_protocol/items/2830.php (accessed July 2016).

79. "COP—What's It All About," Sustainable Innovation Forum, United Nations Environment Programme: Climate Action, 2015, http://www.cop21paris.org/about/cop21 (accessed July 2016).

80. "UNFCCC and the Kyoto Protocol," United Nations, http://www.un.org/wcm/content/site/climatechange/lang/en/pages/gateway/the-negotiations/the-un-climate-change-convention-and-the-kyoto-protocol (accessed July 2016).

81. "COP—What's It All About."

82. Ibid.

83. Ed King, "Kyoto Protocol: 10 Years of the World's First Climate Change Treaty," *Climate Home*, February 16, 2015, http://www.climatechangenews.com/2015/02/16/kyoto-protocol-10-years-of-the-worlds-first-climate-change-treaty/ (accessed July 2016).

84. "UNFCCC and the Kyoto Protocol," United Nations.

85. King, "Kyoto Protocol."

86. Ibid.

87. Anup Shah, "Reactions to Climate Change Negotiations and Action," *Global Issues*, March 5, 2012, http://www.globalissues.org/article/179/reactions-to-climate-change-negotiations-and-action (accessed July 2016).

88. Ibid.

89. Kline, *First Along the River*.

90. Shah, "Reactions to Climate Change Negotiations and Action."

91. Nick Paton Walsh, "Russian Vote Saves Kyoto Protocol," *Guardian*, October 22, 2004, https://www.theguardian.com/world/2004/oct/23/society.russia (accessed August 24, 2016).

92. King, "Kyoto Protocol."

93. "Q&A: The Kyoto Protocol," BBC News, February 16, 2005, http://news.bbc.co.uk/2/hi/science/nature/4269921.stm (accessed November 15, 2015).

94. Ibid.

95. William K. Stevens, "Dead Trees and Shriveling Glaciers as Alaska Melts," *New York Times*, August 18, 1998, http://www.nytimes.com/1998/08/18/science/dead-trees-and-shriveling-glaciers-as-alaska-melts.html?pagewanted=all (accessed July 2016).

96. Ibid.

97. Ibid.

98. Kline, *First Along the River*.

99. Stevens, "Dead Trees."

100. Ibid.

101. "Mission and History," Interfaith Power & Light, 2016, http://www.interfaithpowerandlight.org/about/mission-history/ (accessed February 20, 2016).

102. Ibid.

103. "Our Staff," Interfaith Power & Light, 2016, http://www.interfaithpowerandlight.org/about/staff-and-board-of-directors/ (accessed February 20, 2016).

104. "Letter from Vice President Gore," *Environmental Diplomacy: The Environment*

and U.S. Foreign Policy, US Department of State, http://1997-2001.state.gov/www/global/oes/earth.html#gore (accessed November 15, 2015).

105. Scripps Institute of Oceanography, "Massive Pollution Documented Over Indian Ocean," *ScienceDaily*, June 10, 1999, https://www.sciencedaily.com/releases/1999/06/990610074044.htm (accessed August 25, 2016).

106. Kline, *First Along River*, p. 167.

107. Ibid.

108. Ibid.

109. Ibid.

CHAPTER 19: A SERIES OF UNFORTUNATE EVENTS

1. Todd S. Purdum, "Counting the Vote: The Overview; Bush Is Declared Winner in Florida, But Gore Vows to Contest Results," *New York Times*, November 27, 2000, http://www.nytimes.com/2000/11/27/us/counting-vote-overview-bush-declared-winner-florida-but-gore-vows-contest.html?pagewanted=all (accessed July 2016).

2. Martin Kettle, "Florida 'Recounts' Make Gore Winner," *Guardian*, January 28, 2001, https://www.theguardian.com/world/2001/jan/29/uselections2000.usa (accessed July 2016).

3. Benjamin Kline, *First Along the River: A Brief History of the US Environmental Movement* (Lanham, MD: Rowman & Littlefield, 2011), pp. 191–92.

4. Mark Danner, "How Dick Cheney Became the Most Powerful Vice President in History," *Nation*, February 11, 2014, https://www.thenation.com/article/how-dick-cheney-became-most-powerful-vice-president-history/ (accessed August 26, 2016); "Dick Cheney: Biography," *Biography.com*, January 6, 2016, http://www.biography.com/people/dick-cheney-9246063#vice-presidency (accessed August 26, 2016).

5. Kline, *First Along the River*, p. 172.

6. Joshua Dorner, "Cheney's Culture of Deregulation and Corruption: How the Bush Administration Inaction Created the BP Disaster," Center for American Progress, June 9, 2010, https://www.americanprogress.org/issues/green/news/2010/06/09/7900/cheneys-culture-of-deregulation-and-corruption/ (accessed August 26, 2016).

7. Ibid.

8. Ibid.

9. Ibid.

10. Ibid.

11. "The Clear Skies Initiative," *Now: Science and Health*, PBS, April 15, 2005, http://www.pbs.org/now/science/clearskies05.html (accessed August 27, 2016).

12. "Clear Skies," US Environmental Protection Agency, March 18, 2016, https://archive.epa.gov/clearskies/web/html/ (accessed August 27, 2016).

13. "Clear Skies, R.I.P.," *New York Times*, March 7, 2005, http://www.nytimes

.com/2005/03/07/opinion/clear-skies-rip.html (accessed July 2016).

14. "The Clear Skies Initiative," PBS.

15. "Clear Skies, R.I.P.," *New York Times*.

16. Ibid.

17. Tom Valtin, "Bush Chips Away at Clean Air Act," Sierra Club, http://vault.sierraclub.org/planet/200302/clean_air_act.asp (accessed August 27, 2016).

18. Ibid.

19. Ibid.

20. David Whitman, "Partly Sunny," *Washington Monthly*, December 2004, http://visualprime.com/Pub/WashingtonMonthly-2004dec-00031 (accessed March 31, 2016).

21. Ibid.

22. Shankar Vedantam, "Senate Impasse Stops 'Clear Skies' Measure," *Washington Post*, March 10, 2005, http://www.washingtonpost.com/wp-dyn/articles/A20314-2005Mar9.html (accessed March 31, 2016).

23. "Clear Skies," US Environmental Protection Agency, https://archive.epa.gov/clearskies/web/html/ (accessed August 27, 2016).

24. J. M. Robine et al., "Death Toll Exceeded 70,000 in Europe during Summer of 2003," *Comptes Rendus Biologies* 331, no. 2 (February 2008): 171–78, http://www.ncbi.nlm.nih.gov/pubmed/18241810 (accessed March 31, 2016).

25. Megan Gannon, "Why Heat Waves Can Mean High Death Tolls," *LiveScience*, August 1, 2012, http://www.livescience.com/22050-heat-waves-high-death-tolls.html (accessed July 2016).

26. Ed King, "Kyoto Protocol: 10 Years of the World's First Climate Change Treaty," Climate *Home*, February 16, 2015, http://www.climatechangenews.com/2015/02/16/kyoto-protocol-10-years-of-the-worlds-first-climate-change-treaty/ (accessed July 2016).

27. Alain Bernard, Sergey Paltsev, John M. Reilly, Marc Vielle, and Laurent Viguier, "Russia's Role in the Kyoto Protocol," MIT Joint Program on the Science and Policy of Global Change, report 98, June 2003, http://globalchange.mit.edu/files/document/MITJPSPGC_Rpt98.pdf (accessed August 29, 2016).

28. "Hurricane Katrina," *History*, A&E Networks, 2016, http://www.history.com/topics/hurricane-katrina (accessed July 2016).

29. Ibid.

30. Ibid.

31. Ibid.

32. Ibid.

33. Justin Worland, "Why Climate Change Could Make Hurricane Impact Worse," *Time*, August 27, 2015, http://time.com/4013637/climate-change-hurricanes-impact/ (accessed July 2016).

34. Joseph Romm, *Climate Change: What Everyone Needs to Know* (New York: Oxford University Press, 2016), pp. 55–57.

35. Angie Drobnic Holan, "Halliburton, KBR, and Iraq War Contracting: A History

So Far," *PolitiFact*, June 9, 2010, http://www.politifact.com/truth-o-meter/statements/2010/jun/09/arianna-huffington/halliburton-kbr-and-iraq-war-contracting-history-s/ (accessed August 26, 2016).

36. Wenonah Hauter, "10 Years Later: Fracking and the Halliburton Loophole," *EcoWatch*, August 11, 2015, http://ecowatch.com/2015/08/11/halliburton-loophole-fracking/ (accessed October 11, 2015).

37. Ibid.

38. Ellen Cantarow and Dory Hippauf, "The Federal Agency Behind the Gross Expansion of Fracking Pipelines," *Truthout*, October 17, 2015, http://www.truth-out.org/news/item/33239-the-federal-agency-behind-the-gross-expansion-of-fracking-pipelines (accessed August 29, 2016).

39. Coral Davenport, "New Federal Rules Are Set for Fracking," *New York Times*, March 20, 2015, http://www.nytimes.com/2015/03/21/us/politics/obama-administration-unveils-federal-fracking-regulations.html (accessed March 31, 2016).

40. Hauter, "10 Years Later."

41. Ibid.

42. Ibid.

CHAPTER 20: CLIMATE CHANGE BEGINS A COMEBACK

1. Stephen Lacey, "Interview: 'Chasing Ice' Star James Balog Talks Art, Science, Rationality, and Climate Denial," *ThinkProgress*, December 11, 2012, https://thinkprogress.org/interview-chasing-ice-star-james-balog-talks-art-science-rationality-and-climate-denial-2f3c70b122f6 (accessed August 30, 2016).

2. Ibid.

3. Ibid.

4. Tim Appenzeller and James Balog, "The Big Thaw," *National Geographic* 211, no. 6 (June 2007).

5. *Chasing Ice*, directed by Jeff Orlowski (Los Angeles: Exposure Labs, 2012), DVD, https://chasingice.com (accessed November 16 2015). *Chasing Ice* trailer found at: https://www.youtube.com/watch?v=eIZTMVNBjc4 (accessed July 2016).

6. "About C40," C40 Cities, 2016, http://www.c40.org/about (accessed February 27, 2016).

7. "C40 Fact Sheet," C40 Cities Climate Leadership Group, http://c40-production-images.s3.amazonaws.com/fact_sheets/images/1_About_C40_April_2016.original.pdf?1459877902 (accessed August 29, 2016).

8. "About C40," C40 Cities.

9. Ibid.

10. Fred Pearce, *The Climate Files: The Battle for the Truth about Global Warming*

(London: Penguin Random House, 2010).

11. David Appell, "Behind the Hockey Stick," *Scientific American*, March 2005, http://www.scientificamerican.com/article/behind-the-hockey-stick/ (accessed January 2, 2006).

12. "Temperature Variations in Past Centuries and the So-Called 'Hockey Stick,'" *RealClimate*, December 4, 2004, http://www.realclimate.org/index.php?p=7 (accessed August 30, 2016).

13. David Appell, "Behind the Hockey Stick."

14. Ibid.

15. *An Inconvenient Truth*, directed by David Guggenheim and Al Gore (Los Angeles: Lawrence Bender Productions, 2006), DVD.

16. Andrew Revkin, "'An Inconvenient Truth': Al Gore's Fight against Global Warming," *New York Times*, May 22, 2006, http://www.nytimes.com/2006/05/22/movies/22gore.html?_r=0 (accessed July 2016).

17. Ibid.

18. Ibid.

19. "Documentary: 1982–Present," Box Office Mojo, http://www.boxofficemojo.com/genres/chart/?id=documentary.htm (accessed March 31, 2016).

20. Shane Stacks, "Keynote Conveniently Powers Al Gore's 'An Inconvenient Truth,'" *Ars Technica*, June 20, 2006, http://arstechnica.com/apple/2006/06/4392/ (accessed January 2, 2016).

21. "SOTC: Ice Sheets," *State of the Cryosphere*, National Snow and Ice Data Center, November 9, 2015, http://nsidc.org/cryosphere/sotc/ice_sheets.html (accessed November 20, 2015).

22. Ibid.

23. Ibid.

24. Ibid.

25. Ibid.

26. Ibid.

27. Ibid.

28. Spencer Weart, "The Discovery of Global Warming: Ice Sheets and Rising Seas," *American Institute of Physics*, February 2015, https://www.aip.org/history/climate/floods.htm (accessed October 20, 2015).

29. Ibid.

30. "Our Story," Step It Up 2007, http://www.stepitup2007.org/article.php-list=type&type=48.html (accessed January 2, 2016).

31. "80% by 2050: From 'Radical' to Mainstream," 350.org, http://350.org/80-by-2050-from-radical-to-mainstream/ (accessed August 29, 2016).

32. "Our Story," Step It Up 2007.

33. Bill McKibben, *Oil and Honey: The Education of an Unlikely Activist* (New York: St. Martin's, 2013), p. 10.

34. Ibid.

35. Ibid.

36. Ibid., p. 11.

37. Ibid.

38. Ibid., p. 12.

39. Ibid., p. 12.

40. Van Jones, *The Green Collar Economy: How One Solution Can Fix Our Two Biggest Problems* (New York, NY: Harper Collins, 2008), p. 1.

41. Ibid.

42. Ibid., p. 17.

43. Ibid., p. 14.

44. Ibid., p. 15.

45. Thomas Friedman, "The Green-Collar Solution," *New York Times*, October 17, 2007, http://www.nytimes.com/2007/10/17/opinion/17friedman.html (accessed March 30, 2016).

46. Ibid.

47. "Green for All: Campaign for Green-Collar Jobs," Clinton Global Initiative, https://www.clintonfoundation.org/clinton-global-initiative/commitments/green-all-campaign-green-collar-jobs (accessed August 30, 2016).

48. Friedman, "The Green-Collar Solution."

49. Ibid.

50. Weart, "Discovery of Global Warming."

51. Ibid.

52. Stefanie Spear, "Tim DeChristopher's Peaceful Uprising," *EcoWatch*, March 21, 2013, http://ecowatch.com/2013/04/21/tim-dechristophers-peaceful-uprising/ (accessed February 2016).

53. Ibid.

54. Patty Henetz, "Judge Blocks Disputed BLM Utah Oil, Gas Leases," *Salt Lake Tribune*, January 20, 2009, http://archive.sltrib.com/story.php?ref=/ci_11506451 (accessed August 30, 2016).

55. Spear, "Tim DeChristopher's Peaceful Uprising."

56. Ibid.

57. Ibid.

CHAPTER 21: ONSET OF A SOCIAL REVOLUTION OR JUST BUSINESS AS USUAL?

1. Andrew Lawler, "Face Off," *Audubon*, September 1, 2008, at http://www.andrewlawler.com/face-off/ (accessed September 5, 2016).

2. "The Obama-Biden Plan," Change.gov: The Office of the President-Elect, http://change.gov/agenda/energy_and_environment_agenda/ (accessed August 9, 2016).

3. "Remarks by the President in State of the Union Address," White House, Office of the Press Secretary, January 27, 2010, https://www.whitehouse.gov/the-press-office/remarks-president-state-union-address (accessed September 5, 2016).

4. Ben Jervey, "The Top 10 Environmental Plays of Obama's Rookie Season," *Daily Good*, January 23, 2010, https://www.good.is/articles/the-top-10-environmental-plays-of-obama-s-rookie-season (accessed July 2016).

5. Glenn Kessler, "When Did McConnell Say He Wanted to Make Obama a 'One-Term President,'" *Washington Post*, September 25, 2012, https://www.washingtonpost.com/blogs/fact-checker/post/when-did-mcconnell-say-he-wanted-to-make-obama-a-one-term-president/2012/09/24/79fd5cd8-0696-11e2-afff-d6c7f20a83bf_blog.html (accessed August 31, 2016).

6. "Global Warming and Polar Bears," National Wildlife Federation, http://www.nwf.org/Wildlife/Threats-to-Wildlife/Global-Warming/Effects-on-Wildlife-and-Habitat/Polar-Bears.aspx (accessed July 2016).

7. Ibid.

8. "Wilkins Ice Shelf News," National Snow and Ice Data Center, April 8, 2009, https://nsidc.org/news/newsroom/wilkins/ (accessed July 2016).

9. Ibid.

10. Ibid.

11. "Cyclone Nargis Embodied the 'Perfect Storm,'" NBC News, May 8, 2008, http://www.nbcnews.com/id/24526960/ns/world_news-asia-pacific/t/cyclone-nargis-embodied-perfect-storm/#.VohZHDGxU (accessed January 2, 2016).

12. Ibid.

13. "About," Grantham Research Institute on Climate Change and the Environment, http://www.lse.ac.uk/GranthamInstitute/about/ (accessed July 2016).

14. Ibid.

15. Eric Roston, "Green Websites: Climate Progress," *Time*, April 17, 2008, http://content.time.com/time/specials/2007/article/0,28804,1730759_1731034_1731042,00.html (accessed March 30, 2016).

16. Thomas Friedman, "The Inflection Is Near?" *New York Times*, March 7, 2009, http://www.nytimes.com/2009/03/08/opinion/08friedman.html (accessed January 2, 2016).

17. Steven James Snyder, "Best Blogs: Climate Progress," *Time*, June 28, 2010, http://content.time.com/time/specials/packages/article/0,28804,1999770_1999761_1999752,00.html(accessed January 2, 2016).

18. Rhiannon Edwards and Adam Vaughan, "Top 50 Twitter Climate Accounts to Follow," *Guardian*, May 11, 2010, http://www.theguardian.com/environment/blog/2010/may/11/top-50-twitter-climate-accounts (accessed March 30, 2016).

19. Jess Henig, "'Climategate,'" *FactCheck*, December 10, 2009, http://www.factcheck.org/2009/12/climategate/ (accessed July 2016).

20. Peter Kelemen, "What East Anglia's E-mails Really Tell Us about Climate Change," *Popular Mechanics*, December 17, 2009, http://www.popularmechanics.com/

science/environment/a5154/4338343/ (accessed July 2016).

21. Ibid.

22. Brian Winter, "Scientist: Leaked Climate E-mails a Distraction," ABC News, November 28, 2009, http://abcnews.go.com/Politics/scientist-leaked-climate-mails-distraction/story?id=9178656 (accessed March 27, 2016).

23. Leo Hickman and James Randerson, "Climate Sceptics Claim Leaked E-mails Are Evidence of Collusion Among Scientists," *Guardian*, November 20, 2009, http://www.theguardian.com/environment/2009/nov/20/climate-sceptics-hackers-leaked-emails (accessed January 2, 2016).

24. "Debunking Misinformation about Stolen Climate Emails in the 'Climategate' Manufactured Controversy," Union of Concerned Scientists, http://www.ucsusa.org/global_warming/solutions/fight-misinformation/debunking-misinformation-stolen-emails-climategate.html (accessed July 2016).

25. Jess Henig, "FactCheck: Climategate Doesn't Refute Global Warming," *Newsweek*, December 10, 2009, http://www.newsweek.com/factcheck-climategate-doesnt-refute-global-warming-75749 (accessed January 2, 2016).

26. Benjamin Dolson, "A Climate for Change: An Interview—Katharine Hayhoe," *Plough*, April 9, 2015, http://www.plough.com/en/topics/faith/discipleship/a-climate-for-change (accessed September 1, 2016).

27. Katharine Hayhoe and Andrew Farley, *A Climate for Change: Global Warming Facts for Faith-based Decisions* (New York: Faith Words, 2009).

28. Michael D. Lemonick, "For Katharine Hayhoe, Climate Change Not a Leap of Faith," *Climate Central*, March 29, 2012, http://www.climatecentral.org/news/for-katharine-hayhoe-climate-change-not-a-leap-of-faith (accessed September 1, 2016).

29. "A Climate for Change," http://www.climateforchangethebook.com/ (accessed August 31, 2016).

30. Ibid.

31. "Climate Change in the American Christian Mind," Yale Program on Climate Change Communication, April 1, 2015, http://climatecommunication.yale.edu/publications/climate-change-in-the-american-christian-mind/ (accessed September 1, 2016).

32. "Religious Groups' Views on Global Warming," Pew Research Center: Religion and Public Life, April 16, 2009, http://www.pewforum.org/2009/04/16/religious-groups-views-on-global-warming/ (accessed September 1, 2016).

33. "Copenhagen Climate Change Conference—2009," United Nations Framework Convention on Climate Change, December 2009, http://unfccc.int/meetings/copengagen_dec_2009/meeting/6295.php (accessed October 22, 2015).

34. Ibid.

35. "Options and Considerations for a Federal Carbon Tax," Center for Climate and Energy Solutions (C2ES), February 2013, http://www.c2es.org/publications/options-considerations-federal-carbon-tax (accessed March 22, 2016).

36. Philip Sherwell, "Barack Obama Denies Accusations that He 'Crashed' Secret

Chinese Climate Change Talks," *Telegraph*, December 19, 2009, http://www.telegraph.co.uk/news/earth/copenhagen-climate-change-confe/6845952/Barack-Obama-denies-accusations-that-he-crashed-secret-Chinese-climate-change-talks.html (accessed September 1, 2016).

37. Liz Posner, "Hillary Clinton and Obama 'Crashing' a Secret Chinese Meeting Seriously Happened and She Described It in Detail in 'Hard Choices,'" *Bustle*, October 14, 2015, http://www.bustle.com/articles/117126-hillary-clinton-obama-crashing-a-secret-chinese-meeting-seriously-happened-she-described-it-in (accessed September 1, 2016).

38. Sherwell, "Barack Obama Denies Accusations."

39. Ibid.

40. Mark Lynas, "How Do I Know China Wrecked The Copenhagen Deal? I Was In the Room," *Guardian*, December 22, 2009, https://www.theguardian.com/environment/2009/dec/22/copenhagen-climate-change-mark-lynas (accessed July 2016).

41. Kline, *First Along the River*, p. 182.

42. Tom Eley, "What Caused the Explosion on the Deepwater Horizon?" *World Socialist Web Site*, May 14, 2010, https://www.wsws.org/en/articles/2010/05/spil-m14.html (accessed July 2016).

43. Ibid.

44. Harry Weber, "Blown-Out BP Well Finally Killed at Bottom of Gulf," *Boston Globe*, September 19, 2010, http://archive.boston.com/news/nation/articles/2010/09/19/blown_out_bp_well_finally_killed_at_bottom_of_gulf/ (accessed October 22, 2015).

45. Ibid.

46. Ibid.

47. "A Deadly Toll: The Gulf Oil Spill and the Unfolding Wildlife Disaster," Center for Biological Diversity, April 2011, http://www.biologicaldiversity.org/programs/public_lands/energy/dirty_energy_development/oil_and_gas/gulf_oil_spill/a_deadly_toll.html (accessed September 1, 2016).

48. Ibid.

49. Ibid.

50. Matt Smith, "Empty Nets in Louisiana Three Years After the Spill," CNN, April 29, 2013, http://www.cnn.com/2013/04/27/us/gulf-disaster-fishing-industry/ (accessed July 2016).

51. Robert Mclean and Irene Chapple, "BP Settles Final Gulf Oil Spill Claims for $20 Billion," October 6, 2015, http://money.cnn.com/2015/10/06/news/companies/deepwater-horizon-bp-settlement/ (accessed August 31, 2016).

52. Smith, "Empty Nets."

53. Eley, "What Caused the Explosion."

54. Ibid.

55. *Deep Water: The Gulf Oil Disaster and the Future of Offshore Drilling*, National Commission on the BP Deepwater Horizon Oil Spill and Offshore Drilling: Report to the President, January 2011, http://www.hreonline.com/pdfs/03012011DeepwaterReport.pdf

(accessed August 31, 2016), pp. 131–38.

56. Ibid., p. 133.

57. Jaclyn Lopez, "BP's Well Evaded Environmental Review: Categorical Exclusion Policy Remains Unchanged," Biological Diversity, 2010, http://www.biologicaldiversity.org/publications/papers/Lopez-2010.pdf (accessed August 31, 2016), p. 94.

58. Ibid.

59. Ibid., pp. 93–94.

60. Ibid.

61. Eley, "What Caused the Explosion."

62. Ibid.

63. Naomi Oreskes and Eric Conway, *Merchants of Doubt: How a Handful of Scientists Obscured the Truth on Issues from Tobacco Smoke to Global Warming* (New York: Bloomsbury, 2010).

64. *Gasland*, directed by Josh Fox (New York: HBO Documentary Films, 2010), DVD.

65. "*Gasland*: Synopsis," HBO Documentaries, 2010, http://www.hbo.com/documentaries/gasland/synopsis.html.

66. Ibid.

67. Ibid.

68. Abubakar Sadiq Aliyu et al., "An Overview of Current Knowledge Concerning the Health and Environmental Consequences of the Fukushima Daiichi Nuclear Power Plant (FDNPP) Accident," *Environmental International* 85 (December 2015): 213–28, http://www.sciencedirect.com/science/article/pii/S016041201530060X (accessed October 23, 2015).

69. "Damage Situation and Police Countermeasures Associated with 2011 Tohoku District – Off the Pacific Ocean Earthquake, June 10, 2016," National Police Agency of Japan: Emergency Disaster Countermeasures Headquarters, http://www.npa.go.jp/archive/keibi/biki/higaijokyo_e.pdf (accessed September 4, 2016).

70. James Hansen, "Game Over for the Climate," *New York Times*, May 9, 2012, http://www.nytimes.com/2012/05/10/opinion/game-over-for-the-climate.html?_r=0 (accessed July 2016).

71. Brad Plumer, "9 Questions about the Keystone XL Pipeline You Were Too Embarrassed to Ask," *Vox*, September 22, 2015, http://www.vox.com/2014/11/14/7216751/keystone-pipeline-facts-controversy (accessed October 2015).

72. Ibid.

73. "Facts and Statistics," Alberta Energy, http://www.energy.alberta.ca/oilsands/791.asp (accessed July 2016).

74. John H. Richardson, "Keystone," *Esquire*, August 10, 2012, http://www.esquire.com/news-politics/a15277/keystone-0912/ (accessed July 2016).

75. "Bitumen," *Oxford Dictionaries*, http://www.oxforddictionaries.com/us/definition/american_english/bitumen (accessed September 4, 2016).

76. "About Tar Sands," 2012 Oil Shale & Tar Sands Programmatic EIS, http://ostseis.anl.gov/guide/tarsands/ (accessed March 7, 2016).

77. Plumer, "9 Questions about the Keystone XL Pipeline."

78. Ibid.

79. Kurtis Lee, "Keystone XL Oil Pipeline: What You Need to Know about the Dispute," *Los Angeles Times*, March 6, 2015, http://www.latimes.com/nation/politics/la-na-keystone-qanda-20150307-story.html (accessed July 2016).

80. Plumer, "9 Questions about the Keystone XL Pipeline."

81. Ibid.

82. David Sassoon, "EPA Slams State Department on Proposed Oil Pipeline," *Inside Climate News*, July 27, 2010, https://insideclimatenews.org/news/20100727/epa-slams-state-department-proposed-oil-pipeline. See also, *Final Supplemental Environmental Impact Statement for the Keystone XL Project: Executive Summary*, United States Department of State: Bureau of Oceans and International Environmental and Scientific Affairs, January 2014, https://keystonepipeline-xl.state.gov/documents/organization/221135.pdf.

83. Plumer, "9 Questions about the Keystone XL Pipeline."

84. Elise Labott and Dan Berman, "Obama Rejects Keystone XL Pipeline," CNN, November 6, 2015, http://www.cnn.com/2015/11/06/politics/keystone-xl-pipeline-decision-rejection-kerry/ (accessed November 20, 2015).

85. Ibid.

86. The Solutions Project, http://thesolutionsproject.org/ (accessed July 2016).

87. Ibid.

88. Ibid.

89. Ibid.

90. "Forum: Is Extreme Weather Linked to Global Warming?" *Yale Environment 360*, June 2, 2011, http://e360.yale.edu/feature/forum_is_extreme_weather_linked_to_global_warming/2411/ (accessed October 12, 2015).

91. Spencer Weart, "Impacts of Climate Change," *The Discovery of Global Warming*, February 2016, https://www.aip.org/history/climate/impacts.htm#L_M093 (accessed September 4, 2016).

92. Ibid.

93. "Forum: Is Extreme Weather Linked to Global Warming?"

94. Ibid.

95. Justin Gillis, "Not Even Close: 2012 Was Hottest Ever in US," *New York Times*, January 8, 2013, http://www.nytimes.com/2013/01/09/science/earth/2012-was-hottest-year-ever-in-us.html (accessed July 2016).

96. Alex Sosnowski, "Corn Belt Disaster in Wake of Record Heat Waves," AccuWeather.com, July 10, 2012, http://www.accuweather.com/en/weather-news/record-heat-wave-resulting-in/67651 (accessed July 2016).

97. Gillis, "Not Even Close."

98. Ibid.

99. Ibid.

100. "State of the Climate: Global Analysis—Annual 2015," National Oceanic and Atmospheric Administration: National Centers for Environmental Information, January

2016, https://www.ncdc.noaa.gov/sotc/global/201513 (accessed March 11, 2016).

101. Rob Preece, "The Big Thaw: NASA Scientists Stunned as Satellite Pictures Show 'Unprecedented' Melting of Greenland's Ice Sheet," *Daily Mail*, July 25, 2012, http://www.dailymail.co.uk/sciencetech/article-2178540/Greenland-ice-melting-NASA-stunned-satellite-pictures-unprecedented-melting-ice-sheet.html (accessed July 2016).

102. Ibid.

103. Ibid.

104. Ibid.

105. Chloe Maxmin, in discussion with the authors, May 3, 2015.

106. Ibid.

107. Benjamin Franta, "Three Years Later, Harvard Still Must Divest," *Harvard Crimson*, September 10, 2015, http://www.thecrimson.com/article/2015/9/10/three-years-later-divestment/ (accessed September 4, 2016)..

108. "Divestment Commitments," Fossil Free, http://gofossilfree.org/commitments/ (accessed August 25, 2016).

109. Franta, "Three Years Later."

110. Robert Inglis, in discussion with the authors, August 26, 2015.

111. "H.R. 2380 (111th): Raise Wages, Cut Carbon Act of 2009," GovTrack, https://www.govtrack.us/congress/bills/111/hr2380/text/ih (accessed September 4, 2016).

112. Eric Althoff, "Bob Inglis Breaks from the Republican Party, Advocates Action to Fight Climate Change," *Washington Times*, March 24, 2015, http://www.washingtontimes.com/news/2015/mar/24/bob-inglis-advocates-action-to-fight-climate-chang/.

113. Ibid.

114. "Who We Are," RepublicEn, http://republicen.org/about/index (accessed February 2016).

115. Ibid.

116. Ibid.

117. Joseph Romm, "Superstorm Sandy's Link to Climate Change: 'The Case Has Strengthened' Says Researcher," *ThinkProgress*, October 28, 2013, http://thinkprogress.org/climate/2013/10/28/2843871/superstorm-sandy-climate-change/ (accessed March 16, 2016).

118. Ibid.

119. Lynn Eden et al., "It Is Still Three Minutes to Midnight," *Bulletin of the Atomic Sciences*, January 22, 2016, http://thebulletin.org/it-still-three-minutes-midnight9107 (accessed July 2016).

120. Ibid.

121. Suzanne Goldenberg, "Doomsday Clock Stuck Near Midnight Due to Climate Change and Nuclear War," *Guardian*, January 26, 2016, https://www.theguardian.com/environment/2016/jan/26/doomsday-clock-three-minutes-to-midnight-climate-change-nuclear-war (accessed July 2016).

122. "Doomsday Clock: It is 3 Minute to Midnight," *Bulletin of the Atomic Scientists*, 2015, http://thebulletin.org/clock/2015 (accessed September 4, 2016).

123. Jason Mark, "The Most Important Environmental Stories of 2014," *Earth Island Journal*, December 22, 2014, http://www.earthisland.org/journal/index.php/elist/eListRead/the_most_important_environmental_stories_of_2014/ (accessed October 12, 2015).

124. Ibid.

125. Ibid.

126. Ibid.

127. Jake Thompson, "'Forward on Climate' Rally: More than 35,000 Strong March on Washington for Climate Action," Natural Resources Defense Council, February 17, 2013, http://www.nrdc.org/media/2013/130217.asp (accessed October 12, 2015).

128. Ibid.

129. Ibid.

130. Ibid.

131. Ibid.

132. Ibid.

133. Elizabeth Kolbert, *The Sixth Extinction: An Unnatural History* (New York: Henry Holt, 2014).

134. Elizabeth Kolbert, interview by Terry Gross, "In the World's 'Sixth Extinction,' Are Humans the Asteroid?" *Fresh Air*, WBUR 90.9, NPR, February 12, 2014, http://www.npr.org/2015/01/23/379117018/in-the-worlds-sixth-extinction-are-humans-the-asteroid.

135. Ibid.

136. Mark, "Most Important Environmental Stories."

137. Ker Than, "Causes of California Drought Linked to Climate Change, Stanford Scientists Say," *Stanford Report*, September 30, 2014, http://news.stanford.edu/news/2014/september/drought-climate-change-092914.html (accessed July 2016).

138. Ibid.

139. Mark, "Most Important Environmental Stories."

140. Tom Moore, "California Snowpack and Drought Show Improvement Over Last Year," Weather Channel, May 6, 2016, https://weather.com/news/climate/news/snowpack-sierra-drought-california (accessed September 4, 2016).

141. Mark, "Most Important Environmental Stories."

142. "Current World Population," Worldometers, http://www.worldometers.info/world-population/ (accessed September 4, 2016).

143. Mark, "Most Important Environmental Stories."

144. Patrick J. Kiger, "Carl Sagan and the Cosmos Legacy," National Geographic Channel, February 20, 2014, http://channel.nationalgeographic.com/cosmos-a-spacetime-odyssey/articles/carl-sagan-and-the-cosmos-legacy/ (accessed July 2016).

145. Mark, "Most Important Environmental Stories."

146. Ibid.

147. Ibid.

148. "People's Climate March: Wrap-Up," http://2014.peoplesclimate.org/wrap-up/ (accessed September 4, 2016).

149. Mark, "Most Important Environmental Stories."

150. "Summary of Climate Summit 2014," *Climate Summit Bulletin* 172, no. 18 (September 26, 2014), http://www.iisd.ca/climate/cs/2014/html/crsvol172num18e.html (accessed July 2016).

151. Ibid.

152. Ibid.

153. "Explaining Green Bonds," Climate Bonds Initiative, 2016, https://www.climatebonds.net/market/explaining-green-bonds (accessed September 4, 2016).

154. "Marshall Islands Poet: 'We Deserve To Do More Than Just Survive," *Climate Home*, September 23, 2014, http://www.climatechangenews.com/2014/09/23/marshall-islands-poet-we-deserve-to-do-more-than-just-survive/ (accessed September 4, 2016).

155. "Summary of Climate Summit 2014."

156. "Leonardo DiCaprio at the UN: 'Climate Change Is Not Hysteria—It's a Fact,'" *Guardian*, September 23, 2014, https://www.theguardian.com/environment/2014/sep/23/leonarodo-dicaprio-un-climate-change-speech-new-york (accessed September 4, 2016).

157. Naomi Klein, *This Changes Everything: Capitalism vs. the Climate* (New York: Simon and Schuster, 2014).

158. Ibid., pp. 66–79.

159. Ibid.

160. Ibid.

161. Ibid., pp. 81–82.

162. Ibid., pp. 92–95.

163. Ibid., p. 93.

164. Ibid.

165. Ibid.

166. "This Changes Everything: Capitalism and the Climate," Simon and Schuster, August 2015, http://www.simonandschuster.com/books/This-Changes-Everything/Naomi-Klein/9781451697391 (accessed September 4, 2016).

167. Mark, "Most Important Environmental Stories."

168. Coral Davenport, "Meager Returns for the Democrats' Biggest Donor," *New York Times*, November 6, 2014, http://www.nytimes.com/2014/11/07/us/politics/-meager-returns-for-the-democrats-biggest-donor-tom-steyer.html (accessed September 5, 2016).

169. Mark, "Most Important Environmental Stories."

170. Ibid.

171. Lydia O'Connor, "2015 Has Been a Year of Record-Breaking U.S. Weather Events," *Huffington Post*, October 5, 2015, http://www.huffingtonpost.com/entry/record-breaking-weather-events_us_5612d0e8e4b0dd85030ce995 (accessed November 20, 2015).

172. Ibid.

173. "Running Dry," *San Francisco Chronicle*, 2015, http://www.sfchronicle.com/drought/ (accessed July 2016).

174. Dana Farrington, "California Governor Makes Some Water Restrictions

Permanent," *The Two-Way*, NPR, May 9, 2016, http://www.npr.org/sections/thetwo-way/2016/05/09/477392158/california-governor-makes-some-water-restrictions-permanent (accessed July 2016).

175. "National Overview—May 2015," National Oceanic and Atmospheric Administration: National Centers for Environmental Information, June 2015, https://www.ncdc.noaa.gov/sotc/national/201505 (accessed July 2016).

176. O'Connor, "2015 Has Been a Year of Record-Breaking U.S. Weather Events."

177. Ibid.

178. Ibid.

179. O'Connor, "2015 Has Been a Year of Record-Breaking U.S. Weather Events."

180. Ibid.

181. Kate Connolly, "G7 Leaders Agree to Phase Out Fossil Fuel Use by End of Century," *Guardian*, June 8, 2015, http://www.theguardian.com/world/2015/jun/08/g7-leaders-agree-phase-out-fossil-fuel-use-end-of-century (accessed November 19, 2015).

182. Ibid.

183. Samantha Page, "G7 Leaders Agree on Action to Limit Global Warming to 2 Degrees," *ThinkProgress*, June 8, 2015, https://thinkprogress.org/g7-leaders-agree-on-action-to-limit-global-warming-to-2-degrees-78f7668b8e68#.s7tvrk85e (accessed November 19, 2015).

184. Ibid.

185. Jim Yardley and Laurie Goodstein, "Pope Francis, in Sweeping Encyclical, Calls for Swift Action on Climate Change," *New York Times*, June 18, 2015, http://www.nytimes.com/2015/06/19/world/europe/pope-francis-in-sweeping-encyclical-calls-for-swift-action-on-climate-change.html?_r=0 (accessed November 19, 2015).

186. Max Chafkin, "Elon Musk Powers Up: Inside Tesla's $5 Billion Gigafactory," *Fast Company*, November 17, 2015, http://www.fastcompany.com/3052889/elon-musk-powers-up-inside-teslas-5-billion-gigafactory (accessed July 2016).

187. "Model S," Tesla, https://www.tesla.com/models (accessed September 5, 2016).

188. "Climate Change: Regulatory Initiatives," US Environmental Protection Agency, https://www3.epa.gov/climatechange/EPAactivities/regulatory-initiatives.html (accessed November 19, 2015).

189. David A. Graham, "The Politics of Obama's Greenhouse-Gas Rule," *Atlantic*, August 3, 2015, http://www.theatlantic.com/politics/archive/2015/08/obama-greenhouse-gas-rule/400382/ (accessed November 19, 2015).

190. Ibid.

191. Ibid.

192. "Fact Sheet: Clean Power Plan Benefits of a Clean, More Efficient Power Sector," US Environmental Protection Agency, https://www.epa.gov/cleanpowerplan/fact-sheet-clean-power-plan-benefits-cleaner-more-efficient-power-sector (accessed September 5, 2016).

193. Ibid.

194. Graham, "Politics of Obama's Greenhouse-Gas Rule."

195. "Fact Sheet: Clean Power Plan Benefits."

196. Graham, "Politics of Obama's Greenhouse-Gas Rule."

197. Ibid.

198. Barack Obama, "Remarks by the President at U.N. Climate Change Summit," The White House: Speeches and Remarks, September 23, 2014, https://www.whitehouse.gov/the-press-office/2014/09/23/remarks-president-un-climate-change-summit (accessed July 2016).

199. Graham, "Politics of Obama's Greenhouse Gas Rule."

200. Ibid.

201. Cole Mellino, "24 States Sue Obama Over Clean Power Plan," *EcoWatch*, October 24, 2015, http://ecowatch.com/2015/10/24/clean-power-plan-lawsuits/ (accessed November 19, 2015).

202. Ibid.

203. "Clean Power Plan for Existing Power Plants," US Environmental Protection Agency, February 11, 2016, https://www.epa.gov/cleanpowerplan/clean-power-plan-existing-power-plants (accessed July 2016).

204. "Paris Climate Change Conference—November 2015," United Nations Framework Conventions on Climate Change, http://unfccc.int/meetings/paris_nov_2015/meeting/8926.php (accessed September 5, 2016).

205. "Outcomes of the U.N. Climate Change Conference in Paris," Center for Climate and Energy Solutions, http://www.c2es.org/international/negotiations/cop21-paris/summary (accessed September 5, 2016).

206. Ibid.

207. Joby Warrick, "Bill Gates on Climate Change: 'We Need to Move Faster than the Energy Sector Ever Has,'" *Washington Post*, November 30, 2015, https://www.washingtonpost.com/news/energy-environment/wp/2015/11/30/bill-gates-on-climate-change-we-need-to-move-faster/?utm_term=.5eb305702006.

208. Sheldon Whitehouse, "Time to Wake Up: Moving Toward a Clean-Energy Economy," January 27, 2016, Sheldon Whitehouse: United States Senator for Rhode Island, https://www.whitehouse.senate.gov/news/speeches/time-to-wake-up-moving-toward-a-clean-energy-economy (accessed September 5, 2016).

209. Katie Valentine, "Senator Sheldon Whitehouse's Best Climate Change Burns," *ThinkProgress*, May 18, 2015, https://thinkprogress.org/sen-sheldon-whitehouses-best-climate-change-burns-d545846e1a8e#.jx540gapv (accessed September 5, 2016).

210. Molly Bier, "Watch Leonardo DiCaprio's 2016 Oscar Acceptance Speech for Best Actor," The Oscars, February 29, 2016, http://oscar.go.com/news/winners/watch-leonardo-dicaprios-acceptance-speech-for-best-actor-2016 (accessed March 19, 2016).

CHAPTER 23: SCIENTISTS / RESEARCHERS

1. David L. Chandler, "James Hansen: We Should Look at all Energy Options," *MIT News*, April 16, 2015, http://newsoffice.mit.edu/2015/james-hansen-climate-change-rose-lecture-0416 (accessed June 7, 2015).

2. James Hansen, "James Hansen: Why I Must Speak Out About Climate Change," *TED Talks*, February 2012, https://www.ted.com/talks/james_hansen_why_i_must_speak_out_about_climate_change?language=en (accessed March 19, 2016).

3. Ibid.

4. Naomi Klein, *This Changes Everything: Capitalism vs. the Climate* (New York: Simon and Schuster, 2014), p. 73.

5. "James E. Hansen, cv," The Earth Institute, Columbia University, http://www.columbia.edu/~jeh1/hansencv_201308.pdf (accessed September 5, 2016).

6. Hansen, "James Hansen: Why I Must Speak Out About Climate Change."

7. Ibid.

8. Ibid.

9. Ibid.

10. Ibid.

11. Ibid.

12. Ibid.

13. Ibid.

14. James Hansen, Storms of My Grandchildren: The Truth About the Coming Climate Catastrophe and Our Last Chance to Save Humanity (New York: Bloomsbury, 2009), book jacket.

15. Chandler, "James Hansen: We Should Look."

16. James Hansen, "Tell Barack Obama the Truth—The Whole Truth," Draft Letter to Obama, November 21, 2008, http://www.columbia.edu/~jeh1/mailings/2008/20081121_Obama.pdf (accessed September 5, 2016).

17. Ibid.

18. Chandler, "James Hansen: We Should Look."

19. Ibid.

20. Jim DiPeso, "A Conservative Climate Plan," South Eastman Transition Initiative, http://www.southeasttransition.com/climate/a-conservative-climate-plan (accessed March 19, 2016).

21. Thomas Lovejoy, in discussion with the authors, August 28, 2015.

22. Ibid.

23. Ibid.

24. Ibid.

25. "Thomas Lovejoy: Tropical and Conservation Biologist," *National Geographic*, 2016, http://www.nationalgeographic.com/explorers/bios/thomas-lovejoy/ (accessed March 30, 2016).

26. Elizabeth Kolbert, *The Sixth Extinction: An Unnatural History* (New York: Henry Holt, 2014), pp. 175–76.

27. Lovejoy, in discussion with the authors.

28. Ibid.

29. Ibid.

30. Ibid.

31. Ibid.

32. Ibid.

33. Ibid.

34. Ibid.

35. Ibid.

36. "Michael E. Mann: Biographical Sketch," Penn State: Department of Meteorology, http://www.meteo.psu.edu/holocene/public_html/Mann/about/index.php (accessed March 28, 2016).

37. Ibid.

38. Michael E. Mann and Lee R. Kump, *Dire Predictions: Understanding Climate Change*, 2nd ed. (New York: DK Publishing, 2015).

39. Michael E. Mann, *The Hockey Stick and the Climate Wars: Dispatches from the Front Lines* (New York: Columbia University Press, 2013).

40. "Michael E. Mann: Biographical Sketch."

41. Joshua Holland, "Six Things Michael Mann Wants You to Know about the Science of Global Warming," *Moyers & Company*, June 12, 2014, http://billmoyers.com/2014/06/12/six-things-michael-mann-wants-you-to-know-about-the-science-of-global-warming/ (accessed March 19, 2016).

42. Ibid.

43. Ibid.

44. Michael Mann, "The Wall Street Journal, Climate Change Denial, and the Galileo Gambit," *Huffington Post*, March 28, 2016, http://www.huffingtonpost.com/michael-e-mann/wall-street-journal-climate-change-denial_b_9551482.html (accessed March 30, 2016).

45. Ibid.

46. "About," Biomimicry 3.8, http://biomimicry.net/about/ (accessed December 30, 2015).

47. Janine Benyus, in discussion with the authors, October 26, 2015.

48. Ibid.

49. AskNature, Biomimicry Institute, http://www.asknature.org/.

50. Ibid.

51. Ibid.

52. Ibid.

53. Ibid.

54. Ibid.

55. Ibid.

56. Ibid.

57. Ibid.

58. Ibid.

59. Ibid.

60. Ibid.

61. Ibid.

62. Benyus, in discussion with the authors.

63. "Overview," XPRIZE Foundation, NRG COSIA Carbon XPRIZE, September 2015, http://carbon.xprize.org/about/overview (accessed March 6, 2016).

64. Benyus, in discussion with the authors.

65. XiaoZhi Lim, "How to Make the Most of Carbon Dioxide: Researchers Hope to Show that Using the Gas as a Raw Material Could Make an Impact on Climate Change," *Nature*, October 28, 2015, http://www.nature.com/news/how-to-make-the-most-of-carbon-dioxide-1.18653 (accessed March 6, 2016).

66. Benyus, in discussion with the authors.

67. Mark Jacobson, in discussion with the authors, January 28, 2016.

68. Ibid.

69. Ibid.

70. Mark Z. Jacobson and Mark A Delucchi, "Sustainability: A Plan to Power 100 Percent of the Planet with Renewables," *Scientific American*, November 1, 2009, http://www.scientificamerican.com/article/a-path-to-sustainable-energy-by-2030/ (accessed February 8, 2016).

71. Jacobson, in discussion with the authors.

72. Sarah Shanley Hope, in discussion with the authors, February 22, 2016.

73. Jacobson, in discussion with the authors.

74. Ibid.

75. Ibid.

76. Katharine Hayhoe and Andrew Farley, *A Climate for Change: Global Warming Facts for Faith-Based Decisions* (New York: Faith Words, 2009).

77. Ibid.

78. Katharine Hayhoe, in discussion with the authors, December 7, 2015.

79. Ibid.

80. Ibid.

81. Ibid.

82. Ibid.

83. Ibid.

84. Hayhoe and Farley, *Climate for Change.*

85. Susan Solomon, "Climate Change and Getting Past Gridlock," (presentation, Harvard Club of the North Shore, Hawthorne Hotel, Salem, MA, February 9, 2016).

86. Ibid.

87. Mann and Kump, *Dire Predictions*.

88. Solomon, "Climate Change."

89. Ibid.

90. Ibid.

91. Michael Mann, "Climate Scientist Profile: Susan Solomon," *Dire Predictions: Online Profiles*, 2008, https://media.pearsoncmg.com/ph/esm/esm_mann_dire_2/qr/pntip.html?qr4 (accessed March 28, 2016).

92. Solomon, "Climate Change."

93. Ibid.

94. "Amory B. Lovins: Cofounder, Chief Scientist, and Chairman Emeritus," Rocky Mountain Institute, http://www.rmi.org/Amory+B.+Lovins (accessed February 2016).

95. Amory Lovins, *Reinventing Fire: Bold Business Solutions for the New Energy Era* (White River Junction, VT: Chelsea Green, 2013).

96. Amory Lovins et al., *Winning the Oil Endgame: Innovations for Profits, Jobs, and Security* (Snowmass, CO: Rocky Mountain Institute, 2004).

97. Amory Lovins, *Small Is Profitable: The Hidden Economic Benefits of Making Electrical Resources the Right Size* (Snowmass, CO: Rocky Mountain Institute, 2003).

98. Amory Lovins and L. Hunter Lovins, *Brittle Power: Energy Strategy for National Security* (Baltimore, MD: Brick House, 1982).

99. Paul Hawken and Amory Lovins, *Natural Capitalism: Creating the Next Industrial Revolution* (Boston: Little, Brown, 1999).

100. Patty Kay Mooney, "Amory Lovins and Bob Livingston Chat," Crystal Pyramid Productions, December 4, 2015, http://sandiegovideoproduction.com/amory-lovins-bob-livingston/ (accessed February 9, 2016).

101. "Mission," Carbon War Room, http://carbonwarroom.com/ (accessed September 5, 2016).

102. "Richard Branson Joins Forces with Amory Lovins in Climate Fight," NBC News, December 16, 2014, http://www.nbcnews.com/science/environment/richard-branson-joins-forces-amory-lovins-climate-fight-n269591 (accessed February 9, 2016).

103. Amory Lovins and Richard Branson, "Richard Branson and Amory Lovins Join Forces to Accelerate Clean Energy Revolution," *EcoWatch*, December 17, 2014, http://ecowatch.com/2014/12/17/branson-lovins-clean-energy-revolution/ (accessed February 9, 2016).

104. "Paul J. Crutzen Biography," *Encyclopedia of World Biography*, http://www.notablebiographies.com/supp/Supplement-Ca-Fi/Crutzen-Paul-J.html (accessed February 1, 2016).

105. Ibid.

106. Ibid.

107. Ibid.

108. Ibid.

109. Ibid.

110. Ibid.

111. Charles Q. Choi, "Small Nuclear War Could Reverse Global Warming for Years," *National Geographic News*, February 23, 2011, http://news.nationalgeographic.com/news/2011/02/110223-nuclear-war-winter-global-warming-environment-science-climate-change/ (accessed September 5, 2016).

112. Paul Crutzen and Eugene F. Stoermer, "Have We Entered the Anthropocene?" International Geosphere-Biosphere Programme: Global Change, 2002, http://www.igbp.net/news/opinion/opinion/haveweenteredtheanthropocene.5.d8b4c3c12bf3be638a8000578.html (accessed September 5, 2016).

113. Ibid.

114. Steve Connor, "Scientist Publishes 'Escape Route' from Global Warming," *Independent*, July 30, 2006, http://www.independent.co.uk/environment/scientist-publishes-escape-route-from-global-warming-409981.html (accessed June 4, 2015).

115. "Paul J. Crutzen Biography."

116. Connor, "Scientist Publishes 'Escape Route.'"

117. "Paul J. Crutzen Biography."

118. Paul Crutzen, email message to authors, February 2016.

119. Ibid.

120. "What Is the Global Ocean Conveyor Belt?" National Oceanic and Atmospheric Administration, http://oceanservice.noaa.gov/facts/conveyor.html (accessed September 5, 2016).

121. "Wallace S. Broecker," Earth Institute, Columbia University, http://www.earth.columbia.edu/articles/view/2246 (accessed August 12, 2015).

122. Wallace S. Broecker, "Will Our Ride Into the Greenhouse Gas Future Be a Smooth One?" *GSA Today* 7, no. 5 (May 1997): 1–7, http://faculty.washington.edu/wcalvin/teaching/Broecker97.html (accessed September 5, 2016).

123. "What's in a Name? Global Warming vs. Climate Change," NASA, December 5, 2008, http://www.nasa.gov/topics/earth/features/climate_by_any_other_name.html (accessed September 5, 2016).

124. "Wallace S. Broecker," Earth Institute.

125. Ibid.

126. Ibid.

127. Wallace S. Broecker and Robert Kunzig, *Fixing Climate: What Past Climate Changes Reveal about the Current Threat—and How to Counter It* (New York: Hill and Wang, 2008).

128. Ibid.

129. "Wallace S. Broecker," Earth Institute. Original interview available at: "Professor Wallace Broecker," interview by Stephen Sackur, *Hardtalk*, BBC News, June 5, 2008, http://news.bbc.co.uk/2/hi/programmes/hardtalk/7438039.stm (accessed September 5, 2016).

130. Wallace Broecker, "Does Air Capture Constitute a Viable Backstop against a Bad CO_2 Trip?" *Elementa: Science of the Anthropocene*, December 4, 2013, https://www.elementascience.org/articles/9 (accessed September 5, 2016).

CHAPTER 24: ADVOCATES / AUTHORS

1. Bill McKibben, in discussion with the authors, December 3, 2015.

2. Bill McKibben, *Oil and Honey: The Education of an Unlikely Activist* (New York: St. Martin's Press, 2014), p. 23.

3. McKibben, in discussion with the authors.

4. Ibid.

5. Ibid.

6. Ibid.

7. McKibben, *Oil and Honey*, p. 23.

8. Ibid.

9. Bill McKibben, email message to authors, July 21, 2015.

10. McKibben, in discussion with the authors.

11. Ibid.

12. Naomi Klein, *This Changes Everything: Capitalism vs. the Climate* (New York: Simon and Schuster, 2014), book jacket.

13. Ibid., pp. 1–28.

14. *This Changes Everything*, film, directed by Avi Lewis (2015; Klein Lewis Productions), DVD.

15. Ibid.

16. Ibid.

17. Ibid.

18. Ibid.

19. Klein, *This Changes Everything*, pp. 367–87.

20. Ibid.

21. Ibid., p. 379.

22. Ibid.

23. Naomi Klein, "Don't Shut Post Offices—Reinvent Them," *The Leap* (blog), *This Changes Everything*, February 29, 2016, http://theleap.thischangeseverything.org/dont-shut-post-offices-reinvent-them/ (accessed March 27, 2016).

24. "*This Changes Everything*: The Book," *This Changes Everything*, http://thischangeseverything.org/book/ (accessed March 27, 2016).

25. Elizabeth Kolbert, *Field Notes from a Catastrophe: Man, Nature, and Climate Change* (New York: Bloomsbury, 2006).

26. Doug MacDougall, "Jolting Messages on Climate Change," *Chronicle of Higher Education*, April 14, 2006, http://chronicle.com/article/Jolting-Messages-on-Climate/24634 (accessed March 26, 2016).

27. T. C. Boyle, *Drop City* (London: Penguin, 2004).

28. MacDougall, "Jolting Messages on Climate Change."

29. Jaimie Etkin, "Backgrounder: Elizabeth Kolbert," *Bullpen*, http://journalism.nyu.edu/publishing/archives/bullpen/elizabeth_kolbert/backgrounder/ (accessed September 5, 2016).

30. Ibid.

31. "Contributors: Elizabeth Kolbert," *New Yorker*, http://www.newyorker.com/contributors/elizabeth-kolbert (accessed September 5, 2016).

32. "The Sixth Extinction," Rutgers School of Environmental and Biological Sciences, http://sebsnjaes250.rutgers.edu/anthropocene/elizabeth-kolbert.html (accessed September 5, 2016).

33. Ibid.

34. Elizabeth Kolbert, *The Sixth Extinction: An Unnatural History* (New York: Henry Holt, 2014), pp. 17–18.

35. Christine Dell'Amore, "Species Extinction Happening 1,000 Times Faster Because of Humans?" *National Geographic*, http://news.nationalgeographic.com/news/2014/05/140529-conservation-science-animals-species-endangered-extinction/ (accessed September 5, 2016).

36. Robert Kunzig, "The Sixth Extinction: A Conversation with Elizabeth Kolbert," *National Geographic*, February 19, 2014, http://news.nationalgeographic.com/news/2014/02/140218-kolbert-book-extinction-climate-science-amazon-rain-forest-wilderness/ (accessed September 5, 2016).

37. Kolbert, *Sixth Extinction*, pp. 75–76.

38. Al Gore, "Without a Trace: 'The Sixth Extinction,' by Elizabeth Kolbert," *New York Times*, February 10, 2014, http://www.nytimes.com/2014/02/16/books/review/the-sixth-extinction-by-elizabeth-kolbert.html?_r=0 (accessed September 5, 2016).

39. Marta Bausells, "The 100 Best Nonfiction Books of All Time: What Should Make the List?" *Guardian*, January 27, 2016, https://www.theguardian.com/books/booksblog/2016/jan/27/the-100-best-nonfiction-books-of-all-time-what-should-make-the-list (accessed September 5, 2016).

40. "National Book Critics Circle Announces Finalists; Sandrof Award to Toni Morrison," *Critical Mass*, January 19, 2015, http://bookcritics.org/blog/archive/national-book-critics-circle-announces-its-finalists-for-publishing-year-20 (accessed September 5, 2016).

41. Justin Gillis, "The Lightning Rod," *New York Times: Science Times*, June 16, 2015, D1 and D3.

42. Claudia Dreifus, "A Chronicler of Warnings Denied," *New York Times: Science*, October 27, 2014, http://www.nytimes.com/2014/10/28/science/naomi-oreskes-imagines-the-future-history-of-climate-change.html (accessed August 10, 2015).

43. Claudia Dreifus, "A Chronicler of Warnings Denied," *New York Times: Science*, October 27, 2014, http://www.nytimes.com/2014/10/28/science/naomi-oreskes-imagines-the-future-history-of-climate-change.html (accessed August 10, 2015).

44. Ibid.

45. Ibid.

46. Naomi Oreskes, in discussion with the authors, August 14, 2015.

47. Naomi Oreskes, email message to authors, August 4, 2015.

48. Oreskes, in discussion with the authors.
49. Ibid.
50. Dreifus, "Chronicler of Warnings Denied."
51. Ibid.
52. Gillis, "Lightning Rod."
53. Dreifus, "Chronicler of Warnings Denied."
54. Ibid.
55. Ibid.
56. Oreskes, in discussion with the authors.
57. Ibid.
58. Oreskes, email message.
59. Ibid.
60. Ibid.
61. Ibid.
62. James Gustave Speth, *Angels by the River: A Memoir* (White River Junction, VT: Chelsea Green, 2014).
63. James Gustave Speth, in discussion with the authors, November 17, 2016.
64. Ibid.
65. James Gustave Speth, interview with Steve Curwood, "Gus Speth Calls for a 'New' Environmentalism," *Living on Earth*, Boston, WBUR 90.9, March 7, 2015.
66. McKibben, *Oil and Honey*.
67. Speth, in discussion with the authors.
68. Ibid.
69. Speth, interview with Steve Curwood.
70. Speth, interview with Steve Curwood.
71. Ibid.
72. Speth, in discussion with the authors.
73. Ibid.
74. Speth, interview with Steve Curwood.
75. Speth, in discussion with the authors.
76. Speth, interview with Steve Curwood.
77. Speth, in discussion with the authors.
78. Joseph Romm, *Climate Change: What Everyone Needs to Know* (New York: Oxford University Press, 2016), p. 1.
79. Ibid.
80. Ibid.
81. *Years of Living Dangerously*, http://yearsoflivingdangerously.com/ (accessed March 30, 2016).
82. Joseph Romm, *Language Intelligence: Lessons on Persuasion from Jesus, Shakespeare, Lincoln, and Lady Gaga* (North Charleston, SC: CreateSpace, 2012).
83. Joseph Romm, in discussion with the authors, March 4, 2016.

84. Ibid.

85. Ibid.

86. Ibid.

87. Ibid.

88. Ibid.

89. Ibid.

90. Joel Makower, "Two Steps Forward: Inside Paul Hawken's Audacious Plan to 'Drawdown' Climate Change," GreenBiz, October 22, 2014, https://www.greenbiz.com/blog/2014/10/22/inside-paul-hawkens-audacious-plan-drawdown-climate-change (accessed March 31, 2016).

91. Ibid.

92. Hawken and Lovins, *Natural Capitalism*.

93. Janine Benyus, *Biomimicry: Innovation Inspired By Nature* (New York: HarperCollins, 1997).

94. Paul Hawken, *Blessed Unrest: How the Largest Movement in the World Came into Being and Why No One Saw It Coming* (London: Penguin, 2008).

95. James Sheehan, "Book Review: *Blessed Unrest*," *Electronic Journal of Sustainable Development* 1, no. 2 (2008): 58, http://www.ejsd.co/docs/REVIEW_OF_BLESSED_UNREST.pdf (accessed November 15, 2015).

96. Paul Hawken, email communication with the authors, February 4, 2016.

97. Makower, "Two Steps Forward."

98. Ibid.

99. Bill McKibben, "Global Warming's Terrifying New Math," *Rolling Stone*, July 19, 2012, http://www.rollingstone.com/politics/news/global-warmings-terrifying-new-math-20120719 (accessed September 5, 2016).

100. Makower, "Two Steps Forward."

101. Hawken, email communication with the authors.

102. Ibid.

103. Makower, "Two Steps Forward."

104. Hawken, email communication with the authors.

105. Makower, "Two Steps Forward."

CHAPTER 25: POLITICIANS / ADVOCATES

1. "Al Gore Biography," *Biography.com*, 2016, http://www.biography.com/people/al-gore-9316028 (accessed July 2016).

2. Melinda Henneberger, "On Campus Torn by 60's, Agonizing Over the Path," *New York Times*, June 21, 2000, http://www.nytimes.com/library/politics/camp/062100wh-gore.html (accessed November 15, 2015).

3. Carl Pope, "Heroes of the Environment: Al Gore," *Time*, October 17, 2007, http://content.time.com/time/specials/2007/article/0,28804,1663317_1663319_1669889,00.html (accessed November 14, 2015).

4. "Mr. Albert Arnold 'Al' Gore—Former United States' Vice President," *Environ Business*, http://www.environbusiness.com/ebios/algore (accessed February 29, 2016).

5. Ibid.

6. Ibid.

7. Al Gore, *Earth in the Balance: Ecology and the Human Spirit* (New York: Plume, 1992).

8. "Mr. Albert Arnold 'Al' Gore"; John F. Kennedy, *Profiles in Courage* (New York: Harper, 1956).

9. "Mr. Albert Arnold 'Al' Gore."

10. Ibid.

11. Ibid.

12. "The Climate Reality Project," Al Gore, https://www.algore.com/project/the-climate-reality-project (accessed March 1, 2016).

13. Ibid.

14. Al Gore, *An Inconvenient Truth: The Planetary Emergency of Global Warming and What We Can Do About It* (New York: Rodale, 2006).

15. Pope, "Heroes of the Environment—Al Gore."

16. Carl Pope, "An Inconvenient Truth Is Getting Out," *Huffington Post*, May 25, 2011, http://www.huffingtonpost.com/carl-pope/an-inconvenient-truth-is-_b_19768.html (accessed September 5, 2016).

17. Rachel Flor, "Former U.S. Congressman Bob Inglis to Receive JFK Profile in Courage Award for Stance on Climate Change," John F. Kennedy Presidential Library and Museum, April 13, 2015, http://www.jfklibrary.org/About-Us/News-and-Press/Press-Releases/2015-Profile-in-Courage-Announcement.aspx (accessed March 5, 2016).

18. Robert Inglis, in discussion with the authors, August 26, 2015.

19. Ibid.

20. Rebecca Leber, "Why Republicans Only Grasp Climate Science Behind Closed Doors," *New Republic*, April 15, 2015, https://newrepublic.com/article/121541/scientists-educate-republicans-about-climate-change (accessed March 4, 2016).

21. Inglis, in discussion with the authors.

22. Ibid.

23. Ibid.

24. Ibid.

25. Katie Valentine, "Senator Sheldon Whitehouse's Best Climate Change Burns," *ThinkProgress*, May 18, 2015, http://thinkprogress.org/climate/2015/05/18/3659937/whitehouse-100-climate-speeches/ (accessed March 27, 2016).

26. Ibid.

27. Ibid.

28. Sheldon Whitehouse, "Time to Wake Up: Moving Toward a Clean Energy Economy," Sheldon Whitehouse: US Senator for Rhode Island, January 27, 2016. http://www.whitehouse.senate.gov/news/speeches/time-to-wake-up-moving-toward-a-clean-energy-economy (accessed March 27, 2016).

29. Valentine, "Senator Sheldon Whitehouse's Best Climate Change Burns."

30. Dave Burroughs, "Sen. Whitehouse Aims to 'Wake Up' Congress on Climate for 100th Time," *Morning Consult*, May 17, 2015, https://morningconsult.com/2015/05/17/sen-whitehouse-aims-to-wake-up-congress-for-100th-time/.

CHAPTER 26: ARTISTS / CELEBRITIES

1. James Balog, in discussion with the authors, August 24, 2015.

2. Ibid.

3. *Chasing Ice*, directed by Jeff Orlowski (Los Angeles: Exposure Labs, 2012), DVD, https://chasingice.com (accessed November 16, 2015).

4. Balog, in discussion with the authors.

5. James Balog, "Time Lapse Proof of Extreme Ice Loss," *TED Talk*, July 2009, http://www.ted.com/talks/james_balog_time_lapse_proof_of_extreme_ice_loss/transcript?language=en (accessed March 22, 2015).

6. Ibid.

7. Ibid.

8. Jason Goodyer, "On Thin Ice," *Engineering and Technology*, December 2012: 44–47, http://eandt.theiet.org/magazine/2012/11/on-thin-ice.cfm.

9. Balog, in discussion with the authors.

10. Ibid.

11. Keith Wagstaff, "Q&A: 'Avatar' Director James Cameron Talks Climate Change," NBC News: Science-Environment, September 23, 2014, http://www.nbcnews.com/science/environment/q-avatar-director-james-cameron-talks-climate-change-n209751 (accessed July 18, 2015).

12. Ibid.

13. Ibid.

14. Ibid.

15. James Cameron, "Why I Eat a Vegan Diet," *Men's Journal*, September 15, 2015, http://www.mensjournal.com/health-fitness/nutrition/james-cameron-why-i-eat-a-vegan-diet-20150915 (accessed September 5, 2016).

16. 100isNow, Twitter post, February 10, 2016, https://twitter.com/100isNow/status/697550372655775744.

17. Sarah Shanley Hope, in discussion with the authors, February 22, 2016.

18. Ibid.

19. Ibid.
20. Ibid.
21. Ibid.
22. Ibid.
23. Ibid.
24. Ibid.
25. Suzanne Goldenberg, "How Leonardo DiCaprio Became One of the World's Top Climate Change Champions," *Guardian*, February 29, 2016, http://www.theguardian.com/environment/2016/feb/29/how-leonardo-dicaprio-oscar-climate-change-campaigner (accessed March 27, 2016).
26. Chris White, "DiCaprio Considered Giving Up Acting to Take Up Environmental Activism Full-Time," *Daily Caller*, February 28, 2016, http://dailycaller.com/2016/02/28/dicaprio-considered-giving-up-acting-to-take-up-environmental-activism-full-time/ (accessed March 27, 2016).
27. Leonardo DiCaprio Foundation, http://leonardodicaprio.org/.
28. "About Us," Leonardo DiCaprio Foundation, http://leonardodicaprio.org/ (accessed September 5, 2016).
29. Goldenberg, "How Leonardo DiCaprio Became One of the World's Top Climate Change Champions."
30. Ibid.
31. Ibid.
32. Steven Rodrick, "Inside Leonardo DiCaprio's Crusade to Save the World," *Rolling Stone*, February 18, 2016, http://www.rollingstone.com/movies/features/inside-leonardo-dicaprios-crusade-to-save-the-world-20160218#ixzz41ZKd1Lf7 (accessed September 5, 2016).
33. Ibid.
34. "Board of Directors: Neil deGrasse Tyson," Planetary Society, http://www.planetary.org/about/board-of-directors/neil-tyson.html (accessed March 27, 2016).
35. Neil deGrasse Tyson, *Death by Black Hole: And Other Cosmic Quandaries* (New York: W. W. Norton, 2007).
36. Neil deGrasse Tyson, *Origins: Fourteen Billion Years of Cosmic Evolution* (New York: W. W. Norton, 2014).
37. Neil deGrasse Tyson, *Space Chronicles: Facing the Ultimate Frontier* (New York: W. W. Norton, 2013).
38. Neil deGrasse Tyson, *The Sky Is Not the Limit: Adventures of an Urban Astrophysicist* (Amherst, NY: Prometheus Books, 2004).
39. "Board Directors: Neil deGrasse Tyson."
40. Chris Mooney, "Finally, Neil deGrasse Tyson and 'Cosmos' Take On Climate Change," *Mother Jones*, May 5, 2014, http://www.motherjones.com/environment/2014/05/neil-tyson-cosmos-global-warming-earth-carbon (accessed August 15, 2015).
41. Kristina Bravo, "Watch Neil deGrasse Tyson Debunk Climate Change

Denier Claims in Two Minutes," *Take Part*, June 1, 2014, http://www.takepart.com/video/2014/06/01/watch-neil-degrasse-tyson-debunk-climate-change-denier-claims-two-minutes (accessed August 15, 2015).

42. Sean Illing, "Neil deGrasse Tyson Lets the Science Deniers Have It: 'The Beginning of the End of an Informed Democracy,'" *Salon*, October 20, 2015, http://www.salon.com/2015/10/20/neil_degrasse_tyson_lets_the_science_deniers_have_it_the_beginning_of_the_end_of_an_informed_democracy/ (accessed September 5, 2016).

43. "Environmental Media Association (EMA)," Norman Lear, http://www.normanlear.com/citizenship/environmental-media-association-ema/ (accessed March 26, 2016).

44. Ibid.

45. Ibid.

CHAPTER 27: BUSINESSPEOPLE

1. "The World's Billionaires: Thomas Steyer (#962)," *Forbes*, March 5, 2008, http://www.forbes.com/lists/2008/10/billionaires08_Thomas-Steyer_RUIR.html (accessed March 26, 2016).

2. Richard Valdmanis, "Exclusive: Billionaire Green Activist Steyer Not Ready to Back Clinton, Open to Sanders," Reuters, January 20, 2016, http://www.reuters.com/article/us-usa-election-steyer-idUSMTZSAPEC1K9038AB (March 26, 2016).

3. "Tom Steyer: Co-Founder," Next Generation, 2016, http://thenextgeneration.org/about/people/tom-steyer (accessed July 2016).

4. Ibid.

5. Timothy Cama, "Green Billionaire's '16 Gameplan? Shame GOP on Climate Change," *The Hill*, April 6, 2015, http://thehill.com/policy/energy-environment/238026-steyer-readies-new-2016-attacks (accessed March 26, 2016).

6. Julia Pyper, "Tom Steyer Calls on Activists to Push for Clean Energy Solutions Post-Keystone," *Greentech Media*, November 11, 2015, https://www.greentechmedia.com/articles/read/Tom-Steyer-Calls-On-Activists-to-Push-for-Clean-Energy-Solutions (accessed March 26, 2016).

7. Ibid.

8. Ibid.

9. Cama, "Green Billionaire's '16 Gameplan?"

10. Ibid.

11. Jeremy Grantham, "Be Persuasive. Be Brave. Be Arrested (If Necessary)," *Nature*, November 14, 2012, http://www.nature.com/news/be-persuasive-be-brave-be-arrested-if-necessary-1.11796 (accessed March 21, 2016).

12. Ibid.

13. Ibid.

14. Jeremy Grantham and Ramsay Ravenel, in discussion with the authors, November 30, 2015.

15. Naomi Klein, *This Changes Everything: Capitalism vs. the Climate* (New York: Simon and Schuster, 2014).

16. "The Fossil Fuel Resistance: Meet the New Green Heroes," *Rolling Stone*, April 11, 2013, http://www.rollingstone.com/politics/lists/the-fossil-fuel-resistance-meet-the-new-green-heroes-20130411#ixzz3eOT1RoQb (accessed June 12, 2015).

17. Specifics about the programs funded by the Grantham Foundation can be found on its website, www.granthamfoundation.org.

18. Lawrence Delevingne, "Battle of the Billionaires Erupts over Keystone," CNBC, October 25, 2013, http://www.cnbc.com/2013/10/25/big-hedge-fund-managers-fight-over-keystone-xl-pipeline.html (accessed March 21, 2016).

19. Leah McGrath Goodman, "Jeremy Grantham to Join Keystone Pipeline Protest," *Fortune*, February 13, 2013, http://fortune.com/2013/02/13/jeremy-grantham-to-join-keystone-pipeline-protest/ (accessed March 21, 2016).

20. Ibid.

21. Grantham, "Be Persuasive."

22. Jeremy Grantham, "Conference on Inclusive Capitalism, Speech," Grantham Foundation for the Protection of the Environment, May 27, 2014, http://www.granthamfoundation.org/our-blog (accessed March 27, 2016).

23. Ibid.

24. Grantham and Ravenel, in discussion with the authors.

25. "Jeremy Grantham: In a Climate of Risk," interview by Christopher Lydon, Open Source Radio, October 4, 2014, http://radioopensource.org/jeremy-granthams-seven-year-forecast-climate-risk/ (accessed September 5, 2016).

26. Jeff Goodell, "Q & A: Bill Gates on How to Stop Global Warming," *Rolling Stone*, December 9, 2010, http://www.rollingstone.com/politics/news/the-miracle-seeker-20101028 (accessed July 2016).

27. Robert M. Christie, "Calculating Human Survival," *The Hopeful Realist* (blog), August 16, 2016, https://thehopefulrealist.com/tag/ecological-economy/ (accessed September 5, 2016).

28. Coral Davenport and Nick Wingfield, "Bill Gates Takes On Climate Change with Nudges and a Powerful Rolodex," *New York Times*, December 8, 2015, http://www.nytimes.com/2015/12/09/business/energy-environment/bill-gates-takes-on-climate-change-with-nudges-and-a-powerful-rolodex.html (accessed July 2016).

29. Ibid.

30. Joby Warrick, "Bill Gates on Climate Change: We Need to Move Faster than the Energy Sector Ever Has," *Washington Post*, November 30, 2015, https://www.washingtonpost.com/news/energy-environment/wp/2015/11/30/bill-gates-on-climate-change-we-need-to-move-faster/ (accessed March 26, 2016).

31. Ibid.

32. Ibid.

33. Ibid.

34. Ashlee Vance, *Elon Musk: Tesla, Spacex, and the Quest for a Fantastic Future* (New York: Ecco, 2015).

35. Ibid.

36. Jerry Hirsch, "Elon Musk: Model S Not a Car but a Sophisticated Computer on Wheels," *Los Angeles Times*, March 19, 2015, http://www.latimes.com/business/autos/la-fi-hy-musk-computer-on-wheels-20150319-story.html (accessed March 26, 2016).

37. Pauline Abreu, "The World's Only Electric Sports Car: 2010 Tesla Roadster," *Motor Authority*, April 11, 2010, http://www.motorauthority.com/news/1044161_the-worlds-only-electric-sports-car-2010-tesla-roadster (accessed February 2, 2016).

38. Jeff Cobb, "Tesla Model S Was World's Best-Selling Plug-In Car in 2015," HybridCars.com, January 12, 2016, http://www.hybridcars.com/tesla-model-s-was-worlds-best-selling-plug-in-car-in-2015/ (accessed February 2, 2016).

39. Matthew DeBord, "Elon Musk's Big Announcement: It's Called 'Tesla Energy,'" *Business Insider*, May 1, 2015, http://www.businessinsider.com/here-comes-teslas-missing-piece-battery-announcement-2015-4 (accessed February 2, 2016).

40. "Powerwall: Tesla Home Battery," Tesla, https://www.tesla.com/powerwall (accessed February 2, 2016).

41. Alex Davies, "Elon Musk's Grand Plan to Power the World with Batteries," *Wired*, May 1, 2015, http://www.wired.com/2015/05/tesla-batteries/ (accessed February 2, 2016).

42. "Powerwall: Tesla Home Battery."

43. "About SpaceX," SpaceX, http://www.spacex.com/about (accessed March 26, 2016).

44. Will Nicol, "As Hyperloop Progress Glides Forward, Here's What You Need to Know," *Digital Trends*, February 3, 2016, http://www.digitaltrends.com/cool-tech/hyperloop-news/ (accessed July 2016).

45. Vance, *Elon Musk*.

CHAPTER 28: RELIGIOUS / GRASSROOTS ORGANIZERS

1. "Michael Bloomberg: Mayor, Philanthropist (1942–)," *Biography.com*, http://www.biography.com/people/michael-bloomberg-16466704 (accessed February 2, 2016).

2. Ibid.

3. Robert Frank, "Billionaire Battle: Bloomberg's Worth Versus Trump," *Inside Wealth*, CNBC, January 25, 2016, http://www.cnbc.com/2016/01/25/billionaire-battle-bloombergs-worth-versus-trump.html (accessed July 2016).

4. "Michael Bloomberg: Mayor, Philanthropist."

5. Shannon Brownlee, "Why Americans Need Bloomberg's Big Gulp Ban," *Time*,

June 4, 2012, http://ideas.time.com/2012/06/04/why-americans-need-bloombergs-big-gulp-ban/ (accessed July 2016).

6. Thomas Lueck, "Bloomberg Draws a Blueprint for a Greener City," *New York Times*, April 23, 2007, http://www.nytimes.com/2007/04/23/nyregion/23mayor.html (accessed March 26, 2016).

7. "Mayor Bloomberg, Deputy Mayor Holloway and Office of Long Term Planning and Sustainability Director Sergej Mahnovski Announce Significant Reduction in Greenhouse Gas Emissions and New and Expanded Programs to Continue the Progress," NYC: The Official Website of the City of New York, December 30, 2013, http://www1.nyc.gov/office-of-the-mayor/news/440-13/mayor-bloomberg-deputy-mayor-holloway-office-long-term-planning-sustainibility (accessed November 15, 2015).

8. Lisa Foderaro, "Bronx Planting Caps Off a Drive to Add a Million Trees," *New York Times*, October 20, 2015, http://www.nytimes.com/2015/10/21/nyregion/new-york-city-prepares-to-plant-one-millionth-tree-fulfilling-a-promise.html.

9. Michael Bloomberg, "Forewords to *Climate Change Adaptation in New York City: Building a Risk Management Response*," *Annals of the New York Academy of Sciences* 1196 (May 2010), http://onlinelibrary.wiley.com/doi/10.1111/j.1749-6632.2009.05415.x/epdf (accessed September 6, 2016).

10. Jarrett Murphy, "3 Years after Hurricane Sandy, Is New York Prepared for the Next Great Storm?" *Nation*, October 14, 2015, https://www.thenation.com/article/3-years-after-hurricane-sandy-is-new-york-prepared-for-the-next-great-storm/.

11. Kate Taylor, "New York's Air Is Cleanest in 50 Years, Survey Finds," *New York Times*, September 26, 2013, http://www.nytimes.com/2013/09/27/nyregion/new-yorks-air-is-cleanest-in-50-years-survey-finds.html.

12. Harry Bruinius, "Cleanest Air in 50 Years! How Did New York Do It?" *Christian Science Monitor*, September 27, 2013, http://www.csmonitor.com/Environment/2013/0927/Cleanest-air-in-50-years!-How-did-New-York-do-it (accessed September 6, 2016).

13. Abbey Brown, "Successful NYC Clean Heat Program Wins Award for Outstanding Design," Environmental Defense Fund, February 18, 2005, http://blogs.edf.org/energyexchange/2015/02/18/successful-nyc-clean-heat-program-wins-award-for-outstanding-design/ (accessed September 5, 2016).

14. Andrew Restuccia, "Michael Bloomberg's War on Coal," *Politico*, April 8, 2015, http://www.politico.com/story/2015/04/michael-bloomberg-environment-coal-sierra-club-116793 (accessed March 26, 2016).

15. Mireya Navarro, "Bloomberg Backs 'Responsible' Extraction of Gas and Pays to Help to Set Up Rules," *New York Times*, August 24, 2012, http://www.nytimes.com/2012/08/25/nyregion/bloomberg-backs-gas-drilling-with-rules-to-protect-the-environment.html.

16. Alicia Mundy, "'Risky Business' Report Aims to Frame Climate Change as Economic Issue," *Wall Street Journal*, June 23, 2014, http://www.wsj.com/articles/risky-business-report-aims-to-frame-climate-change-as-economic-issue-1403578637.

17. Maria Gallucci, "States Get $48M Boost from Bloomberg Charity to Help Meet Obama Climate Change Agenda," *International Business Times*, January 21, 2015, http://www.ibtimes.com/states-get-48m-boost-bloomberg-charity-help-meet-obama-climate-change-agenda-1789946.

18. "History of the C40," C40 Cities, http://www.c40.org/history (accessed March 26, 2016).

19. Michael Barbaro, "Bloomberg and Clinton to Merge Climate Groups," *New York Times*, April 13, 2011, http://www.nytimes.com/2011/04/14/nyregion/14bloomberg.html.

20. Jonathan Andrews, "Interview: Michael Bloomberg, Outgoing Chair and Current President C40 Cities," *Cities Today*, February 18, 2014, http://cities-today.com/interview-michael-bloomberg-outgoing-chair-and-current-president-c40-cities/ (accessed February 2, 2016).

21. Greg Scruggs, "Local-Governments Day Announced for Paris Climate Summit," *Citiscope*, July 1, 2015, http://citiscope.org/habitatIII/news/2015/07/local-governments-day-announced-paris-climate-summit (accessed March 26, 2016).

22. Larry Elliott, "Michael Bloomberg to Head Global Taskforce on Climate Change," *Guardian*, December 4, 2015, https://www.theguardian.com/environment/2015/dec/04/mark-carney-unveils-global-taskforce-to-educate-business-on-climate-change (accessed September 6, 2016).

23. Judith Lewis, "Meet the Real Van Jones," *Los Angeles Times*, September 11, 2009, http://articles.latimes.com/2009/sep/11/opinion/oe-lewis11 (accessed March 20, 2016).

24. Van Jones, *The Green Collar Economy: How One Solution Can Fix Our Two Biggest Problems* (New York: Harper Collins, 2008).

25. "About Van Jones," Van Jones, http://www.vanjones.net/about (accessed March 21, 2016).

26. "Our Victories," Ella Baker Center for Human Rights, http://www.ellabakercenter.org/our-victories (accessed March 21, 2016).

27. Sarah Shanley Hope, in discussion with the authors, February 22, 2016.

28. Terrence McNally, "Q&A with *Green Collar Economy* Author Van Jones," *Huffington Post*, January 11, 2009, http://www.huffingtonpost.com/terrence-mcnally/qa-with-van-green-collar_b_150400.html (accessed March 21, 2016).

29. Jones, *Green Collar Economy*, p. 198.

30. Chadwick Matlin, "Van Jones: The Face of Green Jobs," *The Big Money*, April 22, 2009, http://warisacrime.org/node/41867 (accessed March 21, 2016).

31. Jones, *Green Collar Economy*, p. 21.

32. Van Jones, "Beyond Eco-Apartheid," *Truthout*, April 2007, http://truth-out.org/archive/component/k2/item/70209:van-jones--beyond-ecoapartheid (accessed March 20, 2016).

33. Jones, *Green Collar Economy*, p. 50.

34. Jones, "Beyond Eco-Apartheid."

35. Jones, *Green Collar Economy*, p. 55.

36. Jones, "Beyond Eco-Apartheid."
37. Ibid.
38. Jones, *Green Collar Economy*, p. 3.
39. Ibid., p. 5.
40. Ibid., p. 16.
41. Ibid., pp. 16–17.
42. Ibid., p. 64.
43. Ibid., p. 65.
44. Sally Bingham, in discussion with the authors, August 7, 2015.
45. Ibid.
46. Ibid.
47. Ibid.
48. "The Reverend Canon Sally Grover Bingham," Interfaith Power & Light: A Religious Response to Global Warming, http://interfaithpowerandlight.org/wp-content/uploads/2009/11/Sally-Bio-0610.pdf (accessed March 21, 2016).
49. Susan Stephenson, "I Invite You to Swallow an Alarm Clock," *Interfaith Power & Light Blog*, March 26, 2016, http://interfaithpowerandlight.tumblr.com/post/141643165591/i-invite-you-to-swallow-an-alarm-clock (accessed March 31, 2016).
50. Bingham, in discussion with the authors.
51. Ibid.
52. Ibid.
53. Ibid.
54. Ibid.
55. "Bidder 70: About," Tim DeChristopher, 2015, http://www.timdechristopher.org/about (accessed September 6, 2016).
56. Jeff Goodell, "Meet America's Most Creative Climate Criminal," *Rolling Stone*, July 7, 2011, http://www.rollingstone.com/politics/news/meet-america-s-most-creative-climate-criminal-20110707 (accessed September 6, 2016).
57. "Bidder 70," Tim DeChristopher, 2015, http://www.timdechristopher.org/bidder_70 (accessed March 21, 2016).
58. "The Fossil Fuel Resistance: Meet the New Green Heroes," *Rolling Stone*, April 11, 2013, http://www.rollingstone.com/politics/lists/the-fossil-fuel-resistance-meet-the-new-green-heroes-20130411#ixzz3eOT1RoQb (accessed June 12, 2015).
59. James Hansen, *Storms of My Grandchildren: The Truth About the Coming Climate Catastrophe and Our Last Chance to Save Humanity* (New York: Bloomsbury, 2009), p. 248.
60. Ibid.
61. Tim DeChristopher, "Lead, Don't Follow on Climate Justice," *Harvard Divinity Bulletin* 43, no. 1–2 (Winter/Spring 2015), http://bulletin.hds.harvard.edu/articles/winterspring2015/lead-don%E2%80%99t-follow-climate-justice (accessed March 21, 2016).
62. Ibid.
63. Ibid.

64. Ibid.

65. Ibid.

66. Ibid.

67. Ibid.

68. Jason Easley, "Bernie Sanders Files a New Constitutional Amendment to Overturn Citizens United," *PoliticusUSA*, January 21, 2015, http://www.politicususa.com/2015/01/21/bernie-sanders-files-constitutional-amendment-overturn-citizens-united.html (accessed September 5, 2016).

69. Ibid.

70. DeChristopher, "Lead, Don't Follow on Climate Justice."

71. Ibid.

72. "Sierra Club Names New Environmental Justice Award after Dr. Robert Bullard," *The Planet*, Sierra Club, August 5, 2014, http://www.sierraclub.org/planet/2014/08/sierra-club-names-new-environmental-justice-award-after-dr-robert-bullard (accessed June 24, 2015).

73. Brentin Mock, "Robert Bullard, Pioneer in Environmental Justice, Is Honored by the Sierra Club," *Washington Post*, September 24, 2013, http://www.washingtonpost.com/lifestyle/style/robert-bullard-pioneer-in-environmental-justice-is-honored-by-the-sierra-club/2013/09/24/88e0e882-251c-11e3-b3e9-d97fb087acd6_story.html (accessed June 24, 2015).

74. "Bullard: Green Issue Is Black and White," CNN, July 17, 2007, http://www.cnn.com/2007/US/07/17/pysk.bullard/index.html (accessed June 24, 2015).

75. Mock, "Robert Bullard, Pioneer."

76. "Bullard: Green Issue Is Black and White."

77. Mock, "Robert Bullard, Pioneer."

78. Ibid.

79. "Sierra Club Names New Environmental Justice Award."

80. Gregory Dicum, "Meet Robert Bullard, the Father of Environmental Justice," *Grist*, March 15, 2006, http://grist.org/article/dicum/ (accessed June 25, 2015).

81. "Awards: Sierra Club Names New Environmental Justice Award after Dr. Robert D. Bullard," Dr. Robert Bullard: Father of Environmental Justice, 2013, http://drrobertbullard.com/awards-honors/ (accessed July 20, 2015).

82. Ibid.

83. Robert Bullard, *Dumping in Dixie: Race, Class, and Environmental Quality* (Boulder, CO: Westview Press, 1990).

84. Dicum, "Meet Robert Bullard."

85. Ibid.

86. Ibid.

87. Ibid.

CHAPTER 29: HOW THE CLIMATE-CHANGE WAR CAN BE WON

1. Winston Churchill, "The Russian Enigma," The Churchill Society, October 1, 1939, http://www.churchill-society-london.org.uk/RusnEnig.html (accessed September 6, 2016).

2. *This Changes Everything*, film, directed by Avi Lewis (2015; Klein Lewis Productions), DVD.

3. Thomas Lovejoy, in discussion with the authors, August 28, 2015.

4. Alejandero Davila Fragoso, "Americans' Concern about Climate Change Is Growing," *ThinkProgress*, March 18, 2016, http://thinkprogress.org/climate/2016/03/18/3761720/climate-change-worries-increase (accessed March 23, 2016).

5. Ibid.

6. Lydia Saad and Jeffrey Jones, "U.S. Concern about Global Warming at Eight-Year High," Gallup, March 16, 2016, http://www.gallup.com/poll/190010/concern-global-warming-eight-year-high.aspx (accessed March 23, 2016).

7. Ibid.

CHAPTER 30: SOLUTIONS FOR THE UNITED STATES

1. John Cook et al., "Consensus on Consensus: A Synthesis of Consensus Estimates on Human-Caused Global Warming," *Environmental Research Letters* 11, no. 4 (April 13, 2016), http://iopscience.iop.org/article/10.1088/1748-9326/11/4/048002 (accessed September 6, 2016).

2. Naomi Oreskes and Eric Conway, *Merchants of Doubt: How a Handful of Scientists Obscured the Truth on Issues from Tobacco Smoke to Global Warming* (New York: Bloomsbury, 2010).

3. Joseph Romm, in discussion with the authors, March 4, 2016.

4. Andrew Gelman, "All Politics Is Local? The Debate and the Graphs," *FiveThirtyEight* (blog), *New York Times*, January 3, 2011, http://fivethirtyeight.blogs.nytimes.com/2011/01/03/all-politics-is-local-the-debate-and-the-graphs/?_r=0 (accessed July 2016).

5. Joseph Romm, *Climate Change: What Everyone Needs to Know* (New York: Oxford University Press, 2016).

6. Donella Meadows, Dennis Meadows, and Jorgen Randers, *Limits to Growth: The 30-Year Update* (White River Junction, VT: Chelsea Green Publishing, 2004).

7. Naomi Klein, *This Changes Everything: Capitalism vs. the Climate* (New York: Simon and Schuster, 2014).

8. James Gustave Speth, *The Bridge at the Edge of the World: Capitalism, the Environment, and Crossing from Crisis to Sustainability* (New Haven: Yale University Press, 2009).

9. Richard Heinberg, *The End of Growth: Adapting to Our New Economic Reality* (Gabriola Island, BC, Canada: New Society, 2011).

10. Mark Jacobson, in discussion with the authors, January 28, 2016.

11. Mark Jacobson et al., "100% Clean and Renewable Wind, Water, and Sunlight (WWS) All-Sector Energy Roadmaps For the 50 United States," *Energy & Environmental Science* 8 (2015): 2093, http://web.stanford.edu/group/efmh/jacobson/Articles/I/USStatesWWS.pdf.

12. Ibid.

13. "Options and Considerations for a Federal Carbon Tax," Center for Climate and Energy Solutions, February 2013, http://www.c2es.org/publications/options-considerations-federal-carbon-tax (accessed March 22, 2016).

14. Rick Piltz, "Supreme Court Rules Clean Air Act Gives EPA Authority to Regulate Greenhouse Gases," *Climate Science and Policy Watch*, April 2, 2007, http://www.climatesciencewatch.org/2007/04/02/supreme-court-rules-clean-air-act-gives-epa-authority-to-regulate-greenhouse-gases/ (accessed July 2016).

15. "Endangerment and Cause or Contribute Findings for Greenhouse Gases Under Section 202(a) of the Clean Air Act," US Environmental Protection Agency, August 9, 2016, https://www3.epa.gov/climatechange/endangerment/index.html (accessed September 5, 2016).

16. "Fracking by the Numbers, Key Impacts of Dirty Drilling at the State and National Level," *Environment America*, October 3, 2013, http://www.environmentamerica.org/reports/ame/fracking-numbers (accessed September 5, 2016).

17. Elizabeth Bast et al., "Empty Promises: G20 Subsidies to Oil, Gas, and Coal Production," Overseas Development Institute, November 2015, http://www.odi.org/publications/10058-production-subsidies-oil-gas-coal-fossil-fuels-g20-broken-promises (accessed March 26, 2016).

18. Avaneesh Pandey, "US Fossil Fuel Subsidies Increase 'Dramatically' Despite Climate Change Pledge," *International Business Times*, November 12, 2015, http://www.ibtimes.com/us-fossil-fuel-subsidies-increase-dramatically-despite-climate-change-pledge-2180918.

19. Ibid.

20. Ibid.

21. Bast et al., "Empty Promises."

22. Josh Block, "CLF Sues ExxonMobil over Decades-Long Climate Deceit," Conservation Law Foundation, May 17, 2016, http://www.clf.org/newsroom/clf-sues-exxonmobil/ (accessed September 5, 2016).

23. Brian Mansfield, "'We Are the World' at 30: 12 Tales You Might Not Know," *USA Today*, January 28, 2015, http://www.usatoday.com/story/life/music/2015/01/27/we-are-the-world-30th-anniversary/22395455/ (accessed July 2016).

24. "1985: Live Aid Concert," *This Day in History*, *History*, http://www.history.com/this-day-in-history/live-aid-concert (accessed December 15, 2015).

CHAPTER 31: SOLUTIONS FOR THE WORLD

1. David Waskow and Jennifer Morgan, "The Paris Agreement: Turning Point for a Climate Solution," World Resources Institute, December 12, 2015, http://www.wri.org/blog/2015/12/paris-agreement-turning-point-climate-solution (accessed July 2016).

2. Ibid.

3. Eliza Northrup, "Not Just for Paris, but for the Future: How the Paris Agreement Will Keep Accelerating Climate Action," World Resources Institute, December 14, 2015, http://www.wri.org/blog/2015/12/not-just-paris-future-how-paris-agreement-will-keep-accelerating-climate-action (accessed July 2016).

4. Ibid.

5. Waskow and Morgan, "The Paris Agreement.".

6. Elizabeth Cobbs Hoffman, "The Sincerest Form of Flattery: The Peace Corps, the Helsinki Accords, and the Internalization of Social Values," in *Making the American Century: Essays on the Political Culture of Twentieth Century America*, ed. Bruce J. Schulman (Oxford, UK: Oxford University Press, 2014), pp. 124–37.

7. Thomas Lovejoy, in discussion with the authors, August 28, 2015.

8. Roddy Scheer and Doug Moss, "Deforestation and Its Extreme Effect on Global Warming," *Scientific American*, 2016, http://www.scientificamerican.com/article/deforestation-and-global-warming/ (accessed September 6, 2016).

9. Julian Smith, "Survival in the Great Rift," *Nature Conservancy*, December/January 2015, http://www.nature.org/magazine/archives/survival-in-the-great-rift.xml (accessed September 6, 2016).

10. Louis Blumberg, Erin Meyers Madeira, and Rane Cortez, "Nature Is Key to Achieving Governor Brown's Ambitious Climate Goal," *Talk* (blog), *Nature Conservancy*, January 7, 2015, http://blog.nature.org/conservancy/2015/01/07/nature-is-key-to-achieving-governor-browns-ambitious-climate-goal/ (accessed September 7, 2016).

11. E. Schrack et al., *Restoration Works: Highlights from a Decade of Partnership between the Nature Conservancy and the National Oceanic and Atmospheric Administration's Restoration Center* (Arlington, VA: Nature Conservancy, 2012), http://www.habitat.noaa.gov/pdf/restorationworks.pdf (accessed July 2016).

12. Joseph Romm, *Climate Change: What Everyone Needs to Know* (New York: Oxford University Press, 2016), p. 168.

13. Lovejoy, in discussion with the authors.

14. Ibid.

15. Paul Crutzen, email communication with the authors, February 2016.

16. Romm, *Climate Change*, p. 167.

17. Madeline Ostrander, "How One Alaskan Community Is Attempting to Adapt to Climate Change," *Audubon*, January-February 2016: 50–55, http://www.audubon.org/magazine/january-february-2016/how-one-alaskan-community-attempting.

18. Ibid.

19. Ibid.

20. Ibid.

21. Laura Parker, "Climate Change Economics: Treading Water," *National Geographic*, February 2015, http://ngm.nationalgeographic.com/2015/02/climate-change-economics/parker-text (accessed July 2016).

22. Ibid.

23. Ibid.

24. Ibid.

CONCLUSION

1. Bill McKibben, in discussion with authors, December 3, 2015.

ABOUT THE AUTHORS

Budd Titlow is a professional wetland scientist and wildlife ecologist (MS, Virginia Tech), as well as an international and national award-winning nature photographer and widely published writer. He has operated Naturegraphs Freelance Photography and Writing for more than forty years. He has authored three natural history books, most recently *Bird Brains: Inside the Strange Minds of Our Fine Feathered Friends*. He is currently teaching ecology, environmental science, birding, and photography courses at his alma mater Florida State University and the Tallahassee Senior Center, while writing a weekly birding column for the *Tallahassee Democrat Daily* newspaper. He is also serving as president of the Apalachee Audubon Society.

Mariah Tinger has more than twelve years of experience leading individuals and teams in environmental stewardship and education. She has worked in both corporate and nonprofit settings, including several years in Yosemite National Park and as an environmental program manager for Genzyme Corporation. She began her in-depth studies of climate change while pursuing a master's degree in sustainability and environmental management at Harvard University. She is currently a teaching fellow for sustainability and environmental management courses at the university.

INDEX